R00004 13821

22.50
T
12·13·73
509826

D1463361

mcs/

SF997·5 I5S61972
004150304
SPARKS, ALBERT K.
INVERTEBRATE PATHOLOGY

INVERTEBRATE PATHOLOGY

Noncommunicable Diseases

INVERTEBRATE PATHOLOGY

Noncommunicable Diseases

Albert K. Sparks

Director, Gulf Coastal Fisheries Center
National Marine Fisheries Service
Galveston, Texas

and

Professor of Veterinary Microbiology
Texas A & M University
College Station, Texas

ACADEMIC PRESS   1972   *New York and London*

COPYRIGHT © 1972, BY ACADEMIC PRESS, INC.
ALL RIGHTS RESERVED.
NO PART OF THIS PUBLICATION MAY BE REPRODUCED OR
TRANSMITTED IN ANY FORM OR BY ANY MEANS, ELECTRONIC
OR MECHANICAL, INCLUDING PHOTOCOPY, RECORDING, OR ANY
INFORMATION STORAGE AND RETRIEVAL SYSTEM, WITHOUT
PERMISSION IN WRITING FROM THE PUBLISHER.

ACADEMIC PRESS, INC.
111 Fifth Avenue, New York, New York 10003

United Kingdom Edition published by
ACADEMIC PRESS, INC. (LONDON) LTD.
24/28 Oval Road, London NW1

LIBRARY OF CONGRESS CATALOG CARD NUMBER: 72-82649

PRINTED IN THE UNITED STATES OF AMERICA

£4/5030

In memory of
Edward A. Steinhaus

Contents

Chapter III. Physical Injuries

Chapter IV. Chemical Injuries

Chapter V. Venom and Biotoxin Injuries

Chapter VI. Pathological Effects of Ionizing Radiation

Preface

Invertebrate pathology can be approached in two equally valid ways—the noninfectious and infectious diseases of each major group of invertebrates can bè considered in ascending taxonomic or phylogenetic order or it can be approached from the disease point of view, each major type of deviation from the normal state of health being considered. Although the specialist interested in the diseases of a particular species or a restricted group of invertebrates would probably prefer the first approach, the second lends itself more readily to the understanding of the phylogenetic and comparative relationships found in invertebrate pathology.

Despite the predominant interest of many investigators in a specific group of invertebrates such as the mollusks, the challenge of considering invertebrate pathology as a comparative, phylogenetically related entity induced me to adopt the subject-matter approach. The following subjects are discussed: death and postmortem change, reaction to injury and wound repair, physical injury, chemical injury, effects of venoms and biotoxins, effects of ionizing radiation, and tumors and tumorlike growths. With the exception of the chapter on the effects of venoms and biotoxins, the discussion is organized taxonomically or phylogenetically, beginning with protozoans and progressing through the ascending taxonomic order to the chordates.

Because of the dichotomous background of the potential user of this volume, I have attempted to review briefly the normal histology of each animal discussed concomitant with descriptions of pathological deviations from the norm and to define each term at its first use in the text. Most readers, however, will probably have to refer to a standard textbook of invertebrate zoology or a medical dictionary to obtain maximum value from the text. I have endeavored to include sufficient detail from the published literature on each subject to minimize the need for reference to the original publications. This approach has unquestionably led to repetitiveness and perhaps excessively exhaustive treatment of some phenomena and taxonomic groups for which I beg the reader's indulgence.

One of the most persistently perplexing problems in writing this volume, and for that matter in the field of invertebrate pathology, involves terminology. Although much of the current furor is, in my opinion, semantic, some valid questions require answers. Some respected invertebrate pathologists question the propriety of applying terminology developed to describe phenomena in vertebrate pathology to similar but not necessarily identical phenomena in invertebrates. The same rationale is frequently applied to the use of descriptive terminology for cells and tissue components in invertebrates that, in their opinion, should be restricted to vertebrate nomenclature. Others, equally respected, feel that a common terminology, when possible, is desirable. Some usage of vertebrate terminology, appearing in descriptions of pathological phenomena in invertebrates is patently erroneous, particularly in the area of neoplasia. Limitations of space preclude an exhaustive discussion of the divergent viewpoints, but several examples are illustrative.

A salient case in point is the use of the term inflammation. Although most dictionaries and textbooks of human pathology define inflammation as the local reaction of tissues to injury, many invertebrate pathologists object to the use of the term in references to invertebrates because not all the cardinal signs of inflammation (redness, swelling, heat, pain, and loss of function) occur in the reaction to injury by invertebrates. While the cardinal signs are included in most definitions, they are, in my opinion, descriptive of the process in mammals rather than part of a rigid definition. To briefly expand this concept: the redness characteristic of the vertebrate inflammatory response is due to the accumulation of erythrocytes resulting from vascular dilatation and congestion; temperature elevation of a surface inflamed area in birds and mammals results from the transport of large amounts of warm blood from the interior of the body to the cooler body surface; and pain results from pressure on nerves caused by the swelling accompanying the accumulation of exudate. Thus, the cardinal signs of mammalian and avian inflammation accompany the process but are not an integral part of it in that they serve no protective function. Therefore, it seems legitimate to use the term inflammation in reference to the local reactive changes in tissues following injury in any animal, invertebrate as well as vertebrate. One must realize, however, that the process will vary in different taxonomic groups.

A similar controversy has arisen over the use of the term leukocyte. The chief objection to its use lies in the mistaken view, in my opinion, that it should be restricted to vertebrates. The word leukocyte literally means a white or colorless cell, Webster's Unabridged Dictionary defines it as "a white or colorless blood corpuscle, one of the nucleated ameboid cells found in blood or lymph" and as "wandering cells in the tissues of the body and constituting the chief cellular elements in pus." Medical dictionaries define leukocyte identically or similarly.

Most invertebrates possess colorless, amoeboid, nucleated cells that are associated with the tissues and frequently with the circulatory system if one is present. These cells, though probably not homologous throughout the invertebrate phyla or with vertebrate leukocytes, have morphological and functional similarities. They are variously referred to in the literature as amoebocytes (or amebocytes), wandering phagocytes, hemocytes, and leukocytes. Because they are not invariably amoebocytic, phagocytic, or associated with a circulatory system and are frequently more numerous in tissues or cavities, the terms amoebocyte, phagocyte, or hemocyte are not universally applicable. Leukocyte does, however, appear appropriate simply because the cells are colorless and fit the general definition quoted earlier. Despite my opinion that leukocyte is the most nearly correct term for these cells, I have considered all the above names as synonyms and have used them interchangeably. In most instances I have used the name preferred by the original author.

The widespread, perhaps indiscriminate, use of the word collagen to describe connective tissue fibers of invertebrates is another item of concern to many invertebrate pathologists. True, or vertebrate, collagen is an albuminoid protein that is chemically resistant, except to digestion by pepsin and to a slight degree by trypsin, has an axial periodicity of about 640 angstroms (Å), and has a characteristic staining reaction with Mallory triple stain and other connective tissue stains. It has been pointed out that most invertebrates examined appear to possess collagen, but with axial periodicities other than 640 Å. Fibrils of 270 Å period have been described in some invertebrates; fibrils of this periodicity in vertebrates participate in the formation of the mature 640 Å period collagen. Thus the collagenlike material found in many invertebrates probably represents an early stage in collagen formation. I have applied the term collagen to those substances in invertebrates that look like collagen, stain like collagen, appear to have the function of collagen, and are called collagen by the original author even though not proved to be identical to vertebrate collagen by chemical tests and electron microscopy.

Because of the controversy over nomenclature and terminology in invertebrate pathology, a nomenclature committee should be established and charged with the responsibility of standardizing terminology. This would appear to be a valuable activity for the Society for Invertebrate Pathology to sponsor. Rulings by such a committee would eliminate much of the controversy and chaos that characterizes a good deal of the current literature and would ensure a common language of comparative pathology.

I derived great pleasure in the requisite careful study of the vast literature covering a wide spectrum of subject matter published over a time period of more than seventy years. Much of this literature is found in obscure journals or buried in articles or books whose titles give no clue to the presence of sometimes extensive information on some phase of invertebrate pathology. Johnson's "Annotated Bibliography of Pathology in Invertebrates Other than Insects" alerted me to much information I am sure would have otherwise been omitted.

My greatest aim in producing this book was that it would stimulate an acceleration of research effort in this most fascinating field.

Albert K. Sparks

Acknowledgments

Limitations of space preclude individual recognition of the many people who contributed in some way to the writing of this volume. Authors who kindly loaned me original figures ur provided prints from their publications are acknowledged in the text. Many of my colleagues in the Society for Invertebrate Pathology provided encouragement and stimulation, frequently when the task seemed impossible. Students enrolled in my course in invertebrate pathology at both the University of Washington and Texas A & M University, using preliminary drafts of the manuscript as a text, found many errors of omission and commission.

There were some whose contributions were so important that they must be individually recognized. Dr. John G. Mackin, under whom I worked both as a student and a colleague, provided me with my basic training in oyster pathology and kindled my interest in invertebrate disease. The late Dr. Edward A. Steinhaus, to whom this volume is dedicated, first convinced me that I was capable of writing the text, then served as a constant source of inspiration and encouragement until his untimely death. Dr. Phyllis T. Johnson kindly gave me a prepublication copy of the manuscript of "An Annotated Bibliography of Pathology in Invertebrates Other than Insects," saving months of library search time and unquestionably prevented me from missing important studies buried in the scattered literature of this diverse field. Dr. Thomas C. Cheng meticulously read the manuscript and made suggestions which

xv

led to drastic revision, resulting in the elimination of much redundancy and excess verbiage.

Although all my graduate students and my staff at the University of Washington enthusiastically encouraged and helped me during the early phases of the effort, the graduate students working on pathology theses deserve special recognition for their contributions: Dr. Daniel P. Cheney, Mr. David M. DesVoigne, Mr. Richard S. LeGore, Dr. Stanley C. Katkansky, Dr. Stephen G. Martin, Dr. Michael C. Mix, Dr. Gilbert B. Pauley, Dr. Craig Ruddell, and Mr. Donald E. Weitkamp. Dr. Mix was particularly helpful in editing the chapter on the effects of ionizing radiation.

Mr. Gregory Tutmark and Mr. Daniel Patlan warrant my appreciation for the preparation of many of the illustrations.

Since this entire volume was written before eight o'clock in the morning and on weekends and holidays, my wife Pat deserves special thanks for her gracious acceptance of my peculiar work and sleeping habits.

Finally, the authors of the articles from which the material was drawn warrant both my appreciation and that of the reader because without their individual efforts even this first attempt to collect the knowledge of the noninfectious diseases of invertebrates would have been impossible.

CHAPTER I

Death and Postmortem Change

Death in the invertebrates above the Protozoa is difficult to define or determine since somatic death, or death of the entire organism, typically occurs only after necrosis, or cell death, of large areas of the body has taken place. While the same phenomenon is characteristic of the vertebrates (for example, even though nerve cells die within minutes after termination of blood circulation, cartilage cells live for several days), death of the individual ensues within a very few minutes after heartbeat stops because anoxia of vital organs such as the central nervous system leads to irreversible changes. Somatic death in invertebrates is a much more gradual process; for example, the cilia on the gills and mantle epithelium of oysters continue to beat for several days subsequent to massive necrosis and autolysis of much of the rest of the body (Sparks and Pauley, 1964).

Since most invertebrates do not have a determinate life-span, natural death or death from old age is probably rare and occurs, as Steinhaus (1964) pointed out for insects, largely in the laboratory or other controlled situations. There is considerable evidence that there is neither senescence nor natural death in protozoans, and the concept of potential immortality has attracted the attention of numerous investigators (Kudo, 1947). Child (1915), however, in his monograph on senescence and rejuvenescence contended that senescence does occur in most invertebrates; the metabolic rate decreases with advancing age. According to

1

Child, "senescence is a characteristic and necessary feature of life and occurs in all organisms"; It may be balanced by rejuvenescence in many lower animals for an indefinite time, but death is inevitable if rejuvenescence does not occur. Orton (1929) noted that it was difficult for the marine naturalist who observes the fluctuations in populations of sponges, hydroids, worms, mollusks, and echinoderms to accept the theories that natural death is not common to all animals. He advanced the theory that death frequently results from overreproduction in lower animals. Senescence and death in invertebrates were reviewed later by Szabó (1935); he concluded that the primary cause of physiological death from senescence may lie in one of several organ systems, varying among different organisms. He also discussed potential immortality and rejuvenescence and the process of death from senescence.

Necrosis

CAUSES OF NECROSIS

Necrosis describes cell death resulting from disease or injury and differs from necrobiosis, which refers to the physiological death of cells, such as epithelial cells that are replaced by new cells in the normal physiological processes of the animals. The numerous agents causing necrosis can be classified under four major categories: anoxia or ischemia, physical agents, chemical agents, and biological agents. The pattern and degree of necrosis produced by any of these types of etiologic agents varies according to the specific agent, its virulence, duration of exposure to injury, and the susceptibility of individual cells to the particular agent. Unfortunately, the effects of these etiologic agents on invertebrate cells have not been studied to the extent that they have in vertebrates. Some generalizations, drawing primarily on vertebrate pathology, can, however, be made.

Anoxia

Cells die when they are deprived of oxygen. Cell death due to ischemia is known as infarction in vertebrates and produces a characteristic histological appearance which is termed coagulation necrosis. A similar histological appearance has been noted in oysters and has been given the same descriptive term (Sparks and Pauley, 1964; Pauley and Sparks, 1965). Ischemia and resulting coagulation necrosis should be common in invertebrates, even though they do not generally affect specific areas (as the case of vertebrate heart muscle infarctions) because of the typically more open circulatory system and pattern of oxygen transport to tissues.

Physical Agents

Trauma, extremes of heat and cold, desiccation, salinity variation, and radiant and electrical energy all cause somatic death and necrosis of invertebrate cells, even though the patterns of necrosis have not been worked out completely. Mechanical destruction of cells by trauma involves the rupture of cell membranes, destruction of the nucleus, and total disruption of normal relationships of cell elements.

Chemical Agents

Chemical agents may be lethal to cells at their point of entrance, where the concentration is greatest, or they may produce their effects only after systemic absorption and transport to a susceptible tissue or organ system. Others cause necrosis at their points of absorption and excretion.

Biological Agents

Biological agents of great variety cause necrosis of invertebrate cells. These organisms may cause cell death by elaboration of toxins, destructive enzymes, physical trauma, or by other poorly understood mechanisms.

MORPHOLOGICAL ALTERATIONS IN NECROSIS

Most of the morphological changes that occur after cell death are produced by enzymes. Since cell death and subsequent enzymatic action ordinarily take place relatively slowly, some time lag is required before recognizable changes become apparent in the appearance of the necrotic cells. If cell death is sudden and is followed by immediate fixation, or if it results from chemical toxins which have fixative properties, no morphological changes occur. Under normal conditions, however, enzymatic action after cell death initiates a sequence of morphological modifications that terminate in complete dissolution and disappearance of the cell. The typical pattern of necrosis in vertebrate cells is well understood, as are the distinctive features that characterize special types of necrosis, but these processes have been studied to only a limited extent in invertebrates. Therefore, it is not certain that the morphological changes typical of vertebrate necrosis occur in most invertebrates or that they follow the same sequence. It seems appropriate, however, to describe the typical pattern of cellular modification in vertebrate necrosis and then compare what is known of morphological changes in dying cells of invertebrates to the vertebrate pattern.

The most prominent histological changes in vertebrate necrosis occur in the nucleus. Less severe injuries, such as degenerations and infiltrations, that are compatible with survival of the cell tend to affect the cytoplasm. The first recognizable change (pycnosis) in the nucleus of a dying cell is loss of the normal distribution of the chromatin material. The nucleus becomes shrunken, irregular, and wrinkled, and it stains more intensely basophilic. Subsequently, the nuclear membrane may rupture and fragment, releasing the nuclear contents into the cytoplasm, in which case karyorrhexis is said to have occurred; or, following the initial intensified basophilic staining reaction, the nucleus may progressively stain less intensely and eventually lose all its affinity for basophilic dyes until it disappears without rupturing. This process is termed karyolysis. All the nuclear modifications typical of necrosis occur in the cells of dying oysters (Sparks and Pauley, 1964), apparently in the same sequence, and it appears both convenient and technically correct to utilize the terms developed in vertebrate pathology for these phenomena in invertebrates when the processes are similar (if not identical). It may well be that further study of necrosis in invertebrates may reveal variations in the pattern of nuclear changes, but it seems essential that a common language be used if pathology as a science is ever to become truly comparative.

Cytoplasmic changes in necrosis are not so conspicuous as nuclear modifications. Early in the process, the cell may appear somewhat larger than normal, and the cytoplasm may be more granular. Subsequently, the cytoplasm becomes more acidophilic, dense, and opaque, and loses its granularity. Granular coagulation occurs, and fat droplets appear at about the time the nucleus has disappeared by karyorrhexis or karyolysis. Indistinct outlines of the cell membranes may persist for a considerable period of time, but the clarity is lost as they fuse with the opaque cytoplasm to form an amorphous, anuclear mass which eventually disappears through enzymatic dissolution.

The morphological modifications described above represent the typical pattern in vertebrate necrosis and probably represent the usual pattern in invertebrates, but this may vary depending on the characteristics of the agent causing necrosis and the tissue affected. A number of the variations have been given specific descriptive names in vertebrate pathology, and several appear to occur under similar conditions in invertebrates.

Coagulation necrosis occurs when the protoplasm of the dead cell becomes fixed and opaque by coagulation of its protein elements and is recognizable by persistence of the cellular outline for considerable periods of time. Histologically, the coagulated cell mass persists after intracellular detail is lost. It is most commonly observed in vertebrates as a result of ischemia, with the infarcted area sharply demarcated from adjacent normal tissue. Microscopically, the cytoplasm is dense, granular, and opaque, and the nucleus is pycnotic. The nucleus then disappears by karyolysis or karyorrhexis, and the cellular outlines and cytoplasm persist for some time. A number of bacterial toxins also cause coagulation necrosis in vertebrates and presumably in invertebrates, although it has not been described in the latter. Coagulation necrosis has been described in the normal postmortem decomposition of oysters (Sparks and Pauley, 1964), probably as a result of ischemia, and in the digestive epithelium (Pauley and Sparks, 1965) after injection of turpentine into the body stroma. The phenomenon appears to be typical of invertebrate necrosis when the oxygen supply to an organ is terminated.

Liquefactive necrosis is a process of relatively rapid and complete enzymatic dissolution of an entire cell or all of the cells of a tissue. It is, in vertebrates, most common in the brain, where the response to ischemia is rapid, enzymatic autodigestion and liquefaction of the tissues. It is also typical of bacterial infections in which pus is formed. Necrotic tissues similar in histological appearance to liquefactive necrosis are often seen in invertebrates, particularly mollusks, but the term has not been adopted as common usage in invertebrate pathology. It is probably a common phenomenon in tissues with high enzymatic activity.

There are a number of other special types of necrosis in vertebrates, such as caseous necrosis typical of tuberculosis, in which the dead cells are converted into a granular, friable mass of amorphous fat and protein with a total loss of cellular detail grossly resembling soft cheese. Another example is gangrenous necrosis caused by a combination of ischemia and secondary bacterial infection which presents a histological pattern combining coagulation and liquefactive necrosis. Application of these terms to invertebrate necrosis must, however, await further investigation.

Somatic Death

Somatic death is the death of the organism. In the case of protozoans, of course, somatic death and necrosis are synonymous, but in higher animals somatic death occurs only after the necrosis of vital tissues has taken place. In forensic medicine there is a legal definition of somatic death in man, i.e., when heartbeat ceases. This definition is of little value in most invertebrates, many of whom do not possess a heart. It is impossible to establish an exact point at which somatic death has occurred in an invertebrate, but it may be defined as the stage of physiological activity in which necrosis of vital tissues has reached an irreversible point so that repair is no longer possible. Child (1915) noted that, "The individual dies when some tissue or organ which is essential for

its continued existence reaches the point of death, and since the parts
are incapable of dedifferentiation and a new individuation, the other
organs or cells die sooner or later because of lack of nutrition or oxygen,
or because of the toxic products of metabolism."

Gross Postmortem Changes

A sequence of grossly observable events follows somatic death in the
vertebrates. The first of these, in the homeothermic birds and mammals,
is *algor mortis*, or cooling of the body, which takes place at a regular
rate that depends on the ambient temperature. This is followed by
rigor mortis, or rigidity of skeletal muscles, beginning at 6–10 hours
postmortem and lasting for 3–4 days in man. The blood slowly flows by
gravity to the dependent areas of the body, causing a reddish-blue discoloration termed dependent lividity or *rubor mortis*. The entire sequence of events does not occur in the invertebrates. Since the invertebrates are poikilothermic, *algor mortis* is omitted. It is possible that
rigor mortis or a similar phenomenon occurs in some invertebrates with
well-organized skeletal systems and associated skeletal muscles, but no
reports of the gross sequence of events following death are available
to verify the condition. Many of the arthropods, which are flaccid after
somatic death, subsequently become rigid, but whether this is due to
desiccation of the tissues or actual *rigor* of the muscles is not known.
Current research on juvenile penaeid shrimp strongly indicates that
rigor mortis, in the vertebrate sense, does occur in arthropods. Sacrificed
juvenile shrimp placed in jars containing water-saturated air become
rigid. The time of onset of rigidity is, as in vertebrates, temperature dependent, occurring earlier at higher temperatures and progressively delayed at several lower temperatures (D. V. Lightner and C. T. Fontaine,
personal communication). Soft bodied invertebrates, such as the mollusks, remain flaccid after death, and the gross appearance from somatic
death to virtual dissolution as described in a species of oyster (Sparks
and Pauley, 1964) is probably typical.

In the above study, several large groups of Pacific oysters (*Crassostrea
gigas*) were sacrificed by severing the heart; they were observed grossly
and studied histologically over a period of 144 hours at a temperature
of 14°–16°C. Sacrificed oysters are normal in gross appearance for the
first 32 hours; then abnormally large amounts of mucus and silt begin
to accumulate on the surface of the mantle. An odor of decomposition
is first noticeable at 48 hours and becomes more pronounced at 112
hours. Tactile response is normal for the first 48 hours, then gradually

becomes weaker and ceases by 104 hours after sacrifice. During this time, the villi along the margins of the mantle gradually become more contracted and are invariably contracted after 96 hours. The texture of the body gradually changes in appearance after sacrifice, appearing normal through 88 hours but becoming more watery after 96 hours. After 128 hours most oysters are so decomposed that individual organs of the body can no longer be distinguished grossly, and much of the body has often disintegrated by 144 hours. The adductor muscle becomes detached from the shell between 136 and 144 hours.

Child (1915) discussed the problem of death determination and postmortem changes in invertebrates, noting that in the lower forms (protozoans, coelenterates, and flatworms) with which he had worked death is followed by a rapid, often almost instantaneous, complete disintegration. "The body loses its form, swells, breaks down into a shapeless mass, and may finally disappear completely, except for a slight turbidity in the water which results from the minute particles in suspension." When rapid disintegration does not occur, he pointed out that the occurrence of death of small animals can often be determined under the microscope by changes in the animal's appearance, e.g., becoming opaque or changing color. Child also determined the limits of recovery times of animals subjected to narcotic solutions that were then removed; he noted that the time at which recovery ceases is a least approximately the time of death.

Cameron (1932) listed the signs of approaching death in earthworms as immobility, with slight response to tactile stimulus; abnormal body attitude, lying on lateral or dorsal surface rather than ventral; atonicity, the body being flabby or dry; localized constriction and dilation of segments; autotomy of terminal segments, and evagination of the pharynx and anus. Maceration and putrefaction, according to Cameron, are rapidly initiated once death has occurred.

The small, free-swimming tunicates of the class Copelata, more generally known as the Appendicularia, are extremely delicate and soon turn opaque and die when kept in the laboratory or when subjected to sudden environmental changes. The process of progressive disintegration and death in the genus *Oikopleura* was described by Essenberg (1926). As is typical in invertebrates, death is not a sudden process. The animal becomes noticeably less active, the beating of the flagella around the mouth gradually decreases in rate until it ceases entirely, while other organs appear to function normally. Beginning with the anterior end, necrosis progressively advances posteriorly.

Although the body may remain intact and retain its original shape for some time with only the posterior part of the trunk viable, the body often disintegrates as necrosis proceeds posteriorly. The body wall near the mouth and lips is lost first, but the endostyle is still functioning and the branchial cilia are beating vigorously. The endostyle disintegrates next, the branchial cilia cease to function, and the branchial openings and ducts disappear. Organs and tissues continue to be lost until only the esophagus, stomach, and tail remain. The cilia of the esophagus retain their function, conveying food to the stomach where digestion continues, and the remaining digestive system and tail continue to move vigorously about. Eventually, in 2–4 hours, the animal gradually slows down and dies.

Histological Postmortem Changes

After somatic death, or often preceding it as described above, autolysis of the tissues begins, terminating in complete dissolution of the organism. Knowledge of the rate and pattern of postmortem decomposition in any invertebrate is necessary to differentiate antemortem effects of pathological agents from the normal process of autolysis. The rate at which autolysis proceeds depends on a number of factors including, but probably not confined to, the ambient temperature, dissolved oxygen concentration if it is in a liquid medium, bacterial flora of the medium and the dead animal, the histological composition and physiological properties of the dead or dying organism.

Despite similar variations, it is well known that in vertebrates, particularly in man, postmortem changes occur in a regular and irreversible pattern and at a relatively constant rate from one individual to another when known factors causing variations in the rate and pattern are considered.

Although several workers, cited above, have briefly described the gross pattern of postmortem decomposition of dead and dying invertebrates, a detailed investigation of the normal postmortem changes has been made of only one species, the Pacific oyster, *Crassostrea gigas* (Sparks and Pauley, 1964). Since that study presents the bulk of the information available on this phenomenon in the invertebrates it will be discussed in some detail in the hope that the pattern of postmortem changes in the oyster is typical of related invertebrates and that it will encourage investigators to study the phenomenon in other invertebrates.

In attempting to ascertain the sequence of normal postmortem changes, it is essential for the animal to die with a minimum of histological involvement. Chemicals cannot be used because they might "fix" or affect tissues prior to somatic death. Numerous methods of sacrificing the oysters in the study were considered before eventually deciding to surgically sever the heart. It was realized that such treatment was not, technically, normal death, but it was assumed that subsequent post-mortem changes after such deaths would be more nearly normal than from any other means considered since the following advantages were obtained.

1. Death time, or at least the time at which the fatal damage occurred, was accurately known.
2. Tissue distant from the source was not damaged.
3. There was little possibility of tissue repair and survival after such radical treatment.

RATES AND PATTERN OF POSTMORTEM CHANGES IN OYSTERS

Digestive Tubules

The first tissues to be affected in postmortem autolysis of the oyster are in the area of the digestive tubules (Fig. I.1). At about 24 hours postmortem the nuclei of the Leydig cells between the inner tubules become necrotic; the nuclei disappear and the cytoplasm loses its affinity for stains. The tubule epithelium progressively fragments and is sloughed from the basement membrane. Eventually, between 96 and 144 hours postmortem, the basement membranes are lost, leaving only scattered, fragmenting, epithelial linings of tubules (Figs. I.2, I.3, and I.4).

Under unfavorable environmental conditions or with poor fixation, the changes described begin in what are considered to be normal oysters. Autolysis may proceed as far as karyolysis of the Leydig cells, sloughing of the epithelial lining, and initial fragmentation of the epithelium of the deeper tubules (i.e., to approximately the condition of 72 hours normal postmortem). The digestive tubules have been found to be the most sensitive indicator of unfavorable environmental conditions, poor fixation, or postmortem change.

Intestine

The gut appears normal (Fig. I.5) through at least 40 hours, but at 48 hours, postmortem changes are detectable. The epithelium stains densely basophilic, the nuclei become pycnotic with crumpling and con-centration of the chromatin (Fig. I.6), and the epithelium begins to separate from the basement membrane (Fig. I.7) by 64 hours post-mortem.

At about 88 to 104 hours postmortem (Fig. I.8), the gut wall exhibits coagulation necrosis. The epithelium begins to lose its basophilic stain-ing reaction, and stains a light pink with the nuclei becoming karyolytic.

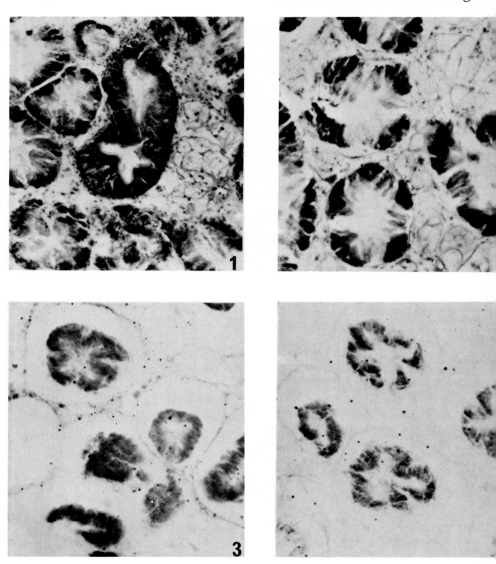

Fig. I.1. Normal digestive tubules and Leydig cells. 366×. (From Sparks and Pauley, 1964.)

Fig. I.2. Digestive tubules and Leydig tissues 48 hours postmortem. 366×. Note fragmenta and sloughing of epithelium and karyolysis of nuclei in Leydig cells. (From Sparks and Pau 1964.)

Fig. I.3. Digestive tubules and Leydig tissues 104 hours postmortem. 366×. Note comp separation of epithelium from basement membrane and complete loss of nuclei in Leydig c (From Sparks and Pauley, 1964.)

Fig. I.4. Digestive tubules and Leydig tissues 128 hours postmortem. 366×. Note fadin epithelium and complete autolysis of Leydig tissue. (From Sparks and Pauley, 1964.)

The cell membranes are lost. The cells fragment, and the secretory granules become scattered and stain less intensely. The greater curvature of the gut seems to exhibit these features prior to the lesser curvature and typhlosole. Bacteria often become conspicuous in the lumen and along the edge of the epithelium. In most instances the cilia begin to slough. At about this time, the gut becomes greatly distended; there is a marked leukocytic infiltration just beneath the basement membrane, and the underlying Leydig cells undergo karyolysis.

The emigration of leukocytes into the area beneath the gut and stomach as the Leydig cells undergo necrosis is particularly conspicuous and is apparently diagnostic of the postmortem breakdown of the digestive system.

It is possible that the gut is a better tissue for determination of time of death of the oyster than the tubule area because it is less sensitive to poor environmental conditions, and the onset and progression of decomposition appear to be more consistent with the time elapsed after death.

Stomach

Postmortem change in the stomach follows the same pattern as the gut, but is usually considerably delayed. Beginning at 64 hours, the epithelium progressively becomes faded, loses its basophilic staining properties, fragments, and, in most cases, pulls away from the basement membrane. There is often a leukocytic infiltration beneath the basement membrane, and the underlying Leydig cells undergo karyolysis. Beginning at approximately 112 hours, portions of the epithelium disappear, but there is a great deal of variation in the time at which this occurs. The epithelium exhibits coagulation necrosis, though very lightly stained, through 144 hours. The Leydig cells beneath the gut begin to lose their nuclei and cytoplasm at about 64 hours postmortem, and this necrosis spreads outward as decomposition progresses.

Mantle

Understanding of the postmortem changes in the mantle is complicated by the changes in the mantle epithelium and connective tissue associated with the gonadal cycle of the oyster (Fig. I.9). The epithelium becomes pycnotic at about 48 hours; then it begins to lose its affinity for stains (Fig. I.10). Extensive sloughing of epithelium from the surface of the mantle does not ordinarily occur before 96 hours and may be delayed for a slightly longer period.

The mantle Leydig area remains relatively normal for approximately 24 hours and then becomes vacuolated by focal necrosis of Leydig cells,

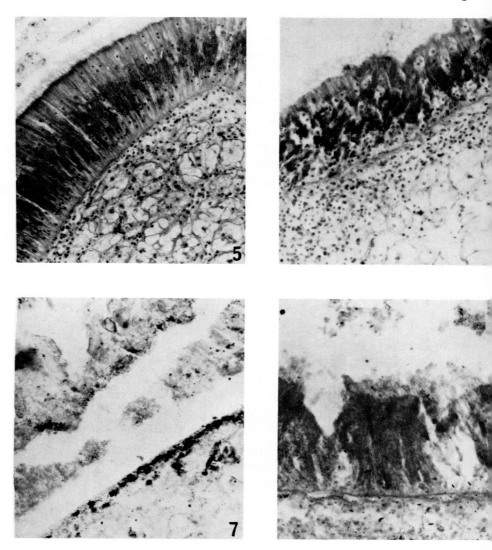

Fig. I.5. Normal gut. 366×. (From Sparks and Pauley, 1964.)

Fig. I.6. Gut 56 hours postmortem. 366×. Note disorganization of epithelium and leuko infiltration beneath basement membrane. (From Sparks and Pauley, 1964.)

Fig. I.7. Gut 104 hours postmortem. 366×. Note separation of epithelium from basement m brane. (From Sparks and Pauley, 1964.)

Fig. I.8. Gut 88 hours postmortem. 366×. Note coagulation necrosis of the epithelium. (F Sparks and Pauley, 1964.)

accompanied by a general leukocyte infiltration. The Leydig cell areas begin to fade at about 72 hours and lose their nuclei and cytoplasm in a progressive pattern from the periphery of the mantle inward (Fig. I.10). There is considerable variation in the time of onset of the general necrosis, with most of the mantle remaining fairly normal in appearance for more than 100 hours.

Beginning usually at 72 hours, numerous multinucleate cells appear in the mantle Leydig areas of most dead oysters, particularly near the periphery (Figs. I.10, I.11, and I.12). They are considerably larger than the typical leukocytes and contain as many as a dozen or more deeply staining nuclei surrounded by faint cell outlines. These "giant cells" are almost certainly uninucleate leukocytes which have engulfed large numbers of necrotic leukocytes and, perhaps, nuclei and cell debris of Leydig cells. The necrotic condition of the phagocytized cells causes them to appear unlike oyster cells. Further indication that these cells are products of decomposition of the oyster is provided by the occasional presence of numerous phagocytized sperm in giant cells from male oysters undergoing postmortem decomposition (Figs. I.11 and I.12).

Gills and Palps

The gills and palps are relatively resistant to postmortem chánge, with no discernible changes occurring until 72 hours postmortem. At this time the epithelium stains much more basophilic and, by 72 to 96 hours, becomes pycnotic. By 96 hours, the epithelium begins to slough, the Leydig cell tissue begins to break down and often contains numerous giant cells. Usually by 120 hours, the epithelium stains less intensely, and by 136 hours postmortem the gills may be fragmented.

Gonadal Region

Often the ova or sperm appear relatively normal (Figs. I.13 and I.15) after all other tissues have undergone extensive autolysis. Also, differences in the gonad because of the normal gonadal cycle must be considered in evaluating the amount of change due to decomposition, particularly in regard to leukocytic infiltration and phagocytosis of the sex products.

The gonads of oysters appear normal through 48 hours postmortem, then, between 56 and 72 hours, the gonadal ducts become slightly pycnotic. Because of phagocytosis of sex products in resorption and the apparent lack of a consistent pattern in the onset and amount of leukocytic infiltration in the gonadal ducts, apparently no significance in postmortem change can be attributed to this condition. Cilia of the gonadal duct epithelium begin to slough between 72 and 88 hours postmortem.

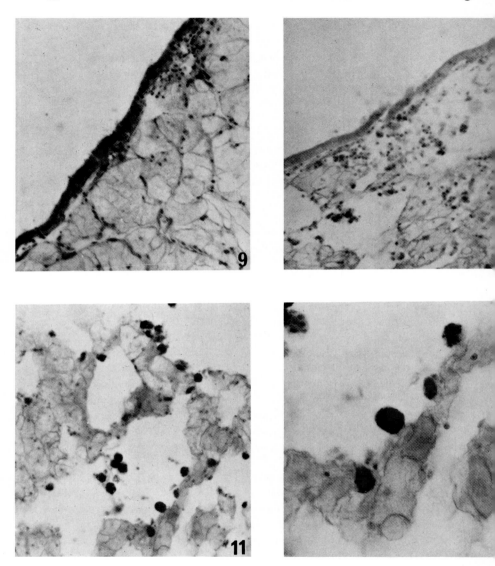

Fig. I.9. Normal mantle epithelium and connective tissue. 366×. (From Sparks and Pa⋯ 1964.)

Fig. I.10. Mantle area 136 hours postmortem. 366×. Note faded epithelium, vacuolated ar⋯ connective tissue, and multinucleate macrophages. (From Sparks and Pauley, 1964.)

Fig. I.11. Leydig cells of mantle area 40 hours postmortem. 366×. Note numerous m⋯ nucleate macrophages, many of which contain phagocytized sperm. (From Sparks and Pa⋯ 1964.)

Fig. I.12. Higher magnification of multinucleate macrophages in Fig. I.11. 915×. (From S⋯ and Pauley, 1964.)

The epithelium itself begins to slough from the basement membrane and starts to lose its staining intensity. The Leydig cells between gonadal follicles are unaffected for a considerable period, but typical karyolysis of these cells is initiated at approximately 112 hours. The sex products themselves are extremely resistant to autolysis, with the first signs of necrosis appearing at 88 hours postmortem. Generally, however, ova (Fig. I.14) and sperm (Fig. I.16) are undergoing phagocytosis, fragmentation, loss of nuclei, and scattering through the surrounding tissues after 96 hours, though normal-appearing gonads have been observed as late as 136 hours postmortem. As breakdown of the gonads proceeds, large concentrations of bacteria typically develop in the area.

Depression and Cellular Degeneration in Hydroids

A phenomenon related to death and postmortem change occurs in freshwater *Hydra* and some marine hydroids. This condition, termed "depression," involves physiological cellular deterioration in response to adverse environmental conditions, and it has been experimentally induced by exposure to toxic substances, fouled media, extremes of pH and temperature, and crowding. It has also been shown to occur occasionally in apparently normal cultures (Chang *et al.*, 1952). The gross and histological aspects of depression were summarized by Zeikus and Steinhaus (1966) from the work of Chang *et al.* (1952) and Rhem (1925). Grossly, when the animal enters into depression, the body and tentacles contract; the animal loses its capacity to react to stimuli, ceases feeding activity, and finally disintegrates. The process may be gradual and may take up to 2 weeks, in which case it can be reversed if the conditions responsible for its initiation are corrected; or the entire process may occur within 24 hours. Rhem (1925), in extensive studies of the histological characteristics of depression, demonstrated that the deterioration begins with the tentacles and proceeds aborally and from the interior outward. The endodermal cells in the region of the proboscis swell, lose their cohesiveness, and pass into the enteron; this is followed by dissolution of the mesoglea; eventually the ectodermal cells disintegrate in the same pattern as the endoderm.

Zeikus and Steinhaus (1966) reported a different type of noninduced, systematic, cellular degeneration in apparently normal cultures of *Hydra;* this degeneration appears to be unrelated to environmental conditions in the culture media since experimental efforts to induce the condition were either ineffective or initiated typical depression. The degenerating animal remains extended and active, is responsive to tactile stimuli,

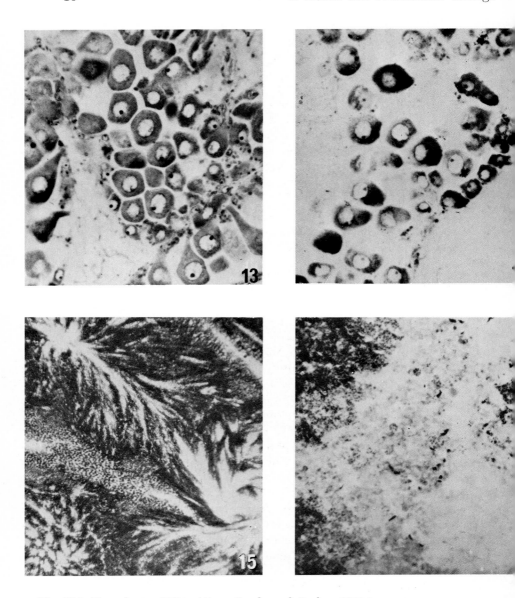

Fig. I.13. Normal ova. 366×. (From Sparks and Pauley, 1964.)

Fig. I.14. Ova 128 hours postmortem. 366×. Note relatively little decomposition of eggs, w
Leydig tissue has undergone almost complete autolysis. (From Sparks and Pauley, 1964.)

Fig. I.15. Normal sperm. 366×. (From Sparks and Pauley, 1964.)

Fig. I.16. Sperm 144 hours postmortem. 366×. Note karyolysis of sperm and large num
of leukocytes in this area. (From Sparks and Pauley, 1964.)

paralyzes and ingests prey, but continually sheds cells. The tentacles, in contrast to depression, remain extended until the final stages when the animal is reduced to 1 mm or less in length. The onset of degeneration is characterized by a coarsely granular appearance of the entire body and cessation of budding; previously formed buds may still be present, and they also exhibit granularity. The above authors termed this condition Stage I of the degenerative process (Table I.1). Subsequently, within 3 to 4 days, any buds present detach and disintegrate, and no new ones are formed. During the following 4 or 5 days, degeneration becomes more obvious, with groups of cells fragments appearing around the animals and an opaque exudate, consisting of bacteria and cell debris, occurring at the point of the animal's attachment to the substrate. Commonly, new tentacles develop along the length of the body. This condition is denoted as Stage II. The degenerating *Hydra* continues to shed cells until it consists of a minute body with long tentacles (Stages III and IV).

TABLE I.1

Ultrastructural Changes Associated with Degeneration[a]

Stage	Days after onset	Ectoderm	Mesoglea	Endoderm
Stage I (granularity; cessation of building)	1	Normal	Normal	General light cytoplasmic swelling; clumped nucleoplasm
Stage II (cell shedding; growth of new tentacles)	3–4	Light cytoplasmic swelling; clumped nucleoplasm; increased number of interstitial cells; cnidoblasts developing at all levels	Fibrous background diminished	Cristae mitochondriales distorted; vesiculation of many cell components; dissolution of cell membrane; appearance of myelinlike figures
Stage III (reduction in size due to cell shedding)	6–10	Same as Stage II endoderm; also vesiculation of endoplasmic reticulum	Cell fragments present	Mitochondria vesiculated; presence of extensively winding membranes
Stage IV (minute body with long tentacles)	10–14	Membranous components replace cytoplasm	Same as Stage III	Cytoplasm replaced by whorls of membranes; no normal structures

[a] From Zeikus and Steinhaus (1966).

Zeikus and Steinhaus studied the ultrastructural modifications associated with cellular degeneration. The first and most dramatic degenerative changes take place in the gastroderm; the ectoderm remains normal in appearance until the entire gastrodermal epithelium is markedly distorted. The onset of the disintegrative process is characterized by general cytoplasmic swelling. As the *Hydra* begins to shed cells, the ectoderm and mesoglea are affected in apparently the same fashion as the endoderm in the previous stage. There is an apparent increase in the number of interstitial cells, and they become markedly swollen. The mesoglea loses much of its fibrillar appearance and staining affinities. Cytoplasmic abnormalities become more pronounced with vesiculation of cell organelles and the appearance of winding membrane components characteristic of this second stage of the degenerative process in the endoderm.

During the third stage, the animal becomes greatly reduced in size because of the shedding of cells. The ectoderm again exhibits changes occurring in the endoderm in the previous stage. Most of the ectoderm becomes conspicuously swollen, with empty vesicles and vacuoles filling the cytoplasm of many cells. The increased extracellular spaces are filled with granules, presumably of a secretory nature, and the surface membrane is represented by only a few short strands above the epitheliomuscular cells. Much of the gastrodermal cytoplasm at this stage has been replaced by myelinlike figures, and the nuclear membranes are markedly distended. Cell borders have largely disappeared and much of the cytoplasmic contents is unidentifiable.

During the fourth and final stage of cellular degeneration, the epithelia of both the gastroderm and ectoderm exhibit severe degenerative changes. Markedly distorted short strands of cristae mitochondriales are the only identifiable organelles, and myelinlike figures are common in both the endoderm and ectoderm.

At 10–14 days after onset of degeneration, only a small remnant of the original *Hydra* remains. Histological examination of a cross section of the body at this time reveals an accumulation of large vesicles, granules of assorted shapes and sizes, and cellular debris in the lumen of the gut and attached to the outer surface. The mesoglea has largely disappeared and, when present, is represented by only a narrow layer, containing tissue debris, between the two epithelia.

The relationships between depression and cellular degeneration and the relationship of both to death and postmortem changes in *Hydra* remain unclear. As Zeikus and Steinhaus pointed out, Ito (1962a,b) demonstrated that vesiculation of swollen cytoplasmic components is the first stage of postmortem change in mammals. Both Ito (1962,a,b) and Freeman (1964) stated that the presence of myelin figures in the cytoplasm is evidence of degenerating membraneous components, except, presumably, tissues such as the nervous system where myelin figures are a normal component of the cell. Therefore, cellular degeneration in *Hydra* seems to fit, at least in part, some of the criteria of postmortem change, even though the animal undergoing the process continues to move about, feed, and respond to tactile stimuli.

Because of the gradual onset of somatic death and the difficulty of defining the exact point at which it occurs in most invertebrates, it is appropriate to consider depression and cellular degeneration as closely related to postmortem changes even though the animal is alive, and in the case of cellular degeneration functional through most of the process.

References

Cameron, G. R. (1932). Inflammation in earthworms. *J. Pathol. Bacteriol.* **35**, 933–972.

Chang, J. T., Hsieh, H. H., and Liu, D. D. (1952). Observations on *Hydra* with special reference to abnormal forms and bud formation. *Physiol. Biol.* **25**, 1–10.

Child, C. M. (1915). "Senescence and Rejuvenescence." Univ. of Chicago Press, Chicago, Illinois.

Essenberg, C. E. (1926). Observations on gradual disintegration and death of Copelata. *Univ. Calif., Berkeley, Publ. Zool.* **28**, 523–525.

Freeman, J. A. (1964). "Cellular Fine Structure," Chapter 11, p. 75. McGraw-Hill, New York.

Ito, S. (1962a). Post-mortem changes in the plasma membrane. *Electron Microsc., Proc. Int. Congr., 5th, 1962* Vol. 2, pp. 1–5.

Ito, S. (1962b). Light and electron microscopic study of membraneous cytoplasmic organelles. *Symp. Int. Soc. Cell Biol.* **1**, 129–147.

Kudo, R. R. (1947). "Protozoology," 3rd ed. Thomas, Springfield, Illinois.

Orton, J. H. (1929). Reproduction and death in invertebrates and fishes. *Nature (London)* **123**, 14–15.

Pauley, G. B., and Sparks, A. K. (1965). Preliminary observations on the acute inflammatory reaction in the Pacific oyster, *Crassostrea gigas* (Thunberg). *J. Invertebr. Pathol.* **7**, 248–256.

Rhem, W. (1925). Über Depression und Reduktion bei *Hydra*. *Z. Morphol. Okol. Tiere* **3**, 358–388.

Sparks, A. K., and Pauley, G. B. (1964). Studies of the normal post-mortem changes in the oyster, *Crassostrea gigas* (Thunberg). *J. Insect Pathol.* **6**, 78–101.

Steinhaus, E. A. (1964). When an insect dies. *Entomol. Soc. Amer.* **10**, 183–189.

Szabó, I. (1935). Senescence and death in invertebrate animals. *Riv. Biol. Firenze* **19**, 377–436.

Zeikus, R. D., and Steinhaus, E. A. (1966). Observations on a previously undescribed type of cellular degeneration in *Hydra*. *J. Invertebr. Pathol.* **8**, 14–34.

CHAPTER II

Reaction to Injury and Wound Repair in Invertebrates

When cells are injured or destroyed, an immediate protective response, termed inflammation, occurs in the tissues of vertebrates and in many invertebrates. The inflammatory response destroys, dilutes, or walls off the injurious agent and the damaged or dead cells. The pattern of response is basically similar, regardless of the nature of the injurious agent, the site of the injury, or the taxonomic classification of the injured organism.

The successful conclusion of the inflammatory response is wound repair, with the dead or damaged cells being replaced by healthy cells. Major features of the reparative process are the removal of the inflammatory exudate, regeneration of parenchymal cells when possible (resulting in an almost perfect reconstruction of the original architecture), and the formation of scar tissue by progressive proliferation of fibroblasts and deposition of collagen.

Although much of the early work on inflammation by Metchnikoff (1893) utilized invertebrates, subsequent use of these forms for experimental study has been relatively superficial. Vertebrate inflammation will be discussed first, since this process is so well documented. Inflammation in invertebrates will then be described, and the two phenomena will be compared.

Injury of vertebrate tissues or cells results in the release of humoral substances, not yet completely understood, that initiate the inflammatory response. Almost immediately, the small blood vessels near the injury contract and then dilate. Vasodilation is accompanied by increased capillary permeability, allowing leukocytes and plasma to escape into the injured area. Erythrocytes become concentrated in the capillaries because of the loss of fluid through transudation. The loss of fluid and packing of erythrocytes increase the viscosity of the blood in the capillaries, resulting in reduced blood flow (stasis). At the same time, the white cells orient along the periphery of the capillary wall (pavementing) in preparation for their emigration into the injured area.

One of the cardinal signs of inflammation is loss of function of the injured tissue. The other cardinal signs are redness, swelling, heat, and pain (the rubor, tumor, calor, and dolor of Celsus). Redness is an early sign of inflammation, and results from accumulation and stasis of erythrocytes in the area of injury. Since invertebrates lack erythrocytes, this sign of inflammation does not occur.

In the vertebrates, swelling is grossly seen as a soft, pliable, tumorous mass; it is seen microscopically as an edematous area, heavily congested with exudate. Although tumorous swellings have been recorded in only a few of the experiments investigating invertebrate inflammation, the histological features in many are quite similar to those of the vertebrates, except for the obvious lack of erythrocyte exudation.

Heat is characteristic of a surface wound in homeothermic vertebrates (Florey, 1958). Since the surface of the body usually has a lower temperature than the interior, blood moving from the interior to a superficial location tends to elevate the temperature over adjacent areas. True inflammation, according to Metchnikoff (1893), occurs without heat in poikilothermic vertebrates, and, of course, no heat is produced by this process in invertebrates.

Pain is most likely initiated by increased pressure on nerve endings caused by the swelling. Since the tissues of many invertebrates do become swollen in the inflammatory process, pain is a possibility after injury. However, since most invertebrates lack a complex central nervous system, it is doubtful that many invertebrates feel pain in the same sense as do the vertebrates.

In vertebrates the ability of an organ or tissue to replace injured or dead cells (wound repair) most often results in fibroblast and collagen infiltration and the subsequent formation of a scar. Insects (Schlumberger, 1952) and numerous other invertebrates respond similarly. On a comparative basis, the processes that occur after injury in many invertebrates are remarkably comparable to vertebrate phenomena, both histologically and sequentially.

Phylum Protozoa

In vertebrates and the higher invertebrates a traumatic lesion invariably invokes an inflammatory response. Metchnikoff and other early workers noted, however, that the results were much simpler in unicellular organisms. When an amoeba is bisected, there is no wound formed along the line of section; the edges unite immediately after passage of the instrument. Two new amoebae are formed; the one containing the nucleus continues to function in a normal fashion, while the other, lacking a nucleus, subsequently dies. Anucleate portions of pseudopods of the shelled amoeboid protozoan *Arcella* that are excised and separated from the mother cell by distances up to 500 μm are reincorporated into the cell. The fragments and the mother cell move toward each other and fuse. This fusion is genuine because contact elicits a "contact shock." This does not occur during feeding; the excised portions flow back into the mother cytoplasm without being encapsulated in a food vacuole (Kepner and Reynolds, 1923; Reynolds, 1924). Fragments will also fuse with pseudopodia of sister cells, but never with cells of another species. Growth of the same race in different media causes incompatibility, resulting in a shattering reaction rather than fusion on contact. Compatibility can be regained by growth in the same media. The heliozoan, *Actenosphaerium*, when squashed under a cover glass, can reconstitute itself within 24 minutes by drawing together and fusing a multitude of dissociated fragments (Kuhl, 1954).

When ciliates (which possess more highly differentiated protoplasm than do the Sarcodina) are artificially bisected, a wound results that destroys the pellicle and bares the inner layer of protoplasm. In a short time the edges grow over the wound and secrete a new pellicle, thus, in the words of Metchnikoff, "securing complete cicatrization." The initial reaction to bisection appears to be identical in both nucleated and enucleated portions, although Balbiani (1888) believed that healing is never completed in the absence of a nucleus which, he felt, exercises a decided influence on secretion of the cuticle. In the wounded ciliate, the portion containing the nucleus is completely repaired within 24 hours, while the portion lacking a nucleus gradually atrophies and dies. In some species, such as the holotrich *Trachelius ovum*, wounds caused by section are immediately covered over by ectoplasm, and the fragments containing a nucleus are completely regenerated in less than 5 hours.

The electron microscope has made possible the ultrastructural study of reaction to injury and wound repair in ciliates. Removal of a portion of the pellicle of *Paramecium caudatum* results in the loss of a small amount of cytoplasm which remains connected to the wound (Janisch, 1964). At 2–5 minutes postinjury, this necrotic extruded cytoplasm is covered by a membrane (probably the precipitation membrane), is coarsely and uniformly granulated, but contains none of the organelles present in normal cytoplasm.

A distinct boundary, consisting of a layer of fine, granulated osmiophilic material, forms between the necrotic and viable cytoplasm within 30 minutes. At 24 hours postinjury, the pellicle covering the wound is smooth and lacks the complicated structure of the normal surface.

The mitochondria in the vicinity of the wound are swollen and the macronucleus is more lobular than in an unwounded *Paramecium*, but Janisch was uncertain as to whether these changes were directly related to the injury or were secondary results of altered physical and chemical conditions after removal of a portion of the pellicle. Subsequently, Janisch (1966) showed that a traumatized *P. caudatum* heals surface injuries by concentrically narrowing the wound and by gradually closing the residual pellicle.

Additional information on the mechanism of wound healing in ciliates has been obtained from Tartar's transplantation studies in *Stentor* recently reviewed in Transplantation Proceedings (1970). If a cell is cut with a glass needle and split open to prevent rapid healing (fusion) of the ectoplasm, the endoplasm is immediately contained within a thin membrane. Tartar points out that this phenomenon may simply be due to the polar organization of cytoplasmic molecules at the endoplasm–water interface. When another cell that is similarly treated is brought into juxtaposition, the tenuous membranes are broken, and the endoplasms fuse, with the ectoplasm joining firmly at the points of contact.

Phylum Mycetozoa

The validity of including the Mycetozoa in a discussion of invertebrate pathology is questionable since there has been considerable debate among biologists as to whether they are plants or animals. The Mycetozoa were long considered to be closely related to the fungi; they were known as Myxomycetes or Myxogasteres, the slime molds. Extensive studies of their development indicate that they are closer to the Protozoa than to the Protophyta, although they are undoubtedly on the borderline between the two groups (Kudo, 1954). Because the plasmodium of Mycetozoa consists of a large syncitial mass of protoplasm that is often a foot or more in diameter, Metchnikoff remarked that the group pre-

sents excellent opportunities for the study of protoplasm in general and
of pathological phenomena in particular. He conducted experiments on
the reaction of *Physarum* to three types of wounds: physical trauma,
heat, and chemical irritants. Insertion of minute glass tubes into the
protoplasm of a *Physarum* results in a puncture wound, allowing some
protoplasmic diffusion into the surrounding media, but the surface
rapidly repairs itself and arrests the loss of protoplasm. The glass tube
is quickly "englobed" as though it were a food particle and is then
ejected in the same manner as any indigestible material. Touching the
central portion of the plasmodium with a small, heated glass rod produces thermal excitation rather than a mechanical lesion. Immediately
upon being touched, the central portion of the plasmodium dies and
may be easily distinguished from the living peripheral portions that are
unaffected by the injury. After remaining motionless for several hours,
the plasmodium "awakes" from its passive condition and moves away,
leaving the necrotic portion behind.

Chemical irritants, such as silver nitrate, invoke rejection of the injured
portion and cause immediate movement away from the source of damage.
In both thermal and chemical injuries, the injured portions and the
foreign material inserted into the protoplasm are rejected. It appears that
the chief reaction to injury in the Mycetozoa is repair of the ectoplasm
and rejection of foreign materials and necrotic protoplasm destroyed by
the irritant.

Phylum Porifera

Reaction to injury, particularly to foreign bodies, in the Protozoa involves the entire mass of protoplasm, with the digestive and excretory
functions playing a prominent part (Metchnikoff, 1893). Defense mechanisms of sponges are somewhat more sophisticated, with certain cells
combining in their activities, although acting independently, to protect
and repair the organism. Although protective functions are concentrated
in the mesenchyme (mesoglea), the endodermal cells play some part in
the process, and the cells of the ectoderm protect the organism by resist-
ing injurious agents.

Defense Mechanisms

If a sharp object (such as a small glass tube or a spicule of asbestos) is
thrust into a sponge, the greater part of the foreign object will lie in the
mesoderm. Usually, numerous amoebocytes accumulate and partially or
completely surround the foreign body, although some materials fail to

elicit this response. When a response occurs, it can be considered to be a primitive inflammatory reaction; amoebocytes infiltrate the area of injury and surround the inserted object. The amoebocytes often become partially fused to form, in the words of Metchnikoff, "small plasmodia." In some members of the phylum, grains of sand and other hard materials are surrounded or encapsulated by masses of spongin, which results in skeletal strengthening. The mesodermal amoebocytes often encapsulate living organisms that have penetrated the mesoderm and eventually digest them. Other organisms have greater resistance to phagocytic activity, and they remain in the sponge, often completely encapsulated, without being altered. Eggs of annelids and crustaceans undergoing apparently normal development may be seen in the mesoderm of *Hiricinia echinata* and *Ceraochalina gibbosa,* even though they are surrounded by masses of amoebocytes forming a follicle around them.

Thus, a foreign body introduced into the parenchyma of a sponge excites the amoebocytes which either engulf it, collect in a large mass around it, or fuse to encapsulate it. If the foreign body is susceptible to digestion, it is soon dissolved; if it is resistant, it remains surrounded by the cells in the mesoderm of the sponge. The latter phenomenon occurs frequently, since the sponge is easily penetrated and provides a favorable abode for many organisms because of the constant stream of water passing through it.

Recently, Cheng and his associates (1968a,b,c) studied the internal defense mechanisms of *Terpios zeteki,* one of the most abundant species of sponges in Hawaii. They began by determining the types and relative abundance of the free cells and agreed with Galtsoff's (1925) and Hyman's (1940) categories of archaeocytes, collencytes, chromatocytes, thesocytes, and scleroblasts (Fig. II.1). These cells can be distinguished, both in histological sections and as free cells, by their dimensions, cytoplasmic inclusions, and other characteristics. In the normal animal, as established by a relatively large number of differential counts, collencytes are most abundant, archaeocytes are the next most common, followed by chromocytes, thesocytes, and scleroblasts, with the latter being found only rarely. An endocytoplasmic symbiotic zooxanthella occurs only in the archaeocytes of *T. zeteki.* Preliminary study (Cheng *et al.,* 1968b) revealed that not only were the collencytes the most abundant of the five cell types, but that they were frequently syncytially arranged *in situ* (Fig. II.2).

Knowing the types and ratios of cells that occur in the parenchyma of *T. zeteki* makes it possible to examine the role of each in phagocytosis and other forms of internal cellular defense mechanisms. An obvious question is whether one or more of these cell types is involved in the

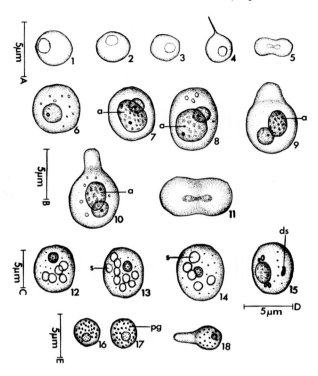

Fig. II.1. Cell types occurring in the parenchyma of the sponge *Terpios zeteki*. Drawn from living material. 1–5, Collencytes—note the fine filamentlike pseudopod in 4 and the dividing collencyte (5); 6–11, archaeocytes—note the symbiotic zo-oxanthella (a) in 7–10, the pseudopods in 9 and 10, and the dividing archaeocyte (11); 12–14, thesocytes—note two types of cytoplasmic inclusions, the larger bodies (s) and extremely small granules; 15, scleroblast—note the intracytoplasmically developing spicule (ds); 16–18, chromocytes—note the pigment granules (pg) in all cells and the pseudopod in 18. (From Cheng *et al.*, 1968b. Courtesy T. C. Cheng.)

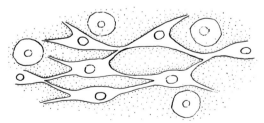

Fig. II.2. Syncytially arranged collencytes in the mesoglea of the sponge *Terpios zeteki*. Drawn from section stained with hematoxylin and eosine. Note isolated collencytes intermingled with syncytially arranged ones. (From Cheng *et al.*, 1968b. Courtesy T. C. Cheng.)

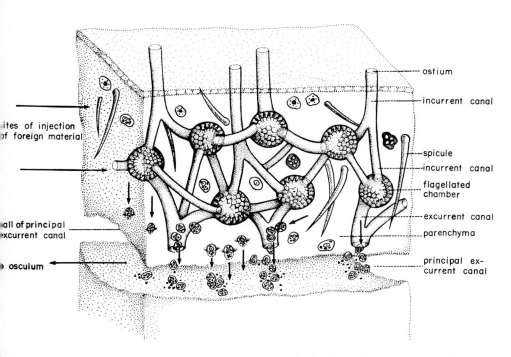

Labels on figure:
ostium
incurrent canal
spicule
incurrent canal
flagellated chamber
excurrent canal
parenchyma
principal ex-current canal

sites of injection of foreign material
all of principal excurrent canal
osculum

Fig. II.3. Diagrammatic representation of a portion of the body wall of *Terpios zeteki* reconstructed from serial sections showing arrangement of incurrent and excurrent canals and flagellated chambers. (From Cheng *et al.*, 1968b. Courtesy T. C. Cheng.)

uptake and elimination of foreign materials, either abiotic or biotic. A second question is whether an increase in the number of one or more cell types occurs in response to the presence of foreign materials. Seawater suspensions of india ink and of carmine were injected into separate series of sponges to answer these questions (Fig. II.3). At six time intervals, the sponges were fixed for sectioning and staining; differential counts of the five types of parenchymal cells were made on an additional series of identically injected sponges at the same time points.

Gross Observations

Upon injection of india ink, the entire subcylindrical trunk of the sponge becomes uniformly black; the coloration extends for approximately 4 cm. Injection of carmine causes a uniformly reddish discoloration of the trunk. In both instances colors are still apparent, although reduced in intensity, after 1 hour. However, by the end of 6 hours only small blotches of black and red are visible to the naked eye, while large amounts of ink and carmine particles collect on the bottom of the bowls in which the sponges are maintained. The discoloration of the sponges diminishes progressively; they return to essentially normal color by 12 hours after injection. At the same time, the amount of ink and carmine particles in the bowls increases as the parenchymal amoebocytes, primarily the archaeocytes, phagocytize the particles (Fig. II.4), migrate into the lumina of the excurrent canals (Figs. II.5 and II.6), and are discharged to the exterior. The collencytes are also involved to a much lesser degree in this activity. With injections of 0.2 ml of 1:10 suspensions of ink and carmine, sponges are essentially cleared of carmine particles by the twenty-fourth hour and of ink particles by the ninety-sixth hour postinjection when maintained at 20°C in seawater with a salinity of 35‰.

Shortly after both ink and carmine particles are injected into the mesoglea, they form extracellular clots (Figs. II.7 and II.8) which are gradually resolved as the particles are phagocytized by archaeocytes and occasionally by collencytes. The authors suggested that some factor or factors in the sponge may be responsible for the clumping that, in turn, serves as a precursor to mass phagocytosis and elimination.

The archaeocytes that harbor zooxanthellae are not actively phagocytic. Interestingly, ink-laden archaeocytes have not been observed to undergo division. Differential counts mentioned previously indicate that, although there were slight fluctuations in the number of each cell type during the course of the experiment (Figs. II.9 and II.10), such increases or decreases are not statistically significant. It appears that even though archaeocytes and occasionally collencytes have been observed to divide their numbers are not significantly increased in response to the introduction of foreign particles. This, of course, may also be interpreted to mean that an increase may occur, but the increased number of cells is offset by the loss of foreign-particle-laden cells, with a dynamic equilibrium established between new and eliminated cells.

In the third study conducted on the internal defense mechanisms of sponges, Cheng et al. (1968c) investigated the cellular reaction of Terpios zeteki to the following implanted heterologous biological materials: human erythrocytes, trematode rediae and cercariae, and molluscan muscle tissue. Human erythrocytes injected or implanted in Terpios zeteki are either phagocytized or encapsulated by archaeocytes (Figs. II.11A and B and II.12A) and are gradually eliminated from the mesoglea via the migration of the host's cells into the excurrent canals (Fig. II.12B). The mesoglea is essentially free of implanted erythrocytes within 48 hours after implantation. Virtually no free erythrocytes occur in the mesoglea even 24 hours postimplantation. Those present are encapsulated, either singly or in small groups, by fused, syncytial archaeo-

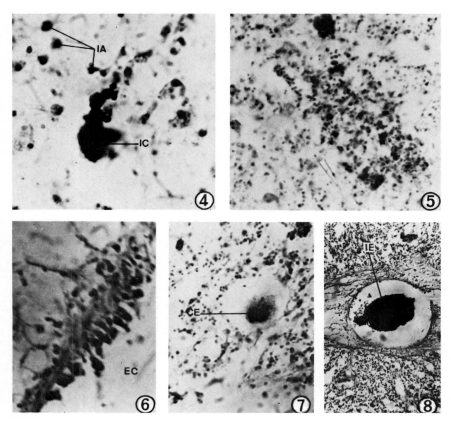

Fig. II.4.* Large aggregation of india ink particles in mesoglea of the sponge *Terpios zeteki*. Note ink-laden archaeocytes in immediate vicinity. One hour post-injection; 90×. (From Cheng *et al.*, 1968a. Courtesy T. C. Cheng.)

Fig. II.5.* Large aggregation of carmine particles in mesoglea of *T. zeteki*. One hour postinjection; 40×. (From Cheng *et al.*, 1968a. Courtesy T. C. Cheng.)

Fig. II.6.* Ink-laden archaeocytes intermingled with pinacocytes lining the wall of an excurrent canal of *T. zeteki*. Note that some archaeocytes have migrated into the lumen of the excurrent canal. Twelve hours postinjection; 90×. (From Cheng *et al.*, 1968a. Courtesy T. C. Cheng.)

Fig. II.7.* Carmine particles in the lumen of an excurrent canal. Twelve hours postinjection; 40×. (From Cheng *et al.*, 1968a. Courtesy T. C. Cheng.)

Fig. II.8.* Large clump of ink particles in lumen of principal excurrent canal. Twenty-four hours postinjection; 10×. (From Cheng *et al.*, 1968a. Courtesy T. C. Cheng.)

* All sections stained with hematoxylin and eosine. CE, carmine particles in excurrent canal; EC, excurrent canal; IA, ink-laden archaeocytes; IC, ink clot; IE, ink particles in excurrent canal.

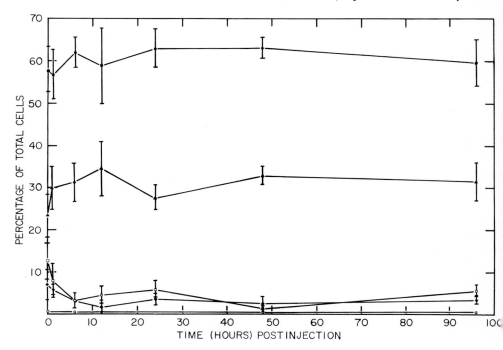

Fig. II.9. Graph showing the percentages of each of the 5 types of parenchymal cells in sponge *Terpios zeteki* at time intervals postinjection of india ink. Each point represents the m of 10 counts and the perpendicular lines represent standard deviations. ●—●, Collency ▲—▲, archaeocytes; ○—○, scleroblasts; ◆—◆, thesocytes; □—□, chromocytes. (From Ch *et al.*, 1968a. Courtesy T. C. Cheng.)

cytes (Fig. II.11C and D). Implants of rediae evoke only slight encapsulation involving archaeocytes and collencytes (Fig. II.12C), with some of the latter forming syncytia. In instances in which cercariae were freed through rupture of the rediae walls, the cercariae became completely encapsulated by a relatively thick layer of archaeocytes and collencytes (Fig. II.12D and E). Although sponge archaeocytes and collencytes do adhere to the outer surfaces of implanted rediae and molluscan muscle fibers (Fig. II.12F), these two foreign materials do not become completely encapsulated. The complete encapsulation of the cercariae by a relatively thick syncytium is of considerable interest since it has been demonstrated (Thakur and Cheng, 1968) that *Philophthalmus gralli* cercariae begin to secrete cytogenous materials from subsurface gland cells prior to their release or escape from the rediae. This material is uniformly deposited on the body surface of each cercaria, thus making it possible that the more spectacular and efficient encapsulation of cercariae,

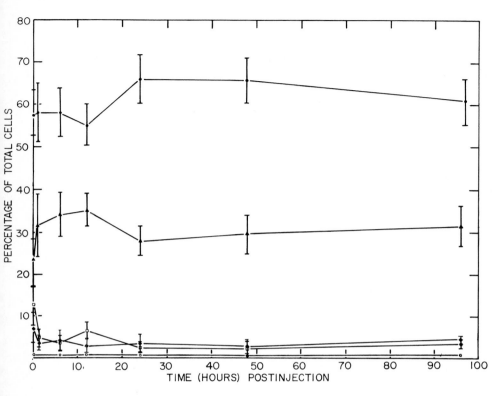

Fig. II.10. Graph showing the percentages of each of the 5 types of parenchymal cells in sponge *Terpios zeteki* at time intervals postinjection of carmine particles. Each point repres the mean of 10 counts and the perpendicular lines represent standard deviations. ●—●, lencytes; ▲—▲, archaeocytes; ○—○, scleroblasts; ◆—◆, thesocytes; □—□, chromocytes. om Cheng *et al.*, 1968a. Courtesy of T. C. Cheng.)

as compared with cellular reactions toward rediae or muscle tissue, represents the response of sponge archaeocytes and collencytes to the proteinaceous cystogenous material (as is the case in mollusks harboring trematode metacercariae) (Cheng *et al.*, 1966a,b).

Archaeocytes are capable of becoming hypertrophied to accommodate foreign materials of slightly greater size than themselves, e.g., human erythrocytes. Both phagocytized and encapsulated erythrocytes are virtually all eliminated from the parenchyma of *T. zeteki* by 38 hours via the migration of erythrocyte-laden or enveloping archaeocytes across pinacocytes into excurrent canals. This indicates that not only are archaeocytes that have phagocytized foreign bodies capable of migration, but small encapsulation complexes, involving several archaeocytes surrounding a

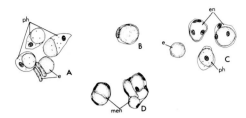

Fig. II.11. Phagocytosis and encapsulation of human erythrocytes by archaeocytes of the sponge *Terpios zeteki*. A, single erythrocytes within single phagocytic archaeocytes and free erythrocytes. B, single erythrocyte within phagocytic archaeocyte, note clumped appearance of cytoplasm of erythrocyte suggesting degradation. C, single erythrocytes encapsulated by 2 or more fused archaeocytes, phagocytic archaeocyte containing single erythrocyte and free erythrocyte. D, 2 or more erythrocytes encapsulated by 2 or more fused archaeocytes. All drawings made from cells observed in sections. e, Erythrocyte; en, single erythrocyte encapsulated by 2 or more sponge archaeocytes; ph, single erythrocyte phagocytized by single archaeocyte. (From Cheng *et al.*, 1968c. Courtesy T. C. Cheng.)

foreign body, are also capable of migration, although less rapidly. The aggregation of sponge cells around implanted foreign bodies should not be totally attributed to heterograft attraction, since examination of control sponges cut similarly but without insertion of foreign bodies reveals there is some migration of cells to the wounded surface, an almost certain indication of wound healing processes. However, the fact that sponge cells adhere to the surfaces of heterografts (and form a complete capsule around them in the case of the cercariae) indicates that the cellular response can be attributed, at least in part, to the presence of the implanted materials.

Phylum Coelenterata

Although coelenterates are more complex organisms than sponges, some forms, such as *Hydra*, have lost much of the mesoderm during evolution. Whereas the mesoderm of sponges plays the primary role in inflammation, those coelenterates lacking amoebocytes and possessing only rudimentary mesoglea are forced to rely on other tissues for response to injury and wound repair. The epidermis of coelenterates consists largely of cuboidal or columnar "supporting" cells; interspersed among them are sensory cells, nematocysts, and small, rounded cells usually found in clumps, called interstitial or indifferent cells. The latter are believed to be persistent, undifferentiated, embryonic cells similar to the archaeocytes of sponges; they give rise to sperm and eggs, and, more importantly in reaction to injury, to any other cell needed. It has been

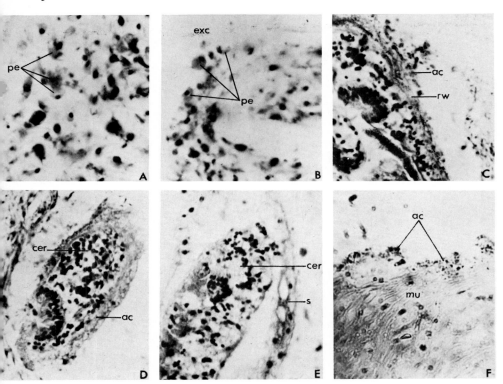

Fig. II.12. Response to tissue implantation in the sponge *Terpios zeteki.* A, erythrocytes phago-
ized by archaeocytes in mesoglea (6 hours postimplantation). B, erythrocyte-laden archaeo-
es migrating into excurrent canal (6 hours postimplantation). C, irregular aggregates of archaeo-
es and collencytes on surface of implanted redia (6 hours postimplantation). D, relatively
iform layer of archaeocytes and collencytes adhering to surface of cercaria (6 hours postim-
ntation). E, formation of syncytial tunic of host cells surrounding cercaria (12 hours postimplan-
ion). F, irregular aggregates of archaeocytes and collencytes on surface of implanted molluscan
scle (12 hours postimplantation). All sections stained with hematoxylin and eosine. ac, Sponge
haeocytes and collencytes; cer, cercaria; exc, excurrent canal; mu, muscle; pe, phagocytized
throcytes; rw, redial wall; s, syncytial capsule of host cells. (From Cheng *et al.*, 1968c. Cour-
y of T. C. Cheng.)

estimated that all cells of a *Hydra* are replaced during a 45-day period
by interstitial cells.

The astonishing powers of regeneration and rapid repair of injuries in
freshwater hydrozoans were studied by several investigators in the last
century. Ishiwaka (1890) found that the oral region of *Hydra* completely
recovered within 20 minutes after infliction of an injury; if one organism
was cut in two and stretched out on a piece of cork, a complete animal
was formed again in 24 hours. When a filament of algae was inserted

between the edges of a cut in one of Ishiwaka's experiments, the *Hydra* changed its position and grew into a closed sac which produced a mouth and tentacles and formed a perfect *Hydra* within 6 days after the injury. Punctures and other superficial lesions heal very quickly without accumulation of amoebocytes at the site of injury. The absence of amoebocytes at the wound site is due to the lack of mesenchyme. This does not mean, however, that phagocytosis does not occur in these forms; the gastrodermis ("entoderm") is lined with large cuboidal or columnar epithelial cells, all of which are phagocytic and can engulf foreign bodies as well as digestible food particles.

In the marine colonial hydrozoans, the epidermis as well as the gastrodermis may contain epithelial cells with phagocytic capabilities; the protective function of this is obvious. These forms, like *Hydra,* have remarkable regenerative powers and are able to rapidly repair injuries. The regenerative capacity of both freshwater and marine hydroids is so rapid and so complete that the possibility of secondary infection is slight, and the rapidity of repair eliminates the necessity for a typical inflammatory process with its classic infiltration of phagocytes.

More recently, Tokin and Yericheva (1959, 1961) demonstrated that the only inflammatory response in *Hydra oligactis* is phagocytosis. When a hair moistened with a suspension of marking ink is introduced into the tissues of *Coryne loveni* (an athecate colonial hydroid) only the cells of the ectoderm phagocytize the ink particles. This phagocytosis is active; almost all ectodermal cells of the limbs of the hydranth contain ink particles after 22 hours. Under conditions of trauma, phagocytosis is greatly accelerated. Trauma-induced intensification of phagocytic activity was also demonstrated in other hydroids, including *Laomedia flexuosa* and *Clovea multicornis,* by use of marking ink and dead *Bacillus subtilis.*

The authors agreed with Metchnikoff that the reaction to injury in the hydroid coelenterates cannot properly be termed inflammation, since phagocytosis is the only phenomenon characteristic of the inflammatory process that occurs. There is no specific orientation of cells in response to the wound, no manifestations of "group phagocytosis," and no tendency to encapsulate the foreign body.

Emigration of phagocytes to the site of an injury is, however, common in those coelenterates with well-developed mesoglea containing wandering amoebocytes. Metchnikoff (1893) observed that the insertion of a splinter or pin into the gelatinous bell of the large medusa, *Rhizostomum cuvieri,* produces a cloudiness (grossly visible by the next day) around the foreign body resulting from the accumulation of numerous amoebocytes. He noted that the same phenomenon occurs under similar conditions in *Aurelia aurita*, and if the foreign body has been previously soaked in carmine, the infiltrating amoebocytes phagocytize the granules. The amoebocytes accumulating at the site of the injury must traverse a gelatinous material without the aid of any vascular system; once there, they may retain their individuality or fuse (according to Metchnikoff) to form minute plasmodia. The mechanism initiating the emigration of amoebocytes to the lesion has not been identified.

Prazdnikov and Mikhailova (1962) studied the early stages of the inflammatory reaction in the coelenterates, *Staurophora mertensii* and *Aurelia aurita*, and a ctenophore, *Beroe cucumis*. These authors introduced a cotton thread saturated with carmine into the mesoglea of the test animals. The mesoglea of the hydromedusa *S. mertensii* contains collagenous fibers but no cellular elements; therefore, no phagocytosis of carmine grains or damaged cells by insertion of thread occurred. Within 12 hours, a layer of homogenized collagenous fibrils had formed around the thread, and epithelial cells had begun to migrate over the edge of the wound. After 3½ days the foreign body was enclosed in a "ring" of homogeneous mesoglea, and the epithelium covered the exposed surface of the wound by production of large generative cells with granular cytoplasm. Numerous mitotic figures appeared in the epithelial cells adjacent to the wound, demonstrating that increased multiplication was initiated by the injury. However, no phagocytic phenomena were exhibited throughout the reaction to injury and early reparative stages.

The mesoglea of *Aurelia* contains numerous amoebocytes as well as collagenous fibers. Within 3 hours after insertion of a carmine-soaked thread into the mesoglea, the traumatized epithelium sloughs and active emigration of amoebocytes is initiated. Frequently the movement of amoebocytes takes place along the collagenous fibers in a manner similar to that of vertebrate fibroblasts. The collagenous fibers are distorted and compressed by the foreign body, and fibers in the vicinity of the thread become homogeneous. During this early period numerous enlarged, vacuolated, dying amoebocytes appear in the vicinity of the foreign body. Little phagocytosis of the carmine by the amoebocytes is noted; this activity is largely restricted to the ectodermal cells.

Six to twelve hours after injury, migration of cells into the wounded mesoglea results in the development of an uneven border between the epithelium and the mesoglea. Spindle-shaped epithelial cells migrate over the wounded surface to close the external wound, while dedifferentiated epithelial cells and amoebocytes form a phagocytic band around the foreign body. Necrotic amoebocytes and epithelial cells appear between the fibers of the thread, and the homogenized substance of the mesoglea near the thread continues to stain intensely for collagen with Mallory's stain.

After 24 hours most of the amoebocytes are out of the traumatized area, and groups of four to six amoebocytes, tightly adherent to one another, may be seen on the surface of the medusa. Infiltration of amoebocytes adjacent to the thread does not occur, but, as in earlier observations, amoebocytes are present in the mesoglea in the area of the trauma.

On the fourth day, large cells appear in the mesoglea and in the disintegrated area of ectodermal epithelium. Mitoses are frequent in both the ectodermally derived cells and in the amoebocytes. Near the ectodermal epithelium in the mesoglea, easily recognizable, regenerating myoepithelial cells appear, and there is an infiltration of a homogeneous collagenous substance, staining brightly with Mallory's, beneath the ectodermal epithelium in the area of the trauma. No live, wandering cell elements are left between the fibers of the thread.

The authors concluded that *Aurelia aurita* does exhibit an inflammatory response, characterized by the movement of amoebocytes into and subsequently out of the traumatized area. Phagocytic activity of the amoebocytes, however, is extremely weak; this function is taken over largely by emigrating ectodermal epithelial cells. The role of the amoebocytes in inflammation is still not clear. The modifications of the collagenous substance and the marked disintegration and dedifferentiation of the epithelial layers, along with the rapid appearance of regeneration phenomena, are characteristic of the inflammatory response and wound repair in these animals. The occurrence of mitotic figures and binucleate amoebocytes indicates that multiplication of amoebocytes contributes to the regeneration of the cell elements of the mesoglea, but the migration of epithelial cells into the mesoglea and their subsequent cell division leave little doubt that the epithelial cells also contribute to these processes.

Phylum Ctenophora

Ctenophores have great powers of regeneration, and the many injuries to which their fragility and watery construction subject them are quickly repaired (Hyman, 1940). Prazdnikov and Mikhailova (1962) reported that the insertion of a thread saturated with carmine into *Beroe* initiates an amoebocyte infiltration into the area, active phagocytosis of carmine particles, and collagenous encapsulation of the thread. The thread is eventually ejected from the organism by muscular contraction in the vicinity of the wound. These reactions are much more intense than those occurring in the coelenterates *Aurelia aurita* and *Staurophora mertensii,* which exhibit only a weak phagocytic response when subjected to identical treatment.

Phylum Platyhelminthes

Studies of reaction to injury and wound repair in Platyhelminthes have, for the most part, been confirmed to the class Turbellaria and, even in that group, have usually been a secondary result of investigations of regeneration. The reaction to injury and regenerative capacities of the parasitic classes have been largely ignored, as might be expected.

CLASS TURBELLARIA

The bulk of the flatworm body consists of a parenchyma or connective tissue of entomesoderm that is made up of contiguous rounded cells or, more commonly, a fixed, netlike syncytium with wandering free cells or amoebocytes. The syncytium is nucleated, usually fibrillar in appearance, and contains numerous fluid-filled interstices (Hyman, 1951). The amoebocytes are rounded cells that migrate freely through the parenchyma; they play an active role in reparative, regenerative, and reproductive processes. Metchnikoff (1893) noted that these cells, which he called mesodermic phagocytes, collect around the lesion if the transparent turbellarian, *Mesostomum ehrenbergi,* is wounded. Hyman (1951) briefly summarized the histological aspects of wound repair from the numerous studies of regeneration in planarians. The wound is partially closed by muscular contraction, unless the organism is in isotonic media; the epithelial cells adjacent to the wound flatten, become syncytial, and creep over the cut surface. If the wound is large, regeneration cells accumulate and participate in the formation of a covering membrane that subsequently differentiates into typical epidermis. A mass of cells, known as regeneration cells, formative cells, or neoblasts, accumulates beneath the covering membrane and differentiates into the missing tissues and organs. Additionally, there appears to be some repair by proliferation of cells adjacent to the injured tissues, e.g., regeneration of cut nerve ends and intestinal branches in land planarians.

The source of the regeneration cells is in considerable dispute, although there is little doubt that they are preexisting cells that migrate to the site of the wound either with or without multiplication by cell division. Some authors believe that they are solely free cells of the mesenchyme, while other investigators describe dedifferentiation of the cells of various tissues, such as gastrodermis, yolk glands, and other gland cells, into amoeboid cells which join the free mesenchymal cells to form the mass of regeneration cells. One view, to which Hyman gives little support, is that the regeneration cells are persistent embryonic cells held in reserve to function in regeneration and in formation of the reproductive system. Planarians with low powers of regeneration, such as *Porocotyla flaviatilis*, have conspicuously fewer free mesenchymal cells than those, like *Dugesia*, which have high powers of regeneration; this indicates that the primary wound repair and regenerative functions lie in the amoebocytes of the mesenchyme.

CLASS TREMATODA

From limited studies of experimentally induced wounds in trematodes and infrequent observations of naturally occurring wounds, some general comments can be made about the pattern and rate of wound repair in the parasitic class Trematoda.

Specimens of *Echinostomum revolutum* injured by incisions in the body wall anterior to the acetabulum or by extensive lacerations of the head crown are expelled by the host within 5 hours with no indications of wound repair (Beaver, 1937). Median dorsoventral, slightly lacerated, septic puncture wounds through the midbody completely heal within 2 days, but Beaver gave no details of the repair process. Worms receiving deep, unilateral incisions through the body wall at the level of the testes exhibit the highest degree of wound repair. These wounds sever the digestive cecum, the vitellaria, and the excretory siphon. Not only is repair initiated (though not completed), but the trematodes continue to grow and produce ova. The wound is closed with a furrowlike scar which is covered by a thick layer of tuberous cuticula. The incised ends of the cecum close, and the posterior portion, although isolated, appears normal. Anterior to the lesion, the excretory siphon becomes greatly enlarged and muscular to form a bladder with an external aperature at the site of the wound. As a result of the injury, the vitellaria posterior to the wound degenerate and the hindbody of the injured side is reduced in size.

Manter (1933) observed a similar condition in a specimen of *Helicometra torta* (Fig. II.13) recovered from a fish, *Epinephelus striatus*. A wedge-shaped portion of the hindbody had been excised, severing the right cecum and extending almost to the middle of the body. The edges of the wound were healed and covered with cuticula, but had not reunited. The isolated posterior portion of the cecum, such as that of *Echinostomum revolutum* in Beaver's studies, was closed but normal in appearance. Vitellaria anterior to the wound were normal, but posterior

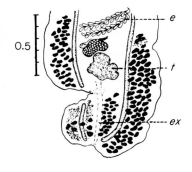

0.5

e

t

ex

Fig. II.13. Posterior portion of the trematode *Helicometra torta* with healed wound. Drawn with aid of camera lucida. (From Manter, 1933.)

to the injury (with the exception of the most posterior), they were degenerated with granules of shell material scattered through the parenchyma. The wound had penetrated the posterior testis, which had completely disappeared. The injury was considered to have been the cause of the degeneration or failure of the testis to develop since more than 100 unwounded specimens examined contained two testes.

Sinitsin (1932) also found that body wounds in *Fasciola hepatica* and *Fascioloides magna* heal without regeneration beyond the secretion of cuticula. Manter noted that, although there was a definite layer of cuticula deposited in the wounded *Helicometra*, the subcuticular musculature had not been regenerated.

Amputation of the hindbody initiates immediate contraction of the body wall to partially close the wound; this is followed by rapid secretion of a layer of cuticula over the cut surface. Both processes are accompanied by a conspicuous aggregation of cells at the site of injury (Beaver, 1937). These aggregating cells, as in all cases in which the body wall is broken, are proliferated primarily from the submuscular layer of the body wall, although infiltration of cells from the surrounding parenchyma may contribute to the aggregate. There is a gradual growth or folding inward of the incised margins of the body wall, and the wound is entirely closed in about 3 days. The cut ends of the gut close, and an outlet for the excretory bladder is established. During the entire healing process the anterior parts of the body remain functional and appear normal.

Beaver pointed out that the speed of major-wound repair in *Echinostomum* is related to its rapid rate of development under normal conditions. Oviposition normally occurs as early as 8 days after infection of the final host; there is an enormous increase in body size and in differentiation and maturation of the gonadal complex. Growth continues after sexual maturation for the 3 or 4 weeks it survives in the host. Since the growth rate is rapid and the life-span short, it is not surprising that wound repair is also accelerated.

One of the few reports of reaction to and subsequent wound repair of a naturally occurring injury in a trematode was provided by Manter (1943) in a description of parasitism of one trematode by another. Larval flukes, *Mesocercaria,* inflicted obvious wounds when they penetrated adult *Neorenifer* (Fig. II.14 *3,4,6*). Most of the effects were limited to the surface area of penetration since the larva seldom migrated far in the host trematode. Once embedded in the parenchyma, the *Mesocercaria* apparently stimulated little or no host reaction (Fig. II.14 *7*), except in two cases where severe damage was done to the reproductive organs (Fig. II.14 *2*).

At the site of attack, the cuticle was scraped or digested away over a considerable area, exposing the circular, longitudinal, and diagonal muscles along the edges of the injury and the parenchyma was usually exposed in the central area (Fig. II.14 *4*). The wounds were large, occasionally almost severing the body, but there was surprisingly little tissue reaction. The exposed parenchyma was slightly more granular than normal, and the stroma was more densely fibrous at the surface with a thin membranelike layer apparently formed by the cell membranes of the parenchymal cells. Manter noted numerous small, light-brown, ovoid particles of uniform texture thought to be either necrotic parenchyma cell nuclei (which they resemble in size) or coalesced particles liberated by yolk cells migrating to the site of the injury (Fig. II.14 *5*). Manter felt confident that yolk-gland cells migrate, in an amoeboid stage, to the affected area. These amoebocytic yolk cells were found in younger wounds in moderate numbers, but were usually lacking in the older wounds, indicating (in my opinion) that they are a part of the acute inflammatory response. It should be pointed out that Cheng and Streisfeld (1963) demonstrated that trematodes (*Megalodiscus temperatus* and *Haematoloechus* sp.) possess phagocytic parenchymal cells. This strengthens the concept that amoebocytic parenchymal cells participate in the reaction to injury and wound repair. Longitudinal and circular muscle layers had regenerated in apparently older wounds. Regeneration or repair of the cuticle was slower and must have occurred (if at all) from the edges of the wound. The small size of some wounds indicated that such repairs had been initiated.

Observations (Fried and Penner, 1964) on the healing of laceration wounds in *Philophthalmus hegneri* verify those of Beaver and Manter. One worm was completely severed behind the acetabulum; another was deeply cut, leaving the posterior third of the body barely attached. Both worms healed the cut surfaces, and the one with the hindbody attached retained it as an appendagelike structure with the included organs functional. The authors noted that wound healing by tissue repair occurred, but there appeared to be no regeneration of amputated parts (Fig. II.15).

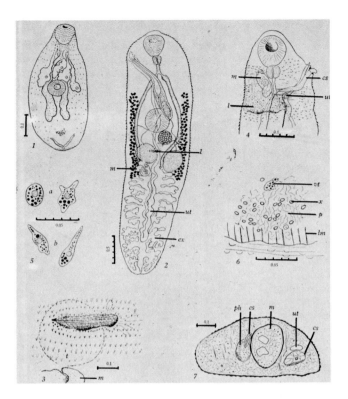

Fig. II.14. Reaction to injury and wound repair in a trematode penetrated by another trematode. (From Manter, 1943.)

1. *Mesocercaria marcianae*—ventral view.

2. *Neorenifer grandispinus*—dorsal view with mesocercaria (m) embedded in parenchyma posterior to left testis and place of entrance dorsal to testis (= lesion or wound). The membrane of the testis was perforated posteriorly, probably during fixation, by the mesocercaria.

3. Enlarged view of wound caused by entrance of mesocercaria shown in 2. Note that the muscle layers of the body wall have apparently regenerated.

4. *Neorenifer grandispinus*—ventral view of anterior portion showing embedded mesocercaria and lesion caused by its entrance.

5. Vitelline cells associated with wounded areas of the parenchyma; a, cells from wound area; b, cells from parenchyma near yolk glands.

6. Enlarged portion of posterior part of wound area shown in 4, showing vitelline cell (vt), ovoid bodies (x) of unknown origin on surface of parenchyma, broken ends of longitudinal muscles (lm), and posterior edge of wound.

7. Cross section through a *Neorenifer* invaded in pharynx region, showing relative size of mesocercaria (m) and absence of tissue reaction. ph, Pharynx; es, esophagus; ut, uterus; cs, cirrus sac. Drawn with aid of camera lucida.

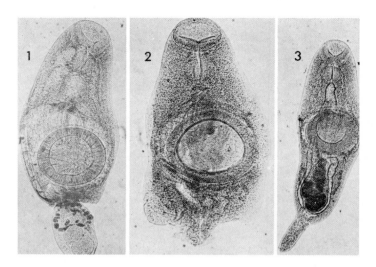

Fig. II.15. Wound healing in a trematode, *Philopthalmus hegneri*. (From Fried and Penner, 1964.)

1. After partial amputation of the posterior third of the body
2. After complete amputation of the posterior third of the body
3. Ten-day-old *P. hegneri* with taillike appendage suggestive of previous mechanical damage.

From the few observations of trematodes wounded under normal conditions and from Beaver's experimental studies, it appears that wound repair consists of rapid closure of the lesion by secretion of cuticula, but that little or no regeneration of damaged or destroyed parts occurs. The degree to which various tissues contribute to the reparative process remains unknown, although Beaver's observation of cellular infiltration and Cheng and Streisfield's discovery of phagocytic parenchymal cells are informative. Manter's assumption that migrating vitelline cells contribute to the process is probably incorrect.

CLASS CESTODA

Information on the reaction to injury and wound repair in cestodes is restricted to a few reports of the regenerative powers and of the teratological results of presumed injuries. Hyman (1951) noted that the scolex can regenerate the neck region, and, subsequently, strobila are produced by normal proliferation. Iwati (1934) studied the regenerative capabilities of bothriocephaloid plerocercoids by making various incisions and transections and then reimplanting the worms into appropriate hosts. He found high regenerative capacities, but gave no details of the tissue responses of the wounded plerocercoids. Thus, we remain uninformed of the mechanism of wound repair in the group, though we can perhaps assume it is similar to the process in related turbellarians and trematodes.

Phylum Rhynchocoela

The nemertine worms possess a circulatory system that contain leukocytes of several types and, in some forms, red blood cells containing hemoglobin (determined by spectroscopic examination). Detailed histological studies have been made of many species of nemertines, and Hyman (1951) presented an excellent summary of the histology of the phylum. The group exhibits remarkable regenerative capabilities, and considerable information on wound healing has been accumulated indirectly in the investigations of regeneration.

When a nemertine is wounded, the wound is closed over by emigration of the deeper cells, presumably interstitial cells, from the adjacent epidermis. Beneath this covering, mesenchyme cells accumulate, consisting of emigrant mesenchyme cells and phagocytic cells, which contain ingested tissue debris that serves a nutritive function. If the worm has been severed, the mass of mesenchymal cells, termed the blastema, regenerates the missing parts. Presumably, they repair damaged parts if the wound does not completely bisect the animal. Hyman notes that *in situ* transformation of old tissue occurs through ingestion by phagocytic cells, which then disintegrate and release their contents as food for mesenchyme cells that regenerate the tissues.

Nemertine epidermis produces copious quantities of mucus; the amount increases when the animal is irritated, injured, or exposed to unfavorable conditions. The role of mucus in reaction to injury or wound repair is unknown.

Under starvation conditions, many of the mesenchyme cells become wandering phagocytes that engulf pigment and body cells; they then disintegrate to furnish food for the starving worm. This illustrates the great phagocytic activity typical of the phylum. Further study would almost certainly demonstrate an active role for these cells in the inflammatory and repair processes. There is also need for additional studies to ascertain the role of the elements of the circulatory system in reaction to injury and wound repair.

Phylum Acanthocephala

As is true of most parasitic worms, no studies have been made of the reaction to injury and wound healing processes in acanthocephalans.

However, the histology of this group has been investigated in considerable detail. The acanthocephalans, like the Aschelminthes, exhibit nuclear constancy or eutely. With the exception of the gonads, the number of nuclei in all tissues remains constant, though, in some families, epidermal nuclei may fragment into many smaller nuclei. No wandering amoebocytes are present in the epidermis or anywhere in the body.

Since the nuclei of most acanthocephalans are constant in number and since amoebocytes are absent, it is difficult to visualize a mechanism of wound repair. Although the syncytial nature of the body may provide some defense mechanisms not yet known, it is possible that wound repair does not occur to any appreciable extent in this phylum.

Phylum Aschelminthes

The Aschelminthes are pseudocoelomate bilateria covered by a resistant cuticle, and, as in the acanthocephalans, nuclear or cell constancy is universal in the phylum. Hyman (1951) includes as classes Rotifera, Gastrotricha, Kinorhyncha, Priapulida, and Nematoda. There is some question as to the correctness of considering these groups as classes rather than separate phyla (as in many modern texts), but there is no question that they are closely related phylogenetically and morphologically.

All possess a tough, resistant, syncytial cuticle which apparently serves as the primary defense against injury. If the cuticle is penetrated or removed, there appears to be little capacity to repair the injury or regenerate injured or lost parts, although virtually no investigations of these phenomena have been undertaken in any members of the group other than the rotifers. Pai (1934) has shown that rotifers are generally incapable of regeneration; they usually die shortly after amputations, often without any evidence of wound healing. He did ascertain that young rotifers were capable of healing wounds and regenerating nonnucleated parts of the corona. According to Hyman (1951), animals with constant cell or nuclear number are theoretically incapable of replacing nucleated parts once their nuclei have ceased to divide.

The rotifers apparently have no wandering amoebocytes or fixed phagocytes, and even digestion is primarily extracellular. The gastrotrichs, like the rotifers, do not have wandering amoebocytes, but the pseudocoel of kinorhynchs is filled with fluid that contains numerous active amoebocytes, apparently originating from the wall of the digestive tract (Hyman, 1951).

The body cavity, the cavity of the caudal appendages, and the spaces in the body-wall musculature of priapulids are also fluid filled and contain numerous rounded cells capable of phagocytizing carbon or carmine particles. Repair of injuries to priapulids occurs under some conditions. The caudal appendages, for example, will regenerate if the sphincter at their attachment to the body is not damaged. Injury of the sphincter allows the escape of body fluid without closure of the wound and the animal dies. The pseudocoel (the space between the body wall and viscera) of nematodes is also filled with fluid and usually contains fixed cells, called pseudocoelocytes, and fibrous strands and membranes. The pseudocoelocytes, however, are not phagocytic, but apparently have an oxidative function. There are no wandering cells reported in this group.

The only report of wound healing in nematodes, to my knowledge, that appears in the literature notes that "destruction accompanied by healing of different organs of free-living marine nematodes seems to be rather rare" (Allgen, 1959). The tail of free-living nematodes is the area of the body most susceptible to injury because of its activity during movement. Allgen described several examples of wound healing of injured tails, mostly from his own observations, but a few were from other authors who failed to recognize the wound healing process. One noted nematodologist, for example, described a new species solely on the basis of the altered appearance of the tail of a single worm that had, in Allgen's opinion, been partially destroyed and then healed. Another author described a greatly thickened cuticle of a female nematode, *Anticoma procera,* that had repaired a wounded tail.

Among the several worms with repaired tail wounds described by Allgen, there were two of particular interest. In one of these, a male *Eurystomatina ornatum* from the French Mediterranean coast, the entire tail had been destroyed along with a short preanal portion of the body. A portion of the spicular apparatus, the posterior supplementary auxiliary organ, was lost, leaving the anterior auxiliary organ as the only intact part of the spicular apparatus. The posterior end of the body was rounded and covered with a thickened cuticle. The second, also a male (*Bathylaimus zostericola*) was from the coast of Norway; it had lost most of its slender tail. The wound had healed, evidently by the cuticle covering the surface at the point of amputation, and, again, the cuticle had thickened considerably, presenting a rounded stump rather than the elongate, slender tail typical of the species.

One can speculate that although the phenomena of reaction to injury and wound repair has largely been lost in the pseudocoelomate animals with well-developed syncytial cuticles those forms, such as kinorhynchs and priapulids, possessing wandering amoebocytes have a greater intrinsic capacity to react to foreign bodies and wounds and ultimately to repair tissue injury. The phenomenon of nuclear or cell constancy, however, is obviously a phylogenetic development detrimental to these processes.

Phylum Annelida

Although Metchnikoff (1893) pioneered the study of inflammation in annelids, a number of earlier workers had laid the groundwork for his investigations by describing the amoeboid wandering cells of the coelomic cavity. The annelids have, in the majority of forms, developed a completely enclosed circulatory system containing oxygen-carrying pigments and amoebocytes, but there appears to be little or no involvement of the elements of the vascular system in reaction to injury. Instead, reaction to trauma, foreign bodies, and parasites appears to be confined to the mesodermic phagocytes of the peritoneal endothelium and to the amoeboid cells suspended in the coelomic fluid. Some forms, such as *Nais proboscidea,* lack perivisceral leukocytes, and foreign body invasion, such as *Gordius* larvae, results in excitation of the phagocytic reaction of the peritoneal cells, but there is no emigration of cells from the adjacent blood vessels.

The most distinguishing anatomical feature of the phylum Annelida is the metameric construction of the body, not only externally but in all internal organs except the digestive system. This metameric arrangement is of considerable value in reacting to and repairing injury; each segment contains a more or less complete protective mechanism, and the wound is generally confined to the injured segment.

Class Polychaeta

Since polychaetes have relatively great powers of regeneration, it is obvious that they rapair wounds rapidly and fairly completely. Attenuated parts such as palps, tentacles, and parapodia are quickly replaced after loss, and some polychaetes display autotomy of certain parts.

Although the healing of surface wounds in polychaetes has not been described, Metchnikoff (1893) studied the cellular reaction to a foreign body inserted into the perivisceral cavity of *Terebella.* Shortly after the introduction of a splinter into the cavity, masses of the wandering, amoeboid, perivascular cells, which he called "lymphatic" cells, surrounded the foreign body. These cells were shown to be actively phagocytic, and they readily engulfed the material present on the surface of the splinter. As noted earlier, reaction to the foreign body is confined to the mesodermic phagocytes, with the vascular system uninvolved in the process.

CLASS OLIGOCHAETA

Metchnikoff (1893) also discussed the reaction to foreign invasion in several oligochaetes; his comments were, for the most part, confined to the response to parasites. He noted that *Nais proboscidea,* which lacks wandering coelomic amoebocytes, combats invasion of *Gordius* larvae into the coelom by surrounding the larvae with protoplasmic processes from the peritoneal cells. A similar reaction occurs when the spores of the "pebrine" organism are introduced into the coelom; the peritoneal endothelium cells actively phagocytize the spores. In commenting on the phagocytic activities of earthworm amoebocytes, which he noted were among the most active of phagocytes, Metchnikoff described active engulfing of any minute foreign bodies encountered and also the encapsulation of larger objects, such as a cotton thread, by groups of amoebocytes.

Cameron (1932) greatly contributed to the understanding of inflammation and wound repair in the annelids with a lengthy and detailed study of these phenomena in earthworms. Earthworms are remarkably resistant to all types of injurious agents. Rather large amounts of foreign particles, cells, and dyes may be injected into the coelom without noticeable effects. Fluid material is rapidly eliminated by the nephridia and dorsal pores of each segment; this efficiently regulates intracoelomic pressure, and massive injections of toxic substances have little effect other than temporary paralysis of the posterior segments and localized dilation of epidermal blood vessels. However, insertion of foreign bodies frequently cause fatalities when introduced into proximal segments. Cameron noted that death occurs earlier when the pharynx or esophagus is injured and suggested that death from insertion of solid foreign bodies probably results from damage to vital organs rather than from reactions to the foreign body.

Cameron systematically investigated the response of the coelomic cells to introduced foreign substances. Phagocytosis of india ink is initiated immediately *in vitro,* with the nongranular coelomic corpuscles reacting first, but all cells (other than chloragogen cells) soon engage actively in the process. Carmine particles, colloidal suspensions of iron, dust, and coal particles elicit the same response, but fat particles, in the form of fine suspensions of olive oil or milk, do not induce phagocytosis *in vitro,* though phagocytosis does occur after 24 hours when fat is introduced into the coelomic cavity.

In vivo, within 5 minutes postinjection, clumping of coelomic cells occurs around groups of particles, and a few coelomic corpuscles, primarily the nongranular varieties, contain ink particles. Most of the ink is found in close apposition to the epithelial cells lining the coelom, some of which contain ink particles. At 15 minutes after injection, more cells contain particles, and some of the phagocytes begin migrating toward the nephridia. By 4 hours, there is a marked increase in the number of coelomic corpuscles, most of which have phagocytized particles, and many of the lining epithelial cells contain ink particles. Proliferation of the lining cells occurs and the cells break off and migrate to the site of the foreign particles. The nephridia are lined with particle-containing phagocytes. The number of coelomic corpuscles continues to increase, and, by 4 days, septal and peripheral epithelial proliferation gives rise to huge masses of cells, packed with ink particles. In 14 days, many free coelomic corpuscles containing ink particles are present, and considerable proliferation of septal and parietal lining cells, especially at the angles between the septum and body wall, is still obvious. Much of the ink is concentrated in the middle tubules of the nephridia, both in the lumen and in the phagocytes. Small areas of proliferating septal cells still occur, but the parietal peritoneal cells are quiescent.

During proliferation, the flattened, coelomic-lining cells swell to a cuboidal shape, and the nuclei become rounded and stain more intensely. Division is rapid; the cells occur in sheets, with the more distal cells moving away. The septal cells and those enclosing the nephridia proliferate most actively. The sheets of developing cells never include acidophilic forms, and differentiation into granular and nongranular types occurs rapidly after migration from the site of origin.

Foreign Cells

Injection of mammalian and oligochaete sperm into the coelomic cavity of earthworms results in delayed phagocytosis, usually beginning after approximately 20 hours. Once initiated, phagocytosis proceeds sluggishly and appears to be confined to the medium- and large-sized basophilic cells. Mammalian spermatozoa apparently are not phagocytized any more rapidly than earthworm spermatozoa of other species, but homologous spermatozoa are left unattacked. Within 48 hours postinjection, many of the ingested sperm are disintegrating.

Mammalian erythrocytes elicit an early, but relatively mild, phagocytic reaction. Again, the medium and large basophilic cells play the dominant role, with a rare acidophilic cell occasionally ingesting an erythrocyte. Between 20 and 40 hours after injection, ingested erythrocytes become pycnotic and subsequently fragment.

Large Foreign Bodies

Insertion of foreign bodies, for example sterilized cotton thread, into various segments of earthworms commonly results in autotomy or death of the worm, apparently because of accidental injury to the alimentary canal. Worms invariably die within a week after insertion of relatively large foreign bodies into the head segments.

Within 24 hours after insertion of a small piece of cotton thread, there is marked infiltration of the thread by all types of coelomic corpuscles. Many of the cells are in close apposition to individual fibers, sometimes partially surrounding the fibers with crescent-shaped cells or completely surrounding the fibers by fusion of several cells. Slight proliferation of septal lining cells may occur when the thread lies near the septum, but this is not a conspicuous phenomenon. Within 3 days, a dense capsule forms around the thread, consisting of flattened and proliferated coelomic cells which have spread toward the foreign body from the portion of the body wall, septum, or nephridial covering nearest the thread. Small blood vessels, growing out from the body wall, invade the area and are surrounded by proliferating cells. Most of the cells within the thread have disappeared by this time. In 7 days, the encapsulating cells are packed closely together, and fibrous tissue appears to be forming around the thread. Blood vessels are still present, but are not as obvious as in the early stages of the reaction.

Since autotomy or death usually occurs at approximately this stage, knowledge of the later stages of the reaction to experimentally introduced foreign bodies is not known. However, the reaction to such parasites as the gregarine *Monocystis* and nematodes is similar in its early stages to that described above. A fibrous capsule is eventually formed completely surrounding the parasite. The capsule consists of closely packed, fine fibers in which few cells can be distinguished. The parasite frequently dies, degenerates, and becomes calcified.

Dorsal Pores and Nephridia. The introduction of large amounts of foreign particles or organisms into the earthworm elicits an immediate phagocytic response, but phagocytosis alone cannot account for the rapid disposal of the foreign material. Cameron conclusively demonstrated that diapedesis occurs through the dorsal pores, with numerous phagocytes containing chloragogen granules, cellular debris from the coelom, and, when foreign particles have been injected, pigment granules and bacteria. The dorsal pores also provide a means of regulating intracoelomic pressure by allowing the discharge of coelomic fluid and by eliminating foreign materials contained in the fluid but not yet phagocytized. Thus, the dorsal pores serve an important role in the defense mechanism of the worm. It should be noted, however, that the dorsal pores may also serve as a portal of entry for bacteria and other foreign materials.

Coelomic fluid is markedly altered in its composition during passage through the nephridia, and the excretory functions of the nephridia undoubtedly play a role in the defense mechanism of the earthworm by eliminating toxins and other foreign materials. Additionally, the cells lining the middle tube have been shown to be phagocytic as well as excretory in function. Liquid foreign materials such as trypan blue and vital red as well as such crystalloids as iron alum, silver nitrate, and potassium ferrocyanide are rapidly excreted by the nephridia. Other foreign substances are often transported to the nephridia and held by the phagocytic cells of the middle tube to be excreted in a more leisurely fashion.

Diffusion. Although the reaction to injury is largely restricted to the injured segment, there is free communication to adjacent segments through septal perforations. Foreign particles or bacteria injected into one segment may diffuse throughout the coelomic cavity in a short while; the extent is dependent on the activity of the worm. The diffusion, though occurring in both directions, is predominantly toward the posterior with the terminal segments containing "brown bodies" consisting of chloragogen granules, cell debris, and foreign materials. The process of diffusion is an important defense mechanism in that it dilutes the foreign substance, allows greater opportunity for phagocytosis and removal via the dorsal pores and nephridia, and concentrates the foreign material in terminal segments in the brown bodies preliminary to discharge by autotomy.

Reaction to Injuries of the Body Wall

Injury of the body wall of earthworms by cauterization with a heated needle or by application of a strong acid results in complete destruction of the superficial epithelial covering and of portions of the underlying subcutaneous connective tissue and muscular layers. The necrotic area protrudes because of muscle contraction of surrounding undamaged tissue. There is an immediate increase in the number of coelomic corpuscles, and they infiltrate the coelom just beneath the injury. The infiltrating cells that approach and adhere to the coelomic lining cells are mainly basophilic cells, though acidophilic cells are also present. Within 3 hours postinjury, there is a marked increase in coelomic corpuscles; these cells virtually fill the coelom and closely adhere to the parietal lining cells beneath the site of injury. The parietal lining cells undergo marked proliferation by this time; they form thick sheets of cells, many of which are elongated and almost spindle shaped. The reaction, however, is purely coelomic at this stage with no invasion of the muscular layer or of the necrotic tissue of the injured area.

By 6 hours, migrating coelomic corpuscles infiltrate the muscle layers and necrotic tissue, and small foci of elongated cells appear at the margin of the healthy muscle tissue. Within 24 hours, there is a marked thickening of the body wall, with an apparent decrease in the size of the coelom due to the formation of a mass of cellular tissue which forms a plug between the necrotic, superficial tissue and the coelom. This plug is bounded on each side by normal tissue, well differentiated into epithelial and muscular layers, but the plug consists of undifferentiated elongated cells and coelomic corpuscles. The great cellularity contrasts markedly with the undamaged, adjacent body wall. By this time, tiny vascular twigs appear at the margin of the plug.

The plug of tissue is still apparent 2 days after injury, but organization of the mass has begun—the cells are arranged in a longitudinal direction, and connective tissue fibers appear between the cells. Muscle fibers overlay and infiltrate the edges of the plug. Desquamation of the necrotic area has occurred, and the epithelium of the skin starts to grow over the denuded area. Coelomic corpuscles, chiefly basophilic, are still numerous, and small blood vessels are fairly well marked at the edges of the wound.

Within 5 days, the denuded surface is completely covered by ingrowth of the epithelial cells of the skin; considerable differentiation into cuboidal and glandular cells occurs, though the new layer is thinner than normal. The inflammatory mass now consists largely of spindle cells and connective tissue fibers arranged in regular longitudinal bands beneath the epithelium, with few coelomic corpuscles. Muscle fibers, growing in from adjacent, undamaged tissue, extend partly across the damaged area, but are not yet arranged in regular groups. Blood vessels (growing inward from the margins) are numerous within the mass.

Repair of the damaged tissue is almost complete within 10 days; the epithelial covering and the body wall are relatively normal in appearance, but some primitive spindle cells, poorly staining fibrils, coelomic corpuscles, and newly formed blood vessels remain, and proliferation continues in the lining coelomic epithelium.

Class Hirudinea

The wound repair process in leeches has received little attention. Mann (1962) noted that phagocytosis by coelomic corpuscles had often been observed, but that subsequent activity of the phagocytes had not been elucidated. Kaestner (1967) reported that particles are engulfed by the coelomocytes and excreted into the gut, connective tissue, or exterior, but he did not describe the mechanism by which this is accomplished. Several authors have postulated that only soluble wastes and foreign materials can be eliminated via the nephridia because of the lack of communication between interior parts and the excretory canals.

In gnathobdellid leeches, foreign or necrotic material is removed from the coelomic fluid by attached phagocytic cells in the botryoidal and vasofibrous tissue (prominent masses of tissue beneath the muscle layers). These tissues, as such, do not occur in rhynchobdellid leeches, and large particles are phagocytized by coelomic epithelial cells, while smaller particles are taken up by pigment cells.

Myers (1935) described some facets of the wound repair process in a glossiphonid leech (*Placobdella parasitica*) as a part of the insemination process. In this species the acting male cements a spermatophore to the body of the acting female, and chemically induced histolysis permits intrusion of the spermatophore contents. In its movement to the "uterus" to fertilize the eggs, the seminal fluid creates a path of tissue destruction and disorganization. The cuticle and epithelium at the site of spermatophore attachment are destroyed; the underlying parenchyma becomes a pale, granular, amorphous mass; the muscle tissue is distorted, and the fibers are partially histolyzed and pigment cells are destroyed.

Immediately after passage of the seminal fluid, the pathway begins to fill with a plug consisting of three types of cells: cells with a small oval nucleus apparently emigrating from adjacent parenchymatous tissue; larger predominant cells believed to be wandering leukocytes; some larger phagocytes of unknown origin. Myers stated that growth and differentiation of these cells results in tissue regeneration, noting that no mitotic figures were identified with certainty. No information was provided on the disposition of the necrotic cellular fragments.

More recently, the reaction to large foreign bodies and the repair process after incision wounds in a rhynchobdellid leech (*Piscicola salmositica*) have been investigated (LeGore, 1970; LeGore and Sparks, 1971).

Large Foreign Body

Cotton thread soaked in aqueous carmine suspension and inserted just dorsal to the middle of the body causes little trauma; the tissues are compressed rather than lacerated by the passage of the needle and the presence of the thread. Wandering cells, which LeGore called oppleocytes, emigrate rapidly into the area, and carmine particles are carried away in the coelomic fluid 2 hours postimplantation. Within 4 hours, coelomic phagocytes engulf carmine particles at the periphery of the thread and migrate, presumably through coelomic channels and epithelial capillaries, to the surface epithelium. They finally traverse the epithelium, causing mechanical damage, carrying the phagocytized carmine particles outside the body (Fig. II.16).

By 12 hours postimplantation, oppleocytes begin to form a thin sheet of connective tissue over the dorsal surface of the thread. By 24 hours postinjury, a thin network of loose connective tissue surrounds the thread, and a few oppleocytes continue to infiltrate the wound (Fig. II.17). Carmine particles can still be seen in interfibrillar spaces, around the wound periphery, and incorporated into the connective tissues. The surrounding tissues, though distorted by compression, appear to be viable.

Beginning approximately 5 days postimplantation, the entire encapsulated mass is transported toward the dorsal surface. Usually, the epithelial surface over the thread is depressed relative to the adjacent body surface, and the epithelium is fragmented, apparently by ejection of debris. The tissues dorsal to the thread apparently histolyze, as indicated by loss of affinity for stain by the muscle tissue and by fragmentation of connective tissue. The thread is eventually (120–150 hours) expelled through the dorsal body surface, leaving behind a plug of newly formed connective tissue (Fig. II.18).

Incision Wounds

Because of the unavoidable variation in severity of experimentally inflicted incision wounds, the time sequence of wound healing is difficult to establish, but the sequence of events is consistent. The leech responds to wounding (incisions, burns, etc.) by rapid movement, apparently in an attempt to escape the source of its injury. The most prominent gross response is local contraction of body segments. The degree of contraction varies with the site of injury; wounds in the posterior portion of the body elicit only contraction of the segments in the area of injury. Midbody injuries initiate contraction of the entire posterior portion of the body from one or two segments anterior to the wound, and wounds in the anterior portion result in only local contraction. Amputation of the posterior body region results in contraction of the three or four adjacent segments anterior to the wound, but the excised posterior portion does not contract. The segment at the cut surface constricts, partially closing the wound, and the wound is effectively sealed within 3–6 minutes.

As mentioned previously, nonamputating incision wounds initiate an almost immediate constriction of the involved segments resulting in an approximation of the adjacent uninjured epithelium (Fig. II.19). The approximated epithelium does not completely close the surface wound, but the underlying tissue compressed by the contraction, together with the epithelium, effectively seals the wound to prevent excess loss of body fluids.

Within 1 hour postinjury, oppleocytes appear in the intermuscular

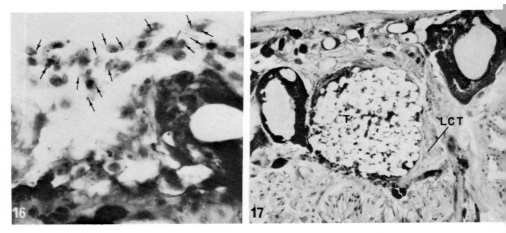

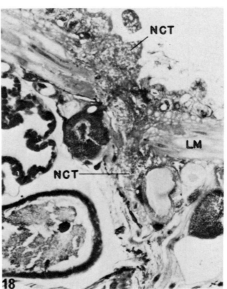

Fig. II.16. Carmine-containing phagocytic cells exterior to body wall of the leech, *Piscic* *salmositica,* 4 hours after insertion of carmine-soaked thread into body. Arrows point to intracell carmine particles. Hematoxylin and eosine. Approximately 119×. (From LeGore, 1970.)

Fig. II.17. Thread (T) insertion in *P. salmositica* 24 hours postimplantation. Note thin laye loose connective tissue (LCT) surrounding thread. Hematoxylin and eosine. Approximately 11′ (From LeGore, 1970.)

Fig. II.18. Thread expulsion channel through dorsal body wall at 120 hours postimplantation. N the depth to which new connective tissue (NCT) extends (LM—longitudinal muscles). Hemat lin and eosine. Approximately 119×. (From LeGore, 1970.)

spaces adjacent to the wound, apparently arising from nearby healthy intermuscular and dermal connective tissue. Between 4 and 8 hours, depending on the severity of the wound, a thin bridge of oppleocytes forms across the wound surface, and a sheet of oppleocytes, 9–10 cells in thickness, grows inward from the periphery of the wound. Concurrently, coelomic phagocytes invade the wound, phagocytize the cellular debris, and migrate away from the wound, leaving the area virtually devoid of debris and coelomic phagocytes by 16 hours postinjury. Within approximately 24 hours in a moderately severe incision, the oppleocyte sheet covers the wound and fills the wound channel with a plug of newly formed connective tissue (Fig. II.20). Larger cell fragments are incorporated into the plug and carried to the body surface where they are sloughed along with the outer portions of the plug. The plug is replenished by oppleocyte emigration to the deeper portions of the wound channel.

Reepithelization begins at about 96 hours postinjury. Uninjured epithelial cells at the periphery of the wound surface flatten somewhat, migrate over the connective-tissue plug, and assume a cuboidal appearance. The process is completed in about 24 hours (120 hours postinjury) without recognizable mitotic activity.

The plug, meanwhile, remains unchanged as an undifferentiated mass of parenchymatous connective tissue histologically resembling a scar. It remains, with no tissue regeneration, for at least 6 additional days (11 days postinjury).

Phylum Sipunculida

The reaction to injury in this group has been studied by several investigators, including Cuénot (1913), who described the "urn cells" as mucociliated scavenger cells that swim through the coelomic fluid accumulating foreign matter and subsequently disposing of it. Cantacuzène (1922a,b,c, 1928) kept several specimens of *Sipunculus nudus* alive in the laboratory for several years and studied their reactions to coelomic injections of bacteria and other foreign materials. He demonstrated that the urn cells collaborate with other phagocytes to eliminate foreign particles. Bang and Bang (1962) elaborated on the morphology of the urn cells and investigated further the immunologic properties of the coelomic fluid suggested by Cantacuzène.

The urn cell is, as shown by Bang and Bang, actually composed of two cells firmly attached to one another with considerable free play at the zone of attachment. The anterior cell has the appearance of a trans-

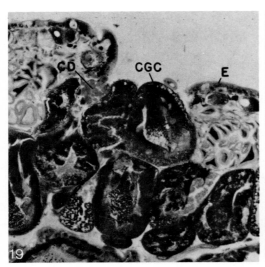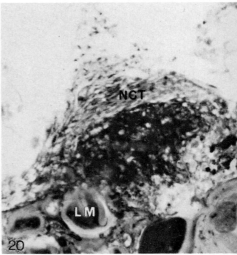

Fig. II.19. Incision wound in *P. salmositica* at 1 hour postinjury. Note approximation of un jured epithelium (E) through constriction of segment and plugging of the wound by coc gland cells (CGC) and associated collecting ducts (CD). Hematoxylin and eosine. Approximat 129×. (From LeGore, 1970.)

Fig. II.20. Incision wound in *P. salmositica* at 24 hours postinjury. Plug of new connective sue (NCT) fills the wound to longitudinal muscle (LM) depth. Note the sloughing of the ou layer of the plug. Hematoxylin and eosine. Approximately 322×. (From LeGore, 1970.)

parent bubble with a flattened base and a short neck. The base and neck fit into a concavity of the saucer-shaped posterior cell, which has a clear secretory central area and is ciliated on its outer convex side (Fig. II.21). The posterior cell is provided with a band of both short and long cilia serving to propel the urn cell through the coelomic fluid and to flick some substances away from the cell and draw others onto the mucous tail. The two rows of cilia are distinct, but it has not been determined whether they alternate or have separate zones of insertion. The long row fends off nonscavenged cells, such as injected erythrocytes, and provides a current for carrying material, such as dead amoebocytes, debris, and bacteria, into the range of the short cilia that propel the particles into the mucous mass or tail that is formed, presumably, by secretory droplets arranged in irregular rows inside the base. It is not known whether the base cell has the form of an open sleeve or whether the mucus is extruded onto the ciliated surface. In either event, the mucous tail accumulates debris until a large mass is attained, then it is dropped off and another tail is gradually secreted.

Cantacuzène and Bang and Bang commented on the ability of sipunculids to eliminate various materials, such as bacteria, ciliate para-

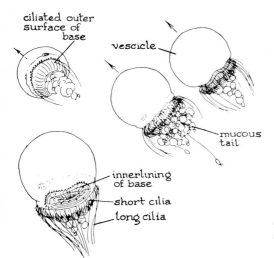

ciliated outer
surface of
base

vescicle

mucous
tail

innerlining
of base

short cilia

long cilia

Fig. II.21. Swimming urn cells of *Sipunculus* drawn while observed under phase microscopy. (From Bang and Bang, 1962.)

sites of crabs, carmine particles, and various blood cells, injected into the coelomic cavity, and they briefly discussed the role of the urn cells in this process. Although wounds were obviously made while injecting materials into or drawing fluid out of the coelom, no information on wound healing was provided. Since some individuals were apparently penetrated a number of times and no comments were made of deleterious effects from the wounding, one must assume that small surface penetrating wounds heal rapidly and without conspicuous pathological effects.

In an investigation of the tissue reactions to homologous and autologous skin grafts and to various materials placed in the coelom of a sipunculid worm, *Dendrostomum zostericolum* (Triplett *et al.*, 1958), no evidence of healing was observed. When materials are inserted into the coelom of *Dendrostomum* they are encapsulated by leukocytes. Bits of severed *Dendrostomum* tentacles placed in the coelom remain alive for up to 70 days, but they are encapsulated. When a second tentacle homograft is inserted, the ability to encapsulate it is decreased, presumably because fewer leukocytes are available, indicating that leukocytes do not proliferate rapidly. Microscopically, the structure of the capsules is rather similar to those described by Cameron (1932) in the earthworm; a layer of compacted cells alternates with a layer of dense, sparsely celled fibrous tissue. Hemerythrocytes, leukocytes, eggs, and sperm are sparsely scattered throughout the capsule wall and in the fluid contained in the capsule. The capsule becomes tightly fused to the severed ends of the tentacle, but it never fuses to the intact epidermis of the implant. Small pieces of cellulose sponge, filter paper, plastic tubing, and sea anemone tentacles are also readily encapsulated; the anemone tentacles are heavily encapsulated, killed, and almost completely digested.

Urn cells do not occur in *D. zostericolum*, but the coelomic fluid contains hemerythrocytes and large numbers of leukocytes of several types, including an eosinophilic cell as a characteristic component. The eosinophils and several other unclassified types are amoeboid and phagocytic; they ingest and break down erythrocytes, bacteria, and yeast injected into the coelom. Phagocytosis *in vitro* is also easily demonstrated. Although phagocytosis obviously plays a major role in reaction to foreign materials (injected cellular material is removed from the coelom within a few hours), there does not appear to be an increase in the number of phagocytes in response to injury.

Phylum Mollusca

There has unquestionably been more investigation of the reactions to injury and wound repair in the mollusks than in any invertebrate group (other than the insects). The economic value of many members of this phylum is a major reason for this. As one might expect, the studies have been confined largely to gastropods and pelecypods, with a few reports of reaction to injury and wound repair in the cephalopods, but virtually nothing is known of these phenomena in the chitons and scaphopods.

CLASS GASTROPODA

The major interest in injury reaction in gastropods has been the response to invasion by larval helminths. Several investigators, however, have studied the response of a number of gastropods to experimental wounds and injections of foreign materials, and they have demonstrated rapid proliferation and repair of the injured tissues.

Tripp (1961) described the tissue reactions initiated by the introduction of several types of particulate material into a pulmonate gastropod, *Biomphalaria glabrata*. Phagocytosis of injected yeast cells in connective tissues begins within 10 minutes after injection and is virtually complete within 8–16 hours. Most amoebocytes in the area contain 1–3 yeast cells, with 6–8 cells per amoebocyte common, and occasionally up to 20 yeast cells may be found in hypertrophied amoebocytes. Subsequent defense mechanisms depend on the prior treatment of the injected cells and involves one or both of the following processes: migration of particle-laden amoebocytes across certain epithelial layers to the exterior and intracellular breakdown of the particles as shown by cellular fragmentation and loss of staining affinities. Response to dried, cultured yeast cells is similar with approximately equal numbers of these untreated cells removed by amoebocyte transport (over a period of 20 days) and intracellular degradation (which is most evident at 2 days). By contrast, moist-heat-killed (held in a water bath at 100°C for 15 minutes) yeast cells are removed by migrating leukocytes during the first 2 days after injection, and the few yeast cells remaining in the tissues at 4 days are degraded intracellularly. Virtually all autoclaved yeast cells are degraded intracellularly within 4 days after injection and are only rarely transported across the epithelial layers to the exterior. Thus, untreated yeast cells persist in the tissues longer than those previously treated.

Injected bacteria are eliminated from the snail in much the same fashion as yeast when pretreated comparably. Some untreated cocci are transported across epithelial borders, beginning at 8–24 hours and lasting through 16 days. Intracellular degradation, however, is a more constant mechanism, occurring from 4 hours through 24 days. Again, like the yeast cells, heat-killed bacteria are readily eliminated by diapedesis, but intracellular breakdown begins relatively late (12 days), when few cocci remain in the tissues. Autoclaved bacteria are only rarely carried to the exterior by amoebocytes, and intracellular degradation occurs from 2 to 23 days.

Formalin-fixed chicken erythrocytes are quickly phagocytized and degraded. Ten minutes postinjection, amoebocytes contain large numbers of normal-appearing red blood cells. Within 2 hours, however, degradation of the ingested erythrocytes has begun, and advanced degeneration is evident at 24–72 hours postinjection. Tissue sections of snails sacrificed at 4–28 days contain no erythrocytes, and there is no indication of transport of red blood cells by amoebocytes through the epithelium.

Carmine grains are phagocytized readily but are not carried through the epithelium; they are retained in clumps near the site of injection for at least 30 days. The reaction to injections of pollen grains and small polystyrene spheres is similar and characteristic. The size of these particles precludes phagocytosis, but amoebocytes surround most particles within 16–24 hours after injection. Within 24 hours, fibroblasts migrate to the site from adjacent connective tissue, and a well-defined circular lesion is formed, consisting of a central mass of densely packed particles and amoebocytes surrounded by one or more layers of fibroblasts. Most pollen grains are enclosed in such nodules within 4 days, or the individual particles are surrounded by a layer of leukocytes. These lesions persist for at least 42 days without any apparent change in the pollen grains. Polystyrene spheres are isolated similarly, but the formation of discrete nodules is delayed, up to 22 days, probably because of the larger size of the spheres.

The cellular reaction of gastropods, as shown by Tripp's studies of *B. glabratus*, to injection of particulate matter depends on the size and nature of the foreign material. Particles small enough to be phagocytized are quickly engulfed and may be carried across epithelial surfaces to the exterior, degraded intracellularly, or retained within the leukocytes, depending on the nature of the particle. Foreign bodies too large to be phagocytized are rapidly encapsulated by fibroblast migration from adjacent connective tissue.

CLASS PELECYPODA

The amoebocytes, leukocytes, or hemocytes of bivalves, as in other mollusks, play a dominant role in defense mechanisms. Accounts of the types, structure, and functions of leukocytes in various mollusks have been presented by a number of authors (Drew, 1910; Cuénot, 1891; Orton, 1922; Yonge, 1926; Haughton, 1934; Takatsuki, 1934; George and Ferguson, 1950; Wagge, 1955; Mackin, 1961).

Under normal conditions, there are large aggregations of leukocytes surrounding the oyster's alimentary tract (Yonge, 1926; Takatsuki, 1934). Takatsuki stated that this phenomenon may erroneously be interpreted as inflammation. The same is certainly true of the phenomenon of resorption of unspawned sex products in which conspicuous leukocytic infiltration of the gonadal areas occurs.

Reaction to Injury

Stauber (1950) demonstrated, in his pioneering and classic study of the role of the leukocyte in the defense mechanisms of oysters that the immediate effect of intravascular injection of agglomerating particles of india ink is virtual embolism of the arterial system, including the branches to the visceral mass, palps, mantle, adductor muscle, and rectum (Fig. II.22). However, the sinuses adjacent to these arteries and to the gill sinuses are unaffected. The arterial emboli consist of only ink particles for several hours, but from 22 hours to 8 days large numbers of leukocytes, most containing ink particles, are associated with the ink mass (Fig. II.23). Between the eighth and seventeenth day, most of the phagocytes migrate away from the arterial system. Although the mass of ink particles may mask the occurrence of phagocytosis, it appears to be initiated somewhere between 2 and 4 hours postinjection. By the end of the first day, large numbers of leukocytes are present in the arterial system, and most of the ink is intracellular, but leukocytes in adjacent sinuses and the gill sinuses are still virtually free of ink particles.

After phagocytosis, ink-laden leukocytes are distributed throughout the oyster (Fig. II.24), both by passage from small arteries to blood sinuses in typical circulatory fashion and by migration through the walls of the blocked arteries. The ink particles are eventually eliminated from the oyster by migration of the ink-laden phagocytes through the epithelia into various lumina (diapedesis). This phenomenon begins at approximately 8 days after injection, and the chief sites of diapedesis are the epithelia of the digestive system (Fig. II.25) (stomach, intestine, digestive diverticula, and rectum), although discharge occurs rarely through the pericardial wall into the pericardium and through the epithelium of the mantle and palps. The epithelia of the external, shell-secreting mantle, the excretory tubules, the gonaducts, and the gill epithelium are almost never involved in diapedesis.

After intracardial injection of rabbit, weakfish, and normal and malarial-parasitized duck erythrocytes, virtual arterial occlusion occurs; but phagocytosis is immediately initiated, beginning within 10 minutes and virtually complete within 6 hours (Tripp, 1958). Blood pressure forces some erythrocyte-laden leukocytes through smaller arteries into sinuses, and they migrate from the vessels and sinuses into the tissues. Diapedesis begins as early as 48 hours postinjection, reaches a peak at 8 days, and involves the same epithelial surfaces noted by Stauber (1950). It is clear from Tripp's experiments that intracellular digestion of the avian erythrocytes regularly occurs. This was demonstrated particularly well when avian red blood cells containing malarial parasites were injected. The pigment associated with the parasite results from the destruction of hemoglobin; the globin is split off and the acid hematin residue remains in the cell. Since the acid hematin is extremely resistant to destruction, it remains in the oyster leukocyte after the erythrocyte has been destroyed. Twelve days after injection of the parasitized erythrocytes, leukocytes containing malaria pigment can be seen readily in the tissues, while intact red cells are difficult to find. Yonge (1926) presented evidence of the presence of intracellular proteases and lipases in oyster leukocytes and showed that shark erythrocytes fed to *Ostrea edulis* are phagocytized and digested to a mass of fat globules within 12 hours. This demonstrates the ability of oyster leukocytes to digest such materials and suggests the mechanism by which it is accomplished. As Stauber (1950) noted, the activity of leukocytes in a poikilothermic animal such as the oyster is correlated

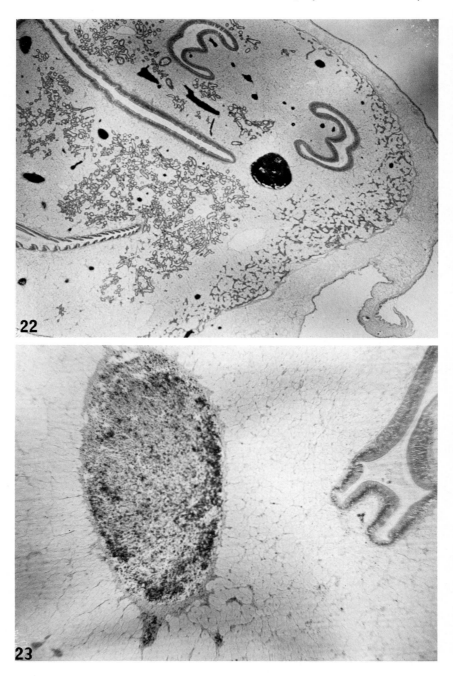

Figs. II.22 and II.23

with the temperature. Numerous studies have shown that physiological processes in the oyster virtually cease when temperatures are depressed below 5°C and are significantly slowed at temperatures of 5°–10°C.

Tripp (1958) summarized the response of leukocytes to particulate matter as engulfment of the particle followed by migration. If the particle is not metabolizable, it is eliminated by diapedeses, but this response is supplemented by intracellular digestion. Some potentially pathogenic microorganisms are probably destroyed in this manner, while others appear to be resistant to the intracellular enzymes of the leukocyte and may even reproduce within them.

Slide preparations of oyster blood showed that freshly introduced bacteria were not phagocitized; frequently an amoebocyte approached a bacterium and then turned away as if to avoid it (Bang, 1961). He was unable to ascertain if certain bacteria were resistant to phagocytosis or if certain oysters were unable to phagocytize a particular strain of bacteria. When phagocytosis occurred in his *in vitro* studies, it was usually preceded by massive sticking of bacteria to the leukocyte. Motile bacteria were shown by electron microscopy to have flagella wrapped around the filamentous pseudopodia of the leukocytes. Bang also observed both extravascular and intravascular clots of leukocytes. He felt that the extravascular clots were not artifacts, since observation of traumatized blood vessels revealed the rapid formation of similar clots that effectively sealed the cut end of the vessel. Similar clumps of leukocytes were noted covering the cut end of the adductor muscle in his preparations, and oysters bled repeatedly developed a shaggy pericarditis consisting of masses of clumped leukocytes.

If oyster leukocytes could be maintained in true tissue culture, their reaction to injury, foreign bodies, and disease organisms and their role in tissue-repair processes could be studied. True cultures of mollusk cells, involving an *in vitro* system in which cells undergo mitosis and from which subcultures can be obtained, have not been established (Perkins and Menzel, 1964), but both Perkins and Menzel and Tripp *et al.* (1966) have been able to maintain tissue explants, utilizing tissue-culture methods, with leukocytes migrating from them for relatively long periods. Perkins and Menzel utilized mantle explants with leukocytes migrating from them from ½ hour to 42 days. A loose monolayer network of cells was formed by contact of pseudopodia of adjacent leukocytes.

Fig. II.22. Cross section of *Crassostrea virginica* shortly after intracardial injection with india ink. Note the virtual embolism of the arterial system. (From Stauber, 1950. Courtesy of L. Stauber.)

Fig. II.23. Arterial emboli several days after intracardial injection of india ink. Note increased numbers of leukocytes present in artery. (From Stauber, 1950. Courtesy of L. Stauber.)

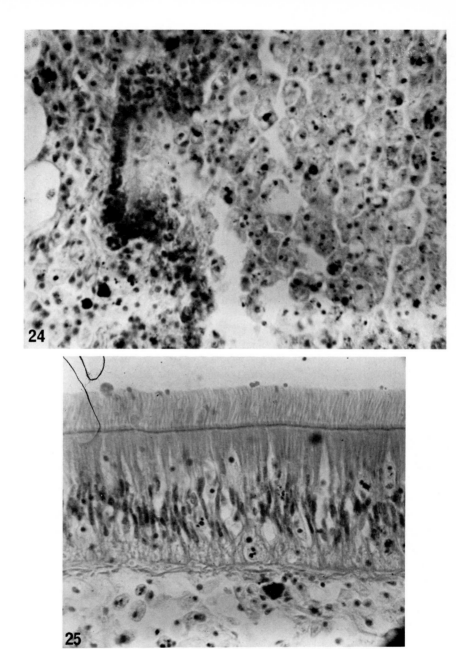

Fig. II.24. Ink-laden leukocytes throughout oyster tissue subsequent to phago-cytosis and migration from the arterial system. (From Stauber, 1950. Courtesy of L. Stauber.)

Fig. II.25. Migration of ink-laden leukocytes to and through intestinal epithelium. (From Stauber, 1950. Courtesy of L. Stauber.)

It was assumed that all cells in the monolayer originated in the explant since no evidence of mitosis was observed. However, since numerous leukocytes were washed away with each change of medium without apparent decrease in numbers in the explant, it was suspected that multiplication occurred in the explant even though it was not demonstrated. Similar results were obtained by Tripp *et al.* (1966) with heart explants that were maintained for periods up to 70 days. The amoebocytes migrating from the heart explant were identical in appearance to those seen in freshly drawn heart blood. They were observed to be actively phagocytic for several weeks, engulfing particles such as yeast, bacteria, algae, and erythrocytes. Multinucleate leukocytes occurring in the medium were assumed to have been mononucleated forms which had phagocytized nuclei released from autolyzing cells, as described by Sparks and Pauley (1964) in decomposing oysters. Although 250 media were tested, mitotic figures were never seen, and attempts to induce mitosis by treating cells with phytohemagglutinin were unsuccessful. Thus it is not known whether these cells can divide or whether they represent fully differentiated cells. Although true cell culture has not been achieved in oysters or other mollusks, the groundwork has been accomplished and detailed, and long-term studies of molluscan leukocytes *in vitro* are now possible.

In situ tissue reactions of bivalve mollusks to the insertion of foreign bodies or materials have been studied in some detail by a number of investigators using a variety of pelecypods as experimental animals. Along with his studies of inflammation in certain gastropods, Messing (1903) investigated the tissue reactions of the freshwater clam *Anodonta* to experimental wounds and injection of foreign material, noting that rapid proliferation of tissues and repair of the injured tissues occurred. Drew (1910) stated that small incisions in the foot of *Cardium norvegicum* healed within 3 weeks by a process remarkably similar to that of mammals. In the same year, Drew and de Morgan (1910) published an excellent histological account of the defense reactions of a scallop, *Pecten maximus*, to the insertion of three different foreign substances into the adductor muscle. Sterile agar and gill-tissue implants elicited a fibrous encapsulation of the foreign materials, and fibroblasts formed around digestive-gland implants in the adductor muscle after 6 days, but the authors were unable to maintain the experimental animals for longer periods. Somewhat surprisingly, Kedrovsky (1925) failed to find fibrous-tissue formation around celluloid implants in the foot of *Anodonta*, but he described, in detail, the nuclear changes occurring in the numerous leukocytes infiltrating the wound area and surrounding the celluloid. However, when Zawarzin (1927) introduced the same material into the mantle of *Anodonta*, a secreting epithelium formed around the irritant and the encapsulated foreign body was rejected from the body, a phenomenon described in the oyster by Mackin (1961). Labbé's (1929) experiments with celluloid-stylet implants in the mantle of *Anodonta* were largely in agreement with the findings of Kedrovsky and Zawarzin. The wounded area is infiltrated with masses of leukocytes that form an agglutination of degenerative cells around the inclusion. Beneath this layer, in 8 days, a layer of small, parallel, fusiform fibroblasts is formed, with a band of nondegenerative, actively anastomosing leukocytes which Labbé believed corresponded to the epithelial layer observed by Zawarzin in *Anodonta* and in the gastropod *Doris* by himself. Labbé noted that the fibroblast layer was inconspicuous and that there was no indication of connective tissue encapsulation.

The reactions of the mussel, *Mytilus edulis*, to the introduction of a carmine-soaked thread or to bacteria were studied by Mikhailova and Prazdnikov (1961, 1962). They noted that the main component of the inflammatory process in mussels is phagocytosis, and they expressed the opinion that the phagocytic reaction plays the decisive role in the protective mechanisms. Reaction of the mussel tissue to injuries varies greatly, depending on the nature of the foreign body, the rate of infiltration of cellular elements to the injury site, the rate of phagocytosis, and the intensity of phagocytic activity. For example, *Bacillus subtilis* accelerates the initiation of the inflammatory response; while carmine grains and microorganisms collected from seawater delay amoebocyte infiltration and phagocytic reactions.

The morphological features of inflammation in mussel mantle tissue occur in three stages. The first stage, developing during the first 12 hours postinjury, is characterized by the deposition of damaged connective tissue and homogenized collagenous tissue around the introduced foreign body; this is accompanied by migration of numerous leukocytes to nearby blood spaces. Encapsulation of the foreign body is incomplete at this stage, and part of the foreign materials, if particle size, enter adjacent blood spaces. This initial inflammatory stage may be prolonged for 24 hours, depending on the nature of the introduced foreign agent, and the authors speculated that antibiotic substances of the tissue play an important role in immunologic reactions during this period.

The second stage begins after 12–24 hours and continues for 5–10 days after the injury. It is characterized by intensive phagocytosis and both intra- and extracellular digestion of the foreign particles. After approximately 48 hours, eosinophilic amoebocytes infiltrating between the fibers of the introduced thread and into the periphery of the wound channel stick together to form symplastic structures (which probably can readily separate) surrounding the foreign bodies in rings or layers. Digestion of bacteria occurs partly outside the amoebocytes in the symplastic loops, probably to compensate for the lack of encapsulating capacity.

The third stage of the inflammatory response of *Mytilus* is characterized by the removal of the particles of the foreign body from the wound area. Amoebocytes containing engulfed particles enter the blood spaces and migrate to the surface epithelium where they are discharged. In the process of removing foreign particles and cellular detritus, basophilic amoebocytes also participate, and desmoblasts and mioblasts are formed near the foreign body to provide supporting tissue for repair. This third stage of the inflammatory process can be superimposed on the second stage, and occurs between 5 and 10 days after the injury.

Pauley and Sparks (1965, 1966, 1967) studied the acute inflammatory response and wound repair process in the Pacific oyster, *Crassostrea gigas*. In separate studies, turpentine or talc was injected into the connective tissue above the palps and into the adductor muscle. Turpentine is an irritating substance that elicits a typical inflammatory reaction, and it is frequently used in studies of vertebrate inflammation. Talc characteristically invokes a postoperative inflammation resulting in granulomatous scar-tissue formation in vertebrates, and it was used in a study of the wound repair process in the cockroach, *Periplaneta americana*, by Schlumberger (1952). The use of these two irritants thus provides a comparison of inflammation and wound repair between the oyster, vertebrates, and insects.

The gross appearance of oysters injected with either talc or turpentine differs both in pattern and time of change. Talc injected into the connective tissue initiates the development of a greenish discoloration around the wound within 16 hours; it persists for 72 hours, then changes to a yellowish-green color, and appears normal after 96 hours postinjection. Adductor muscle injections of talc produce a greenish color along the line of injection within 16 hours, which persists until 168 hours; then it becomes yellow or opaque white. After 216 hours, virtually all oysters exhibit a white scar at the injection site. Turpentine injections into the connective tissue elicit no grossly recognizable response for approximately 32 hours, at which time the adductor muscle becomes swollen and is incapable of holding the valves of the shell closed. Abnormal accumulations of silt on the mantle and gills accompany the loss of function of the adductor muscle and persist for approximately 350 hours. Occasionally, white, tumorous swellings, surrounded

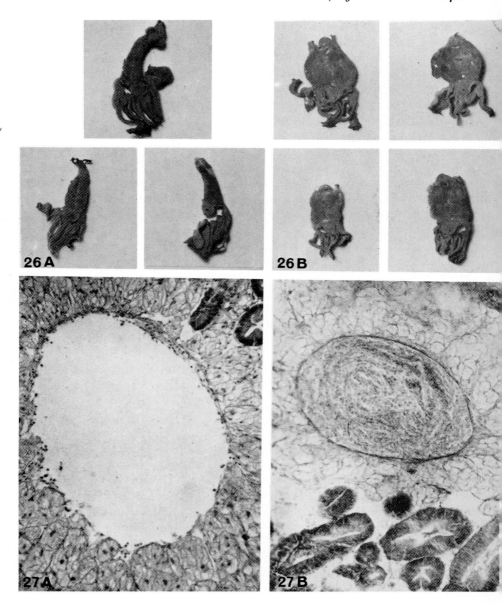

Fig. II.26. Inflammation in the oyster, *Crassostrea gigas*. (From Pauley and Sparks, 1966.)
A. Cross section of 3 oysters 120–128 hours after injection of seawater into the adductor muscle
B. Cross section of 4 oysters 128 hours after injection with turpentine. Note the swollen appearance of the adductor muscle after injection directly into the muscle (top two) or in
   the connective tissue.

by pus, develop on the adductor muscle; they are shown histologically to consist of disorganized, hypertrophied muscle fibers with karyolytic nuclei.

Injection of turpentine directly into the adductor muscle elicits a similar, but earlier and more marked, gross response (Fig. II.26 *A* and *B*). The muscle becomes noticeably swollen and loses its functional capacity within 16 hours. After 32 hours, the attachment of the muscle to the shell is often lost and abnormally large amounts of silt accumulate on the gills and mantle. Pus appears in the area of injection within 48 hours, and grossly identifiable cellulitis (seen histologically as a diffuse area of inflammation) appears by 176 hours and persists for long periods.

A leukocytic infiltration accompanied by edema is the first histological change that occurs about 16 hours after oysters are injected with talc. Blood sinuses adjacent to the injury become congested and contain talc particles after 24 hours. Congestion also occurs in many, but not all, the large vessels some distance from the injured area (Fig. II.27 *A* and *B*). At 24 hours, granulomatous tissue, consisting of elongate cells with round nuclei, begins forming around the talc in the larger blood vessels (Fig. II.28). The nuclei of these cells become elongated after 160 hours, at which time the granulomatous tissue is well organized into thrombi that partially occlude the vessels. The thrombi appear to increase in size and, although never observed, may eventually completely occlude blood vessels causing anoxia and infarction. Lesions begin to form in the connective tissue as early as 24 hours after talc injection and consist of talc particles and loosely aggregated leukocytes surrounded by a compacted, peripheral band of leukocytes (Fig. II.29). The talc particles cause extensive mechanical damage to cells in their immediate vicinity, while cells surrounding, but not involved in, the lesion appear normal. After 40 hours, the leukocytes at the periphery of the lesion become elongated and arranged parallel to one another. By 88 hours, the leukocytes in the center of the lesion are necrotic with light staining cytoplasm and karyolytic nuclei, and aggregations of compact, apparently viable leukocytes occur between the center of the lesion and the elongated cells at the periphery. The nuclei in the peripheral band begin to elongate between 128 and 200 hours (Fig. II.30), and virtually all nuclei are elongated by 250 hours. The central area is infiltrated by elongate cells at approximately 300 hours, and well-formed granulomas

Fig. II.27. Inflammation in the oyster, *Crassostrea gigas.* (From Pauley and Sparks, 1966.)
A. Normal blood vessel. Hematoxylin and eosine. 180×.
B. Blood vessel 32 hours after injection with talc, completely filled with leukocytes and talc particles. Hematoxylin and eosine, 120×.

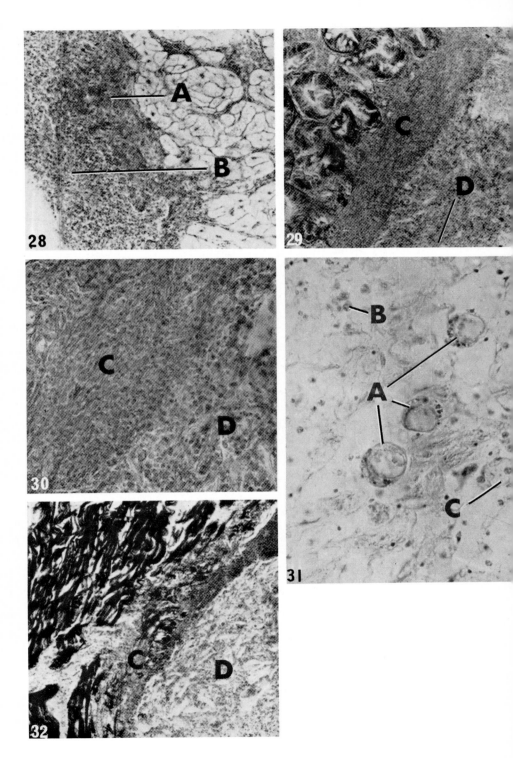

Figs. II.28–II.32

develop, consisting of a band of elongated cells 25–100 cells wide surrounding a mass of talc granules and necrotic leukocytes heavily infiltrated by elongated leukocytes. The lesions also contain large numbers of brown pigment cells after 56 hours, while adjacent normal tissue lack such cells (Fig. II.31). The origin and function of these pigment cells in oyster tissue is unknown, but they occur more abundantly in diseased oyster tissue than in the normal oyster (Stein and Mackin, 1955). They may be a special type of phagocyte with the specific function of removing remnants of necrotic cells, or they may be cells undergoing degeneration.

Talc injection into the adductor muscle elicits a similar response; congestion of blood vessels occurs within 8 hours, and a light leukocytic infiltration appears among the talc particles after 16 hours. Thrombi begin to form in the large vessels in the Leydig tissue adjacent to the adductor muscle by 32 hours, following a pattern similar to the reaction in the connective tissue. Infarcts, characterized by coagulation necrosis of the muscle fibers, occur frequently after 200 hours, probably as a result of anoxia caused by partial thrombic occlusion of blood vessels (Fig. II.32). Granulomas are formed in the same manner as in connective tissue; all muscle fibers in the lesion are necrotic.

Stauber (1950, 1961), Tripp (1958), and Pauley and Sparks (1967) have shown that oysters have an effective defense mechanism against nontoxic, solid particles; they localize the particles by forming granulomas similar to those observed in insects (Schlumberger, 1952) and

---

Fig. II.28. Inflammation in the oyster, *Crassostrea gigas*. Blood vessel 48 hours after talc injection. Note the darker staining granulomatous tissue (A) beginning to form with the lighter staining leukocytes (B) adhering to it. Lumen of vessel at lower left. Hematoxylin and eosine. 180×. (From Pauley and Sparks, 1967.)

Fig. II.29. Inflammation in the oyster, *Crassostrea gigas*. Talc lesion in connective tissue 200 hours after injection. Note the thick band of leukocytes (C) that encapsulates the entire lesion. The central area of the lesion (D) consists of talc particles and necrotic and viable appearing leukocytes. Hematoxylin and eosine. 120×. (From Pauley and Sparks, 1967.)

Fig. II.30. Inflammation in the oyster, *Crassostrea gigas*. Higher magnification of leukocytes encapsulating (C) 200-hour talc lesion. Note the elongated leukocytes at periphery (C) and the round leukocytes in the central portion of the lesion (D). Hematoxylin and eosine. 300×. (From Pauley and Sparks, 1967.)

Fig. II.31. Inflammation in the oyster, *Crassostrea gigas*. Vacuolated brown pigment cells (A); for size comparison, note the leukocyte (B) and Leydig cell (C). Hematoxylin and eosine. (From Pauley and Sparks, 1966.)

Fig. II.32. Inflammation in *Crassostrea gigas*. Infarct in adductor muscle 200 hours after injection with talc. Note the coagulation necrosis in the center of the lesion (D) and the forming granuloma (C) similar to formation in Fig. II.29. Mallory's trichrome. 120×. (From Pauley and Sparks, 1967.)

vertebrates. As demonstrated by Pauley and Sparks (1965, 1967), however, oysters do not possess the ability to successfully combat a caustic liquid such as turpentine. It may spread throughout the body, causing widespread necrosis of vital tissues.

When turpentine is injected into the Leydig tissue of an oyster, microscopic examination reveals edema and general leukocytic infiltration in the area of injury within 8 hours, accompanied by congestion of the adjacent smaller blood vessels and sinusoids. Necrosis of the gut epithelium also occurs at this time, apparently as a result of the turpentine, since the gut does not exhibit postmortem change until 48 hours after fatal injury (Sparks and Pauley, 1964). Marked vascular dilation develops by 16 hours with apparent pavementing of large vessel walls by leukocytes in the vicinity of the injury. Increased numbers of leukocytes appear in the blood vessels and sinuses and begin to move from the blood channels toward the injury site. Edematous areas in the mantle and gonadal regions appear at 24 hours, with marked distention of the gonadal ducts. At the same time, the digestive tubules and Leydig cells in the area of injection undergo massive necrosis, with the cells characterized by faded cytoplasm and pycnotic nuclei. The mantle epithelium over the area of injection is necrotic, with only shadowy outlines of the cell membranes remaining. The wound is heavily infiltrated by leukocytes by 40 hours, and 8 hours later a conspicuous band of leukocytes surrounds the necrotic area (Fig. II.33). Multinucleate giant cells (Fig. II.34), described as normal products of postmortem change by Sparks and Pauley (1964), appear at about 64 hours and are common thereafter. They undoubtedly function as macrophages, phagocytizing necrotic cells and tissue debris. Mass migration of leukocytes across epithelial borders is a constant feature after 64 hours, although it is not known whether the leukocytes involved in this diapedesis contain ingested turpentine. Coagulation necrosis of the epithelia of the gut, stomach, and digestive tubules is conspicuous at 64 hours. By 72 hours postinjection, necrosis of the gonadal area is initiated and is characterized by edema, heavy leukocytic infiltration, distension of the gonadal ducts, and the formation of walls of leukocytes around the necrotic gonadal area. Associated with these conditions is a conspicuous edema in adjacent mantle areas. At approximately the same time, the several types of epithelial cells of the digestive tubules become reduced to an identical-appearing, low, cuboidal epithelium, and the crypt structure is lost (Fig. II.35). However, sloughing or fragmentation of the epithelium, which is characteristic of postmortem decomposition, does not occur. Distinct abscesses develop in the digestive tubule area after 88 hours (Fig. II.36) and appear to be well confined, although there is no indica-

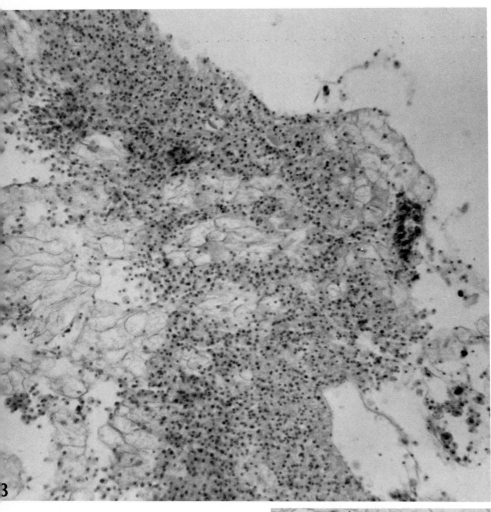

3

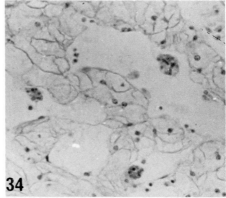

34

Fig. II.33. Site of injection of turpentine in the
ydig tissue of the Pacific oyster (*Crassostrea*
as) 48 hours postinjection. Note the prominent
nd of leukocytes around the necrotic area at right.
maatoxylin and eosine. 225×. (From Pauley and
arks, 1965.)

Fig. II.34. Multinucleated giant cells phagocytiz-
necrotic cells and tissue debris. Seventy-two
urs postinjection. (From Pauley and Sparks,
65.)

73

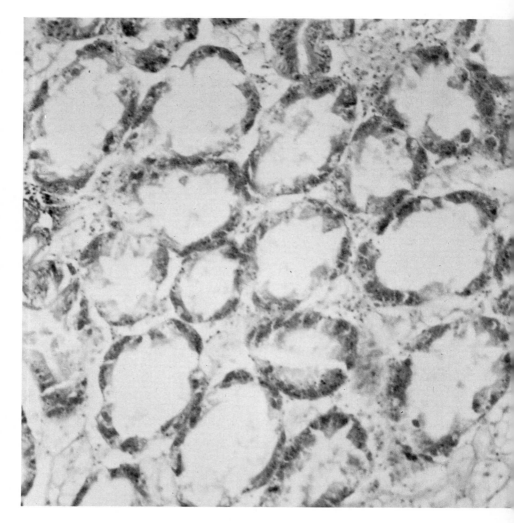

Fig. II.35. Digestive tubules of *Crassostrea gigas* 72 hours after injection with turpentine. N. that the lining epithelium has been reduced to low cuboidal epithelium. Hematoxylin and eosi 230×. (From Pauley and Sparks, 1965.)

tion of fibrous encapsulation. An individual oyster may develop a single abscess or multiple large or small abscesses. The abscesses consist of a central area of liquefaction necrosis, surrounded by a band of necrotic leukocytes and a band of viable-appearing leukocytes at the periphery of the lesion.

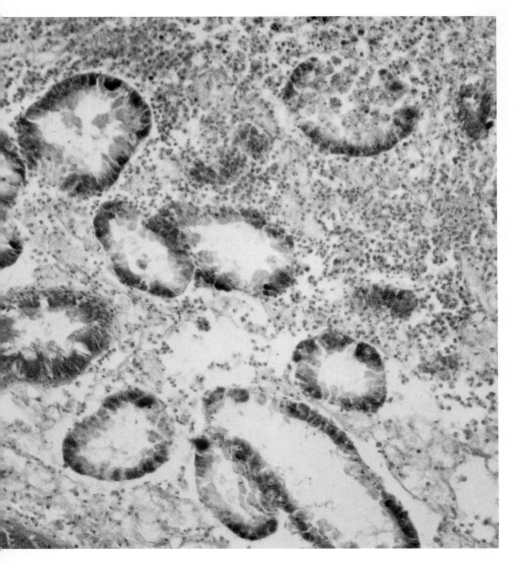

Fig. II.36. Periphery of an abscess in the digestive tubule area of *Crassostrea gigas* 88 hours post-ection with turpentine. Note the progressive necrosis of leukocytes and digestive tubule epithelial s from the periphery of the abscess inward. Hematoxylin and eosine. 250×. (From Pauley and rks, 1965.)

Thirty-two hours after injection of turpentine into the connective tissue, at the time the adductor muscle is grossly observed to be swollen,

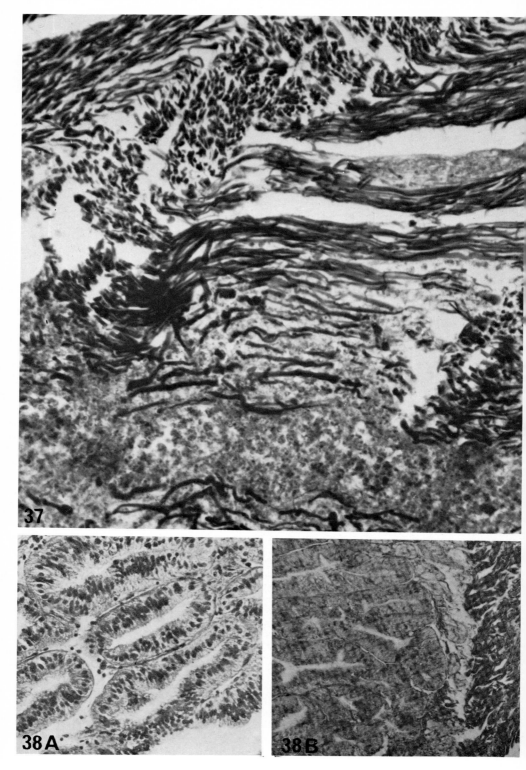

Figs. II.37 and II.38

microscopic examination shows the muscle to be highly edematous. Within a short time, the blood channels become congested, a light leukocytic infiltrate appears in the muscle tissue, and the muscle fibers are necrotic with pycnotic nuclei. Small localized areas of coagulation necrosis, or infarcts, develop in the adductor muscle after 100 hours. The vessels in the tissues adjacent to the adductor muscle become congested by approximately 120 hours, and accumulation of pus, characterized by aggregations of necrotic leukocytes and tissue debris, appears in the adductor muscle. An extensive, spreading cellulitis commonly develops within the muscle after 160 hours, and after 224 hours postinjection, most of the muscle tissue is edematous and heavily infiltrated by leukocytes. The blood channels are congested and the majority of the muscle fibers are necrotic and hypertrophied at this time (Fig. II.37).

The turpentine apparently spreads from the site of injection and causes necrosis of adjacent tissues. By 200 hours, the visceral ganglion becomes necrotic; the nuclei undergo karyorrhexis and often become liquefied by 300 hours. It is not known with certainty whether necrosis of the visceral ganglion results from direct action of the turpentine or from autolysis induced by anoxia resulting from massive damage and loss of function in the adjacent tissue. It seems clear, however, that the spread of turpentine affects other tissues. The epithelium of the kidney becomes hypertrophied, and the nuclei are karyolytic after 80 hours (Fig. II.38); the epithelium begins to slough away from the basement membrane at approximately 240 hours and may be liquefied a short time later. Invasion of the digestive system also occurs with ulceration of the epithelium of the gut, style sac, digestive tubules, and rectum after 200 hours.

Injection of turpentine directly into the adductor muscle initiates inflammatory reactions in the organ similar to, but developing earlier than, those occurring as a result of connective tissue injections. Histologically, necrotic muscle fibers with pycnotic nuclei may be seen within 8 hours after injection. The turpentine remains more localized in the adductor muscle than in the Leydig tissue; the muscle fibers and collagenous fibers retain the material more efficiently than do the Leydig cells. It does, however, spread eventually and causes necrosis of gill, nervous, and kidney tissues.

Fig. II.37. Adductor muscle of *Crassostrea gigas* 168 hours postinjection with turpentine. Note edema, heavy leukocytic infiltrate, and the disorganized, hypertrophied appearance of the muscle fibers. Mallory's trichrome. 400×.

Fig. II.38. (A) Normal kidney epithelium of *C. gigas*. Hematoxylin and eosine. 300×. (B) Hypertrophied kidney epithelium of an oyster (*C. gigas*) sacrificed 120 hours after turpentine injection. Cells which apparently lack nuclei possess karyolytic nuclei. Mallory's trichrome. 120×. (From Pauley and Sparks, 1966.)

Lamellibranchs, as shown by the numerous studies cited previously, possess an effective defense mechanism against nontoxic, solid, foreign materials. Foreign matter is localized, often encapsulated, and removed by diapedesis; then the damaged tissue is repaired. Caustic substances such as turpentine, though eliciting similar tissue responses, are not efficiently localized. They spread to adjacent tissues and cause massive necrosis of vital tissues, possibly resulting in death of the organism.

Wound Repair in Oysters

Several oysters in Pauley and Sparks' experiments developed ulcerated lesions on the surface of the mantle, probably resulting from wounds inflicted during removal of the shell. The wounds were generally characterized by an open ulcer usually lined with low cuboidal epithelium; although they were occasionally lined by relatively disorganized tall columnar epithelium that was confluent with the adjacent, normal, mantle epithelium. Compacted, elongated fibroblasts occurred beneath the epithelium along with collagenous fibers and numerous brown pigment cells. The lower part of the wounds consisted of a thick band of compacted, apparently normal leukocytes, and the blood spaces in the connective tissue below the lesions were congested by apparent migration of leukocytes toward the injury site.

The rather casual observations by Pauley and Sparks (1967) of the repair of the wounds induced by the injection of either seawater or talc into the Leydig cell area or into the adductor muscle led to a more detailed investigation of the wound repair process in the Pacific oyster, C. gigas (Des Voigne and Sparks, 1968).

The tissues surrounding a surface wound in an oyster begin to darken at approximately 16 hours after injury; this is probably analogous to the redness of vertebrate inflammation. Within 24 hours postinjury, the tissues in the vicinity of the lesion acquire a yellow-green coloration which subsequently (48–96 hours) becomes dark green. This dark coloration encircles the wound and persists for approximately 9 days, then begins to fade to yellow-green. However, all specimens sacrificed at the termination of the experiment (during the twenty-eighth day) retained much of this discoloration. Tissues distant from the wound site retain the typical creamy coloration of healthy oysters. The gross response is more pronounced and more persistent in the larger wounds caused by incision with a knife blade than in those induced by a hypodermic needle.

Histologically, the first recognizable response to wounding occurs after about 4 hours. Small blood vessels in the vicinity of the lesion become pavemented by leukocytes, and infiltrating leukocytes begin to form a band underlying the mantle epithelium adjacent to the lesion (Fig. II.39). Small numbers of leukocytes infiltrate the injured area within 4 hours, and by 16 hours postinjury, the band of leukocytes underlying the mantle in the vicinity of the lesion thickens. Infiltration of the wound area by leukocytes continues, and by 24–48 hours after wounding, a thick band of normal, round leukocytes surrounds the entire lesion. By 40 hours, blood vessels adjacent to the wound are heavily packed with leukocytes, and a heavy infiltration into the region of the wound is well established.

Healing proceeds from the interior of the lesion toward the surface. The round leukocytes, after delineating the margin of the wound, become fusiform and line up parallel to the wound (Fig. II.41). The nuclei of these cells retain their normal, round shape until approximately 160 hours postinjury, but subsequently they assume a fusiform appearance. As the band of fusiform leukocytes thickens around the periphery of the lesion, the wound channel is heavily infiltrated by round leukocytes and fibroblasts (Figs. II.40 and II.41); the wound channel is effectively plugged by 144 hours postwounding.

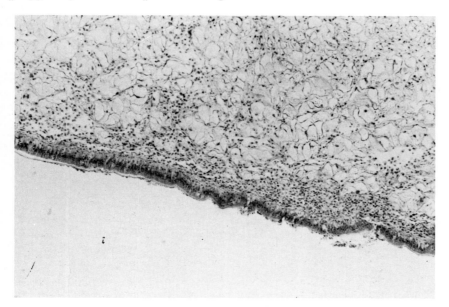

Fig. II.39. Wound healing in the Pacific oyster, *Crassostrea gigas*. Infiltrating leukocytes forming a band underlying the mantle epithelium adjacent to the wound. Four hours postwounding. Hematoxylin and eosine. 120×. (From Des Voigne and Sparks, 1968.)

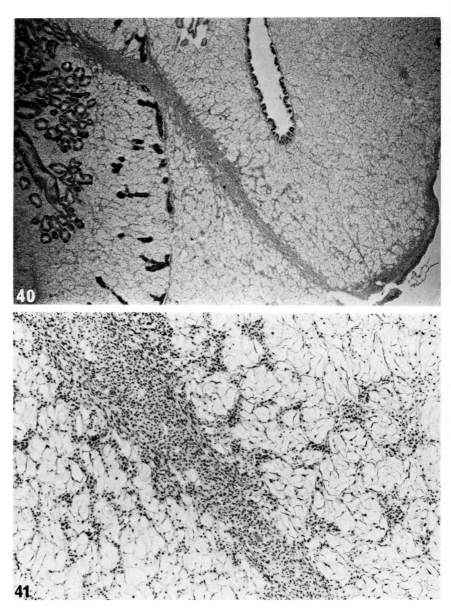

Fig. II.40. Wound healing in *Crassostrea gigas*. Infiltration of leukocytes into the wound channel. Ninety-six hours postwounding. Hematoxylin and eosine. 30×. (From Des Voigne and Sparks, 1968.)

Fig. II.41. Wound healing in *Crassostrea gigas*. Wound channel in 94-hour wound, higher magnification. Note the elongate leukocytes at the periphery of the lesion and round leukocytes in the center. Hematoxylin and eosine. 120×. (From Des Voigne and Sparks, 1968.)

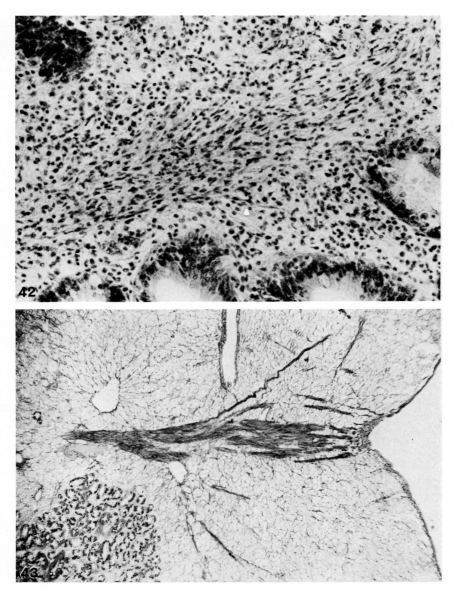

Fig. II.42. Wound healing in *Crassostrea gigas*. Leukocytic plug in 144-hour wound. Note increased elongation of leukocytes and beginning of collagen deposition. Hematoxylin and eosine. 300×. (From Des Voigne and Sparks, 1968.)

Fig. II.43. Wound healing in *Crassostrea gigas*. Increased collagen deposition along wound channel in 144-hour wound. Mallory's trichrome. 30×. (From Des Voigne and Sparks, 1968.)

After infiltration into the wound channel, these round leukocytes also become elongated and align along the axis of the lesion; collagen deposition also increases markedly (Figs. II.42 and II.43). Normal, round leukocytes and fibroblasts continue to infiltrate along the periphery of the wound until 120–160 hours postinjury. Groups of the fusiform leukocytes, filling the wound channel and arranged along the axis of the lesion, form randomly arranged whorls (Fig. II.44). At this point, varying in time from 88 to 448 hours postwounding, the tissue resembles a vertebrate scar. However, the original architecture is eventually restored; the whorls of leukocytes and collagen deposits are replaced by Leydig cells that are indistinguishable from normal tissue. It is not yet known whether this is accomplished by invasion of adjacent Leydig cells into the lesion or by differentiation of the infiltrated, fusiform leukocytes into Leydig cells.

Injured digestive epithelium becomes necrotic, with the nuclei undergoing pycnosis and finally karyolysis. Leukocytes accumulate in the traumatized area and phagocytize the necrotic cell debris (Fig. II.45). Other leukocytes, not involved in the phagocytosis, form a band along

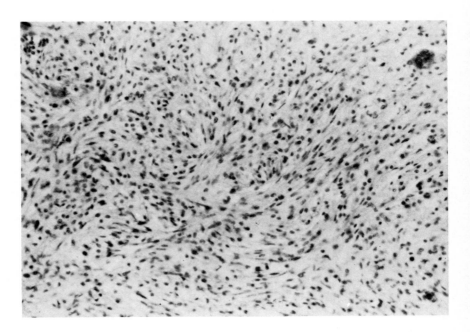

Fig. II.44. Wound healing in *Crassostrea gigas*. Eighty-eight-hour wound. Formation of randomly arranged groups of whorls by the fusiform leukocytes. Hematoxylin and eosine. 300×. (From Des Voigne and Sparks, 1968.)

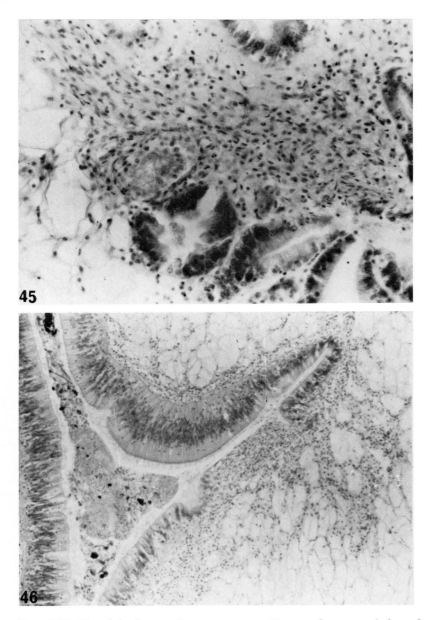

Fig. II.45. Wound healing in *Crassostrea gigas*. Necrotic digestive tubule undergoing phagocytosis. Eighty-eight-hour wound. Hematoxylin and eosine. 300×. (From
Des Voigne and Sparks, 1968.)

Fig. II.46. Wound healing in *Crassostrea gigas*. Band of leukocytes replacing
necrotic epithelium in the intestine. Eighty-eight-hour wound. Hematoxylin and
eosine. 120×. (From Des Voigne and Sparks, 1968.)

the edge of the digestive tissue (Fig. II.46), replace the necrotic epithelium, and elongate. Des Voigne and Sparks (1968) believed that these
cells were totipotent and that they differentiated into epitheloid tissue
or pseudoepithelium. Ruddell (1969), however, questioned this theory
and presented evidence indicating that surface epithelial replacement is
primarily, if not entirely, accomplished by migration of adjacent epithelial cells. Mix and Sparks (1971) have clearly shown that digestive-
epithelium repair in oysters wounded by ionizing radiation is by mitosis
and by subsequent migration of adjacent uninjured epithelial cells.

At the surface of the wound, within 24–32 hours, the band of fusiform
leukocytes underlying the mantle epithelium closes the lesion in the
areas where the epithelium has been injured or removed. The band of
leukocytes gradually thickens and forms a laminated border between
the underlying Leydig tissue and the exterior. The external lamina consists of fusiform leukocytes arranged parallel to the surface; the middle
lamina consists of cells perpendicular to the surface and the overlying
layer and is infiltrated with round leukocytes; the inner lamina contains
normal, round leukocytes and elongate cells that are perpendicular to
the surface or randomly directed. Des Voigne and Sparks reported that
fusiform cells in the outer lamina gradually turn perpendicular to the
surface and apparently form a loose, vacuolated pseudoepithelium in a
manner similar to that previously described in the digestive epithelium
(Figs. II.47 and II.48). At this point, 40–120 hours after wounding, the
underlying lamina is heavily infiltrated with round leukocytes that
eventually become fusiform and align themselves parallel to the surface.
The surface layer frequently becomes stratified and is always deeply
convoluted (Fig. II.49). By 168 hours, the cells appear to be altered
in appearance in a loose, cuboidal and finally into a stratified, tall
columnar form remarkably similar in appearance to the normal adjacent
epithelium; the cells even develop cilia (Fig. II.50).

Ruddell (1969), utilizing histochemical and electron microscopic
techniques, continued the investigation of wound healing in the Pacific
oyster. He divided the process of wound repair into five periods or
stages.

1. Lag period
2. Early inflammatory period
3. Late inflammatory period
4. Early wound-healing period
5. Late wound-healing period

The lag period extends from 1 to 2 hours postwounding during which
time there is no histological or histochemical response evidenced. It

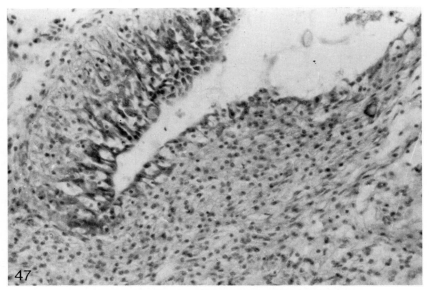

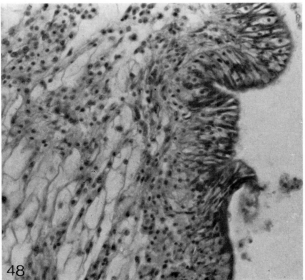

Fig. II.47. Wound healing in *Crassostrea gigas.* Newly formed epithelial covering of the mantle over the surface of the wound. Note the vacuolated nature. Eighty-eight-hour wound. Hematoxylin and eosine. 300×. (·From Des Voigne and Sparks, 1968.)

Fig. II.48. Wound healing in *Crassostrea gigas.* Newly formed epithelial covering over the surface of the wound. Note the more highly organized nature of the mantle covering and the irregular, convoluted border. Eighty-eight-hour wound. 300×. Hematoxylin and eosine. (From Des Voigne and Sparks, 1968.)

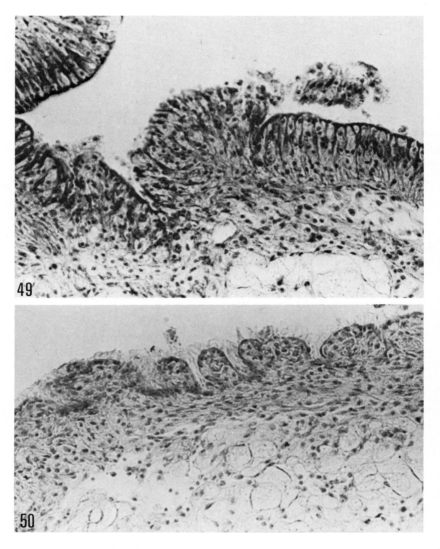

Fig. II.49. Wound healing in *Crassostrea gigas*. Stratified, hyperplastic epithelial covering of 168-hour wound. Note the deep crypt and varying height of the tissue. Hematoxylin and eosine. 300×. (From Des Voigne and Sparks, 1968.)

Fig. II.50. Wound healing in *Crassostrea gigas*. Mantle surface of repaired wound. Note the presence of cilia. One hundred and sixty-eight hours after wounding. 300×. Hematoxylin and eosine. (From Des Voigne and Sparks, 1968.)

must be assumed that the stimuli which initiate the wound repair diffuse from the traumatized area during this period.

The early inflammatory period extends from 2 to 48 hours after wounding. During this period, the Leydig, nerve, and muscle cells in the wound area often appear swollen, and the nuclei of these cells occasionally undergo pycnosis and karyolysis.

Ruddell identified three types of leukocytes (or amoebocytes) (Fig. II.51) involved in the inflammatory response and wound repair process: an agranular amoebocyte (Fig. II.52); an acidophilic granular amoebocyte (AG-Fig. II.53); a basophilic granular amoebocyte (BG-Fig. II.54). Two to 4 hours after wounding, BG amoebocytes in the wound swell to 3–6 times their normal size and burst. At the same time, according to Ruddell, the AG amoebocytes begin releasing copper and a diazotized *p*-nitroaniline-positive substance into the wound area. Four hours after wounding, copper can be detected in small foci of intact BG amoebocytes near the wound area. By 18 hours after wounding, the copper

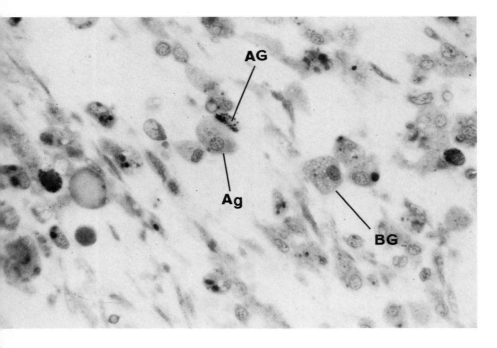

Fig. II.51. Light micrograph of an area adjacent to a 144-hour-old wound showing an agranular ɔebocyte-type cell containing a large glycogen deposit. Granular and agranular amoebocytes are also seen in this section. AG, acidophilic granular amoebocyte; BG, basophilic granular amoebo- ɘ; Ag, agranular amoebocyte. Aldehyde-OsO₄ fixation; embedded in epon; stained in 0.5% toluidine e O, pH 9.0. 990×. (From Ruddell, 1969.)

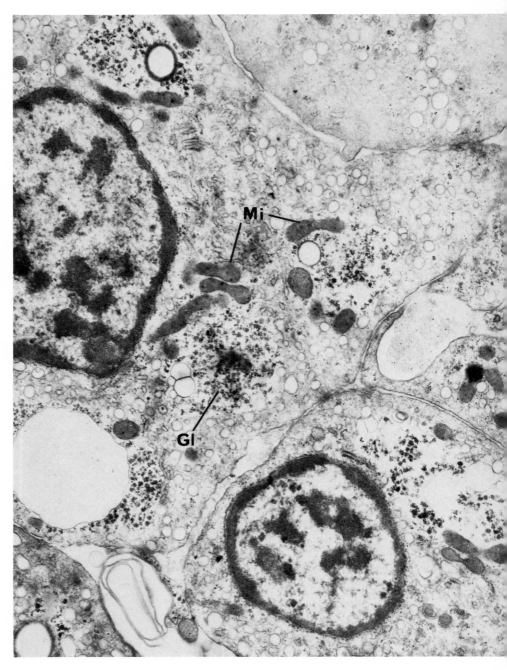

Fig. II.52. Agranular amoebocytes from an area adjacent to a 144-hour-old wound. Gl, glycog
Mi, mitochondria. 18,900×. (From Ruddell, 1969.)

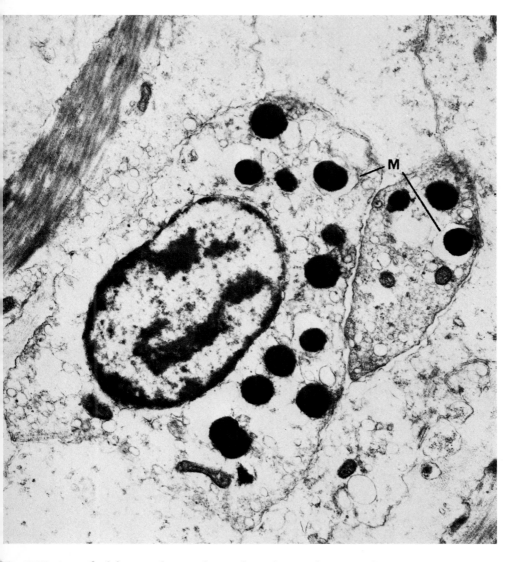

Fig. II.53. An acidophilic granular amoebocyte from the mantle. A membrane (M) can be seen rounding the acidophil granules. 18,900×. (From Ruddell, 1969.)

spreads from the small foci of BG amoebocytes and can be demonstrated bound to the cytoplasm and nucleus of Leydig, muscle, and nerve cells and to the remnants of exploded BG amoebocytes in the wound area (Fig. II.55). Copper is bound to most of the cells in the wounded area including epithelial cells by 24 hours. Maximum dispersal of copper

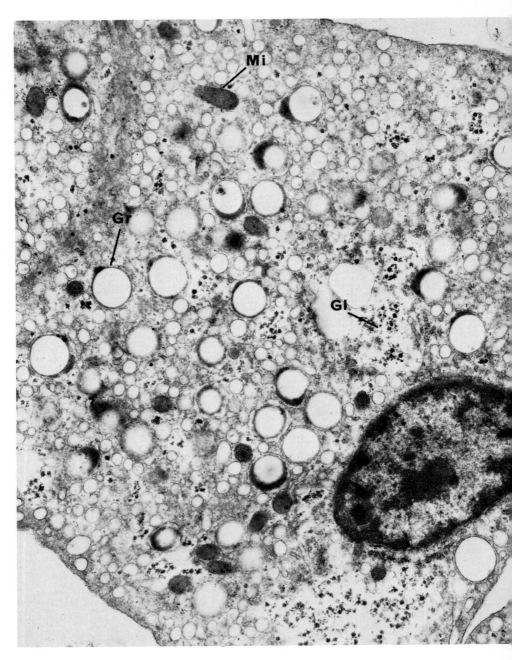

Fig. II.54. A basophilic granular amoebocyte from an oyster mantle. This cell is characterized the presence of large, "hollow" cytoplasmic granules (Gr). Mi, mitochondria; Gl, glycogen. 12,18((From Ruddell, 1969.)

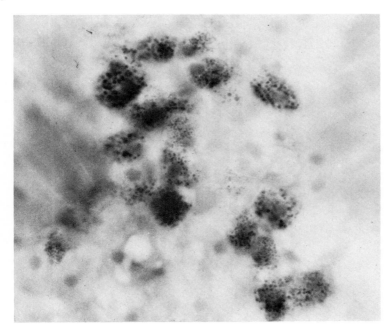

Fig. II.55. Section of an area adjacent to an 18-hour-old wound, stained with aqueous hematoxylin, showing copper bound to virtually every cell in the field. Aldehyde fixation; embedded in HEMA; treated with 0.05% hematoxylin for 20 minutes. 990×. (From Ruddell, 1969.)

appears to be attained within 48 hours after wounding, and it is largely confined to the immediate area of trauma, although it may occasionally be present in BG amoebocytes 1 mm from the wound. The period of copper release is not known, but experimental evidence indicates that it is brief.

The agranular amoebocytes perform an important role in the early inflammatory period by invading the wound and phagocytizing and sequestering cellular debris and particulate foreign matter. Brown (or pigment) cells do not invade the traumatized area, and those in the wound area become vacuolated by 24 hours postwounding. Ruddell noted that all mitotic activity within 1 mm of the wound is suppressed for 120 hours, after which cell division begins in the wound area.

The third stage, the late inflammatory period, extends from 48 to 72 hours after wounding and is characterized by an influx of both granular and agranular amoebocytes into the wound (with the agranular amoebocytes containing phagocytized cellular debris) and the beginning of epithelial migration over the external terminus of the wound.

The fourth stage, early-wound healing, extends from 72 to 144 hours after wounding. There is a continuing emigration of leukocytes into the traumatized area. Agranular amoebocytes and small numbers of granular amoebocytes concentrate and elongate to form a plug that fills the wound channel; this corroborates the findings of Des Voigne and Sparks (1968). Electron microscopy revealed that the leukocytes forming the plug are typical, undifferentiated, agranular amoebocytes (Fig. II.56 and II.57), except that numerous small fibrils approximately 70 Å in diameter and arranged parallel to the axis develop under the cell membranes (Fig. II.58). The agranular amoebocytes infiltrating the wound travel considerable distances, as Ruddell experimentally demonstrated by injecting oysters with carmine particles 2 cm from the wound. He found the carmine in agranular amoebocytes incorporated into leukocytic plugs in oysters sacrificed at 48, 72, and 120 hours after wounding.

During this fourth stage, epithelial cells continue to migrate across the leukocyte plug at the external surface of the body. Epithelial cells adjacent to the edge of the wound begin mitotic division to provide new epithelial cells to the migrating epithelium.

Stage 5, the late wound-healing period, extends from approximately 120 hours postinjury until healing is complete, normally 2–3 weeks after wounding. This terminal period is characterized by the following sequence of events.

1. The amoebocytic cells filling the wound channel synthesize fibrous extracellular material. The deposited material consists of an acidophilic, fast green-positive substance that can be demonstrated to accumulate extracellularly in hydroxyethyl methacrylate-fixed sections of 120- and 144-hour-old wounds stained with the biebrich scarlet and fast green technique.

2. Examination of electron micrographs of 144-hour-old wounds suggests that fibrous material is being deposited extracellularly among many, but not all, of the amoebocytes in the wound plug (Fig. II.59). This material is composed of poorly defined, 100 Å- to 250 Å-diameter fibrils of fuzzy appearance. The amoebocytes or, more appropriately, the fibroblasts which produce these fibrils resemble agranular amoebocytes in every respect, except that the cytoplasmic matrix, cytoplasmic organelles, and cytoplasmic membranes are markedly granular in appearance.

3. At approximately 120 hours postinjury, cells in the wound area begin dividing.

The migrating epithelial cells completely cover the wound between 144 and 200 hours after wounding. On completion of epithelization, the torpedo-shaped epithelial cells of the actively migrating epithelium round up and become oriented perpendicular to the wound surface. Immediately upon termination of migration, the cells of the newly restored

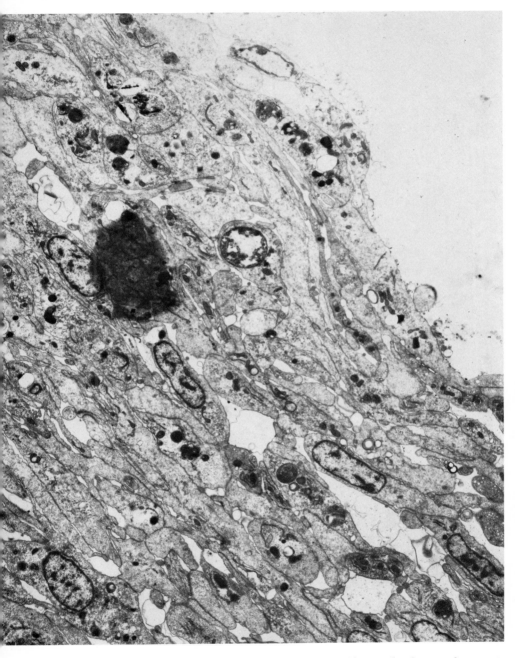

Fig. II.56. Cross section of an amoebocyte plug from a 72-hour-old wound. The amoebocytes in is portion of the plug have elongated. 3910×. (From Ruddell, 1969.)

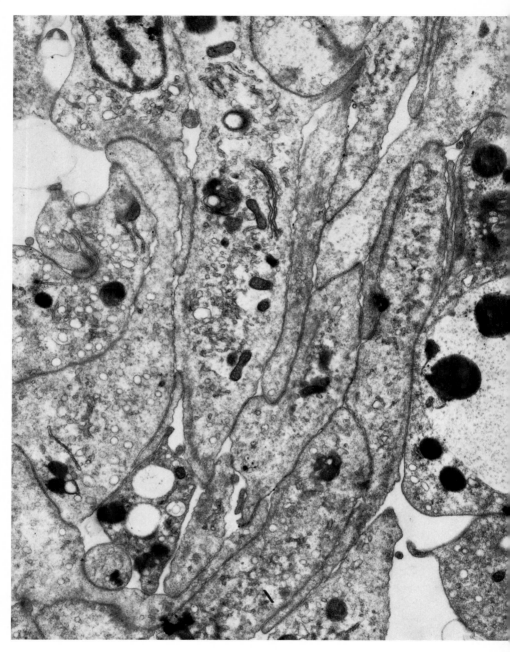

Fig. II.57. Cross section of an amoebocyte plug from a 72-hour-old wound. The cells in the p are typically undifferentiated and lack conspicuous arrays of both endoplasmic reticulum and go membranes. 12,180×. (From Ruddell, 1969.)

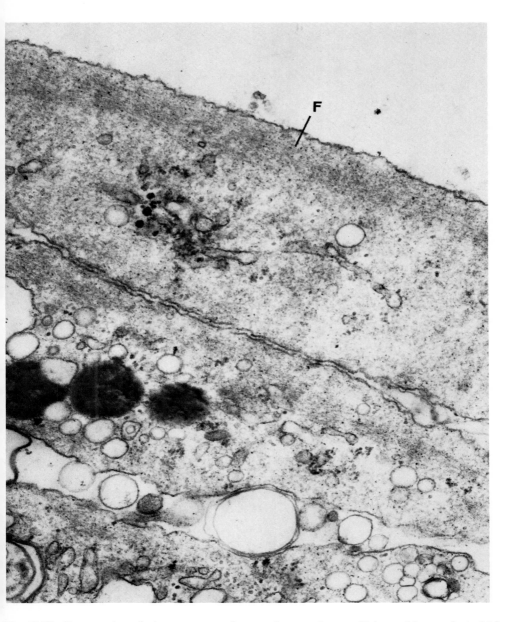

Fig. II.58. Cross section of elongate, agranular amoebocytes from a 72-hour-old wound. A thick nd of fibrils (F), approximately 70 Å in diameter, parallels the plasma membrane. 43,260×. (From ddell, 1969.)

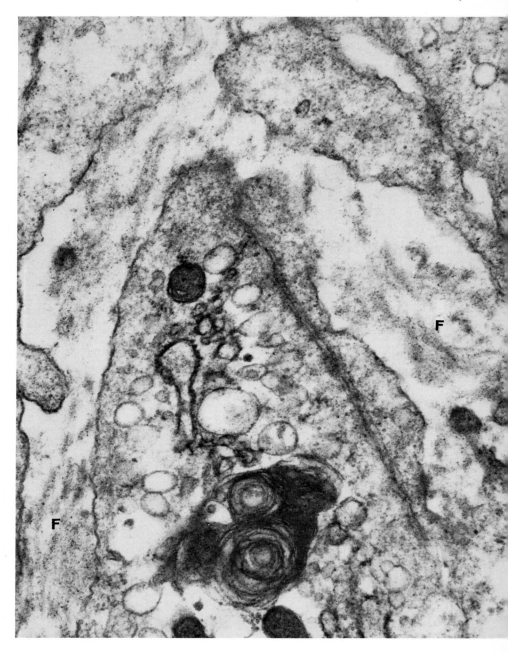

Fig. II.59. Amoebocytelike cells from the amoebocyte plug of a 144-hour-old wound. Poorly defined fibrils (F) are seen in the spaces between cells. 43,260×. (From Ruddell, 1969.)

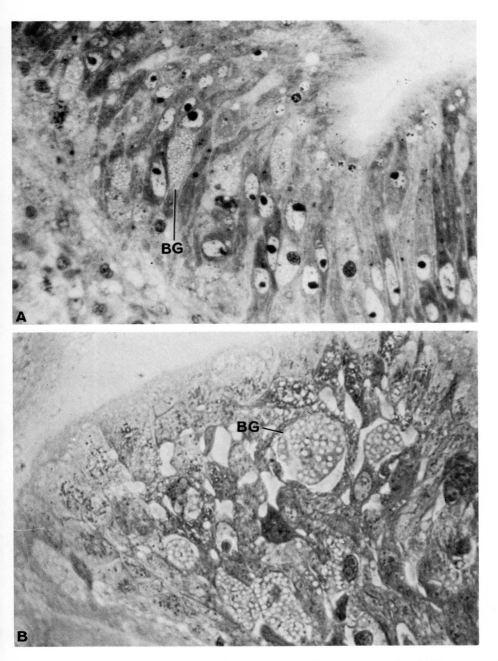

Fig. II.60. Cross sections of epithelial cells covering wounds. A, Section of a 144-hour wound.
⦿0×. B, Section of a 400-hour wound. 1550×. Many basophilic granular amoebocytes (BG) are ob-
ꜱrved incorporated into the wound epithelium. Aldehyde-OsO_4 fixation; embedded in epon; stained
ᵢth 0.05% toluidine blue O, pH 9.0. (From Ruddell, 1969.)

97

epithelium begin dividing mitotically and continue to divide for as long as 720 hours after wounding. This mitotic activity, as well as the epithelial migration, was not observed by Des Voigne and Sparks (1968). There is no explanation for the lack of mitotic activity and for evidence of epithelial cell migration in the earlier work by Des Voigne and Sparks. Extensive independent search by both investigators for evidence of these two phenomena was unrewarding, leading them to conclude, apparently erroneously, that reepithelization is accomplished by differentiation of leukocytes. A partial explanation for these conflicting results may lie in the higher magnifications with which Ruddell worked. However, Mix and Sparks (1971) clearly demonstrated mitotic activity and epithelial migration in the repair of digestive diverticula of oysters damaged by ionizing radiation with the same histological techniques and light microscope examination used by Des Voigne and Sparks.

During this terminal phase of wound repair, large numbers of basophilic granular amoebocytes and acidophilic granular amoebocytes become incorporated into the newly formed wound epithelium (Fig. II.60 A and B), often outnumbering the epithelial cells. This massive infiltration of leukocytes into the newly formed epithelium may have masked the epithelial migration and subsequent mitotic activity and may have caused Des Voigne and Sparks to assume that differentiation of leukocytes occurred. The newly regenerated epithelium typically exhibits a characteristic vacuolated appearance (Figs. II.47 and II.60 B), apparently because of large lacunae between the epithelial cells and leukocytes.

The outcome of wound repair in oysters is an almost perfect reconstruction of the original architecture, rather than the formation of an acellular, collagenous scar typical of wound healing in higher vertebrates. It should be noted, however, that small wounds in vertebrates are repaired without scar formation; so that the process in oysters is of a lesser degree and does not represent a totally different phenomenon. Approximately half the wounded oysters in Ruddell's experiments and many of those studied by Des Voigne and Sparks exhibited hyperplasia of the newly formed epithelium in older wounds (Figs. II.49 and II.61). The epithelium becomes thickened and frequently develops fingerlike projections. The ultimate outcome of the hyperplastic response is not known.

Ruddell has shown that environmental conditions have a marked effect on the rate and degree of wound repair. Oysters held in cold water or out of water do not heal wounds as rapidly or to the same extent as do those held in relatively warm water.

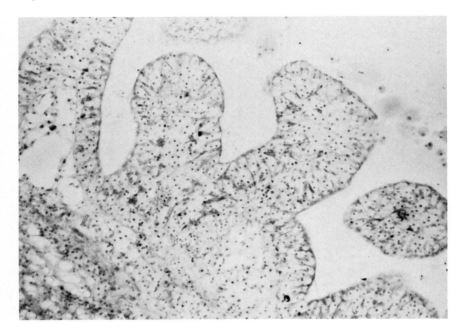

Fig. II.61. Fingerlike epithelial projections from the wound epithelium of a 408-hour-old wound. Aldehyde fixation; embedded in HEMA; stained in toluidine blue O, pH 4.0. 155×. (From Ruddell, 1969.)

Wound Repair in Freshwater Mussels

Wound repair in the freshwater mussel *Anodonta oregonensis*, unlike most pelecypods, is characterized by the lack of a pronounced cellular response (Pauley and Heaton, 1969). Surgical incisions 1 inch long and approximately ¼ inch deep essentially heal by 6 weeks after wounding. No grossly discernible abnormal color (as occurs in oysters—Pauley and Sparks, 1965; Des Voigne and Sparks, 1968; Ruddell, 1969) is associated with the wound.

The first recognizable histological response to surface wounds is invagination of one edge of the wound (Fig. II.62) in an apparent attempt to close the incision. According to Pauley and Heaton, this is often the only response noted during the first 8 days, but it frequently results in complete occlusion of the superficial portion of the wound channel within 1 day postinjury. Infrequently, a "clot" is formed in the narrow wound channel by infiltration and aggregation of hemocytes, and an extremely weak cellular response occasionally develops in surround-

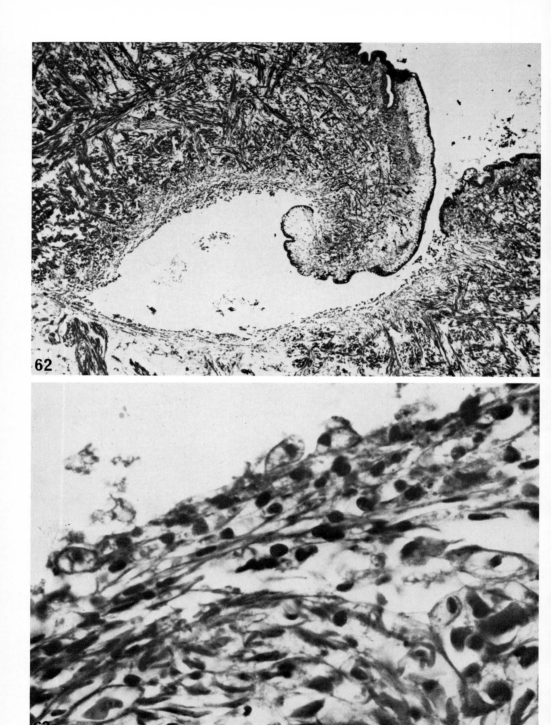

Figs. II.62 and II.63

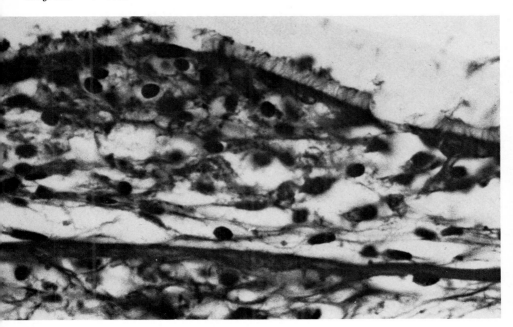

Fig. II.64. Wound repair in *Anodonta oregonensis*. Fibroblasts producing a thin layer of collagen er muscles. Note ciliated cuboidal epithelium. Eight-day wound. Modified Mallory's trichrome. 0×. (From Pauley and Heaton, 1969. Courtesy of G. B. Pauley.)

ing tissues. Usually, however, the cellular response is confined to a slight leukocytic infiltration into the deepest portion of the wound. The aggregating leukocytes often become spindle-shaped and present a layered appearance similar to that found in oyster wounds. Soon after wounding, the spindle-shaped leukocytes extend up the sides of the wound (Fig. II.63), indicating that healing proceeds from the interior to the external surface as in the oyster. Within approximately 1 week, a thin layer of leukocytes, 2–5 cells in thickness, completely lines the wound channel but does not plug the channel as in the oyster.

When the first spindle-shaped leukocytes appear, they are accompanied by a few fibroblasts; by the eighth day, fibroblasts are prominent among the leukocytes, and a few fine strands of collagenlike material have been deposited by the fibroblasts (Fig. II.64). By this time, prom-

Fig. II.62. Invagination of 1 edge of wound in the freshwater mussel *Anodonta oregonensis*. Note the leukocytes lining the wound cavity. Eight-day wound. Hematoxylin and eosine. 21×. (From Pauley and Heaton, 1969. Courtesy of G. B. Pauley.)

Fig. II.63. Layer of spindle-shaped leukocytes (round nuclei) on surface of cut muscles. Note the presence of a few fibroblasts (elongate nuclei) between muscle and leukocytes. Two-day wound. (From Pauley and Heaton, 1969. Courtesy of G. B. Pauley.)

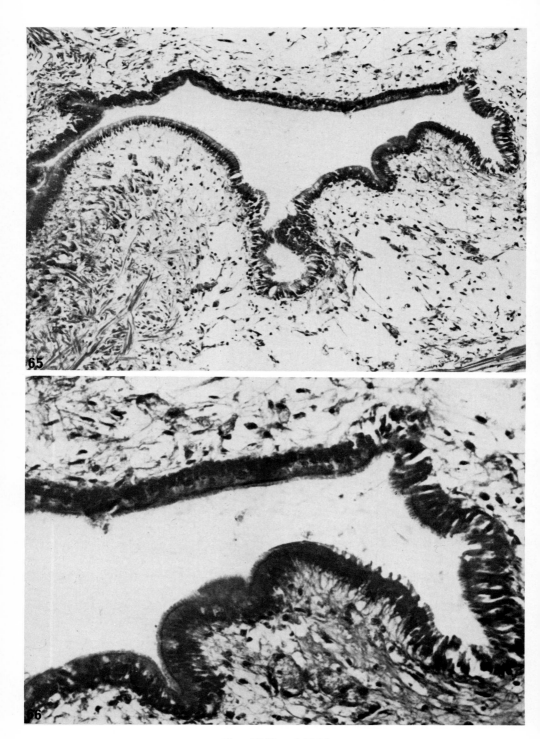

Figs. II.65 and II.66

inent growth of ciliated epithelium from the margins of the wound down the sides is apparent; it consists of cuboidal epithelium (Fig. II.64) near its confluence with the undamaged, ciliated, columnar epithelium of the foot and flattened, squamouslike epithelium in the deeper portion of the wound. New connective tissue fibers are deposited beneath the incipient epithelium and become more prominent by the sixteenth day postwounding. Typically, ciliated, columnar epithelium continues to grow inward from the original foot epithelium and almost completely lines the wound by 4 weeks; there is sometimes a combination of cuboidal and columnar cells. Rarely, regrowth of epithelium does not occur, and, as late as 4 weeks, the entire wound channel is lined by a layer, 8–10 cells thick, of spindle-shaped leukocytes. A network of smooth muscle is formed beneath the columnar epithelium and regeneration of the basophilic mucous glands is initiated.

After 8 weeks, the entire wound channel is lined by ciliated, columnar epithelium (Figs. II.65 and II.66), while regeneration of smooth muscle and mucous glands continues beneath the columnar epithelium. Subsequent to complete reepithelization by columnar epithelium, the free apical borders of the cells become apposed (Fig. II.67). The juxtaposition of the epithelium apparently initiates necrosis of the deeper portions, with the removal of the necrotic tissue by phagocytosis (Fig. II.68), resulting in a shallower, essentially healed wound (Fig. II.69).

The initial lack of a pronounced cellular response and subsequent lack of granulomatous or fibrous reaction in response to and repair of wounds in *Anodonta* is atypical of what is known of these phenomena in mollusks (Pauley and Heaton, 1969); however, it has been reported in other invertebrates such as some insects (Day and Bennetts, 1953) and a holothurian (Cowden, 1968). Reepithelization of the wound channel prior to its being plugged with leukocytes and, subsequently, with connective tissue also differs from the typical molluscan pattern.

In a study primarily designed to determine the morphogenesis of leukocytes, the adult Manila clam, *Tapes semidecussata*, was injected with the DNA precursor, [³H]thymidine (Cheney and Sparks, 1969); Cheney (1969) investigated the response to and repair of incision wounds. A light leukocytic infiltration into the wound develops within 14 hours. The nuclei of the infiltrated cells elongate and typically become oriented parallel to the axis of the wound. At 1 day postwounding,

Fig. II.65. Wound healing in *Anodonta*. Wound completely lined by ciliated columnar epithelium. Fifty-eight-day wound. Hematoxylin and eosine. 50×. (From Pauley and Heaton, 1969. Courtesy of G. B. Pauley.)

Fig. II.66. Higher magnification of 58-day wound. Note regenerating mucous glands beneath epithelium (at 6 o'clock). Hematoxylin and eosine. 133×. (From Pauley and Heaton, 1969. Courtesy of G. B. Pauley.)

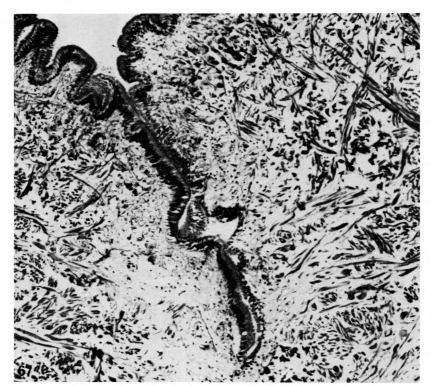

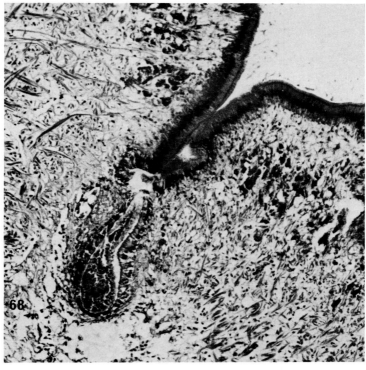

Figs. II.67 and II.68

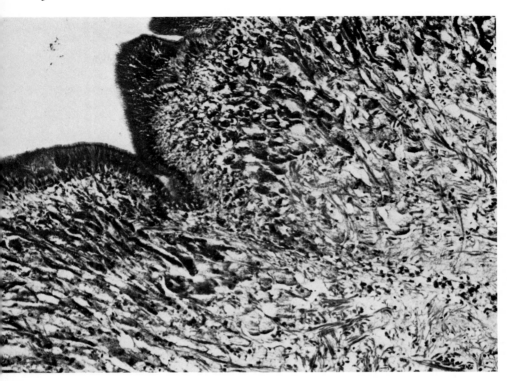

Fig. II.69. Shallow, healed wound in *Anodonta*. Note the still recognizable original line of in-
on. Fifty-eight-day wound. Hematoxylin and eosine. 60×. (From Pauley and Heaton, 1969.
urtesy of G. B. Pauley.)

the infiltrate is much heavier, but similar in appearance to that occurring
at 14 hours. The epithelium adjacent to the incision begins to migrate
into the wound. The leukocytic infiltrate begins to recede and, at 5 days
after wounding, remains only at the base of the wound. The migrating
epithelium continues to move along the surface of the wound channel
in a manner similar to that in the freshwater mussel, *Anodonta*, and
covers most of the cut surface by 5 days postinjury.

The leukocytic infiltrate may still be heavy at 12 days after wounding,
but epithelial pavementing of the wound channel is almost always complete. Within 20–30 days, the wound channel is bridged by connective-

Fig. II.67. Wound healing in *Anodonta*. Epithelium in apposition closing the
wound channel. Forty-four-day wound. Hematoxylin and eosine. 51×. (From Pauley
and Heaton, 1969. Courtesy of G. B. Pauley.)

Fig. II.68. Wound healing in *Anodonta*. Necrosis of epithelium deep in wound.
Hematoxylin and eosine. 51×. (From Pauley and Heaton, 1969. Courtesy of G.
B. Pauley.)

tissue extensions, and the outer surface of the wound is covered with cuboidal epithelium.

The wound is completely closed and covered by thin epithelium by 52 days after wounding, presumably isolating the epithelium lining the wound channel as in *Anodonta*. Cheney did not discuss the fate of the epithelium that migrates down the surface of the wound channel, but it probably follows the same sequence of necrosis and subsequent removal by phagocytosis that occurs in *Anodonta*.

CLASS CEPHALOPODA

The inflammatory response of cephalopods is generally similar to that of the lamellibranchs. When a sterilized thread is aseptically inserted into the ventral portion of the mantle of the cuttlefish, *Sepia*, the following sequence of events is initiated (Jullien, 1928). Within 6 hours, the dermal zone containing the thread is heavily congested with leukocytes that after 12 hours infiltrate the connective tissue in the area of the thread in great numbers. Many of the leukocytes retain the form occurring in the circulating blood—round or slightly oval cells of 7–8 μm containing an excentrically placed nucleus and cytoplasm filled with small eosinophilic granules; however, some become modified in appearance. They become elongate and sluglike, with the long axis measuring 16–20 μm; the nucleus tends to become rectolinear and terminates in an apical thread. Fusion and perhaps enlargement of the small eosinophilic granules occurs, resulting in several large granules. Within 18–24 hours, some cells of both types make contact with the thread, and, after 2–4 days, some cells penetrate the thread and degenerate. Other leukocytes assume a fusiform shape, 20–28 μm in length, and align themselves in 2 or 3 concentric layers to form a cystlike structure around the thread, similar to the granuloma formation around talc particles in oysters.

After 5 or 6 days, the inflammatory region consists of a central zone of necrotic leukocytes in the interior of the thread, an intermediate zone adjacent to the thread forming a cystic crown of four or five layers of cells, and a peripheral zone of transition between the wound and the adjacent normal tissue. The cystic crown becomes thicker between 9 and 14 days, and the leukocytes between the crown and the epidermis diminish in number.

Morphological modifications of the leukocytes surrounding the cyst continue, and, at 18–23 days, the cells in the cystic crown become more elongated, 30–35 μm by 1 μm; they lose their eosinophilic granules and become basophilic, and they closely resemble typical fibroblasts of normal connective tissue. The cells in the peripheral zone approach the form of those found in cystic crown; they are arranged in parallel layers but not as tightly compacted. The collagenous fibers between the peripheral zone and the body surface begin to fragment after 30 days, and 5 days later the thread and the encapsulating cellular elements are eliminated through the body surface, which subsequently heals.

Jullien (1940) subsequently studied the reaction of several cephalopods to subcutaneous injections of tar, a substance known to be carcinogenic in some animals. The tar was found to have no carcinogenic effect on cephalopods, and the reaction to it was the typical inflammatory response to a foreign substance, involving leukocytic infiltration, encapsulation through modification of leukocytes into fibroblastlike cells, and rejection of the cyst through the external epithelium. This work is of interest primarily because it demonstrated that the reaction to tar was the same in all the cephalopods studied, but the rate of response varied considerably with the different species. The inflammatory response is rapid in the cuttlefish, *Sepia*, somewhat slower in the squid, and much slower in ectopod, *Eledone*. The variation in speed of response is not confined to tar, but is typical of all foreign materials injected into the three cephalopods.

Jacquemain *et al.* (1947) further investigated the effect of carcinogens on cephalopods with the introduction of crystals of 1:2–5:6 dibenzanthracene enclosed in Vaseline beneath the skin of cuttlefish. The chemical elicited a much different cellular reaction, perhaps somewhat comparable in certain respects to the caustic effect of turpentine in oysters. The initial reaction is a rapidly expanding, markedly elevated swelling with an oval outline. Unlike the normal inflammatory reaction, there is no tendency for formation of an elimination tract and a scar. The injured tissues fragment, autolyze, putrefy, and are shredded from the animal. There is no cellular encapsulation reaction to the injurious agent, only a diffuse and weakly reactive infiltration; this is an example of the failure of the cephalopod defense mechanisms to successfully combat one specific foreign material.

Phylum Arthropoda

Despite Metchnikoff's pioneering work on phagocytosis and reaction to injury in the crustacean *Daphnia* and despite the tremendous growth in insect pathology, there is surprisingly little known about the reactions to injury and the wound-healing processes in the remainder of this vast group of animals. As Metchnikoff pointed out, the development of a hard exoskeleton, found in most arthropods, protects the animal from most surface wounds and prevents the entrance of many microorganisms. This may result, at least in some arthropods, in a loss or diminution of the animal's capacity to react to injury. However, Metchnikoff stated that "the extremely rapid regeneration of the epidermis in Arthropoda causes their wounds to heal very quickly."

CLASS MEROSTOMATA

The most primitive arthropod whose reaction to injury has been studied is the horseshoe crab, *Limulus polyphemus* (class Merostomata, subclass Xiphosura). The fossil record of Xiphosura extends back to the Ordovician period, and only 3 genera and 5 species have survived to the present. Loeb (1902) studied the inflammatory reactions of *Limulus,* and subsequently, in a series of papers both alone and in collaboration with other workers (see Johnson, 1968), described the role of the amoebocytes in the process and the effects of various substances on the amoebocytes. The primary reaction to trauma and infection is leukocytic infiltration of the injured area, followed by extensive intravascular clotting (Bang, 1956); the initial reaction is cellular and is followed by extracellular gelation. Bang (1956) became interested in a bacterial disease of *Limulus* and the reaction of the blood to it and to nonpathogenic bacteria. These studies are mentioned here because they probably demonstrate the typical reaction to injury by *Limulus. Limulus* amoebocytes were maintained *in vitro* in a siliconized glass culture system in an intact state resembling the *in vivo* state for up to 30 days without replenishing the culture media (undiluted *Limulus* serum) or use of antibiotics, but no cell multiplication was noted. When a pathogenic *Vibrio,* taken from obviously sick *Limulus,* was added to the system, persistent gelation of the media occurred, accompanied by bacterial growth and pathological changes in the amoebocytes and granules. Although gelation is characteristic of shed *Limulus* blood, uninfected cultures and cultures challenged with nonpathogenic bacteria had little or no gelation in the media in siliconized glass containers. Such cultures contained normal granules and consistently contained large clumps of agglutinated dead bacteria, suggesting (in Bang's opinion) involvement of the gelation process in immobilization and killing of invading organisms. Only the siliconized culture system containing intact, granular cultures was effective against nonpathogenic bacteria, with rapid and permanent elimination of large numbers of several strains of marine bacteria. The pathogenic *Vibrio,* on the other hand, or its thermostable toxin even in minute quantities caused prompt changes in the morphology of the intact amoebocytes, culminating in cytolisis.

Loeb (1902), in addition to doing the *in vitro* studies of clotting characteristics of *Limulus* blood, also investigated the inflammatory process *in vivo* by injecting various substances (such as carmine suspension) and inserting foreign bodies (including silk thread, gauze, agar, and heat-coagulated egg white) into the body. After injection of carmine, tissue sections were made from the area of injection. The blood cells, which demonstrate no amoeboid movement, also show only weak phagocytic activity, leading Loeb to surmise that amoeboid movement is necessary for phagocytosis. The carmine particles are collected in large masses surrounded by coagulum in the tissues and blood vessels, demonstrating that the *in vitro* coagulation of *Limulus* blood is a true defense mechanism.

The insertion of foreign bodies was through a joint or directly into one of the anterior segments, with the wound closed by suture. At periods from a few hours up to 11 days, tissue in the wound area was removed, sectioned, and stained. The tissue surrounding a wound or foreign body undergoes necrotic changes, and the muscles degenerate in the first day. Associated with the degenerating muscle are masses of blood cells; the muscle fibers surrounded by them become indistinct and then liquefied. Loeb was not sure whether the liquefication is caused by the pressure of the blood clot or by a proteolytic element in the blood.

A foreign body inserted into *Limulus* is surrounded by a coagulum within a very few hours. Subsequently, the portion of the coagulum nearest to the foreign body becomes necrotic; the nuclei and cells break down and disappear. The typical heavy infiltration of amoebocytes into the area of the foreign body found in vertebrates and in most invertebrates apparently does not occur in *Limulus*, nor do those amoebocytes in contact with the foreign body phagocytize particles. The sole reaction appears to be the formation of a coagulum containing fibers and necrotic cells surrounding the foreign body.

CLASS CRUSTACEA

As noted by Metchnikoff, masses of leukocytes accumulate beneath surface wounds in *Daphnia* and other transparent crustaceans and remain there until the wound is completely healed. Metchnikoff demonstrated the process experimentally by inflicting minute wounds on the caudal appendages of *Argulus* and observing the emigration of leukocytes to the wound site.

Apparently all the primitive crustaceans exhibit marked leukocytic reactions to invasion by disease organisms, and much of Metchnikoff's classic treatise on comparative inflammation was based on his studies of diseases of *Daphnia*. Since the sequence of events occurring in the reaction to parasitic infection are remarkably similar to the reaction to traumatic lesions, it is appropriate to consider them briefly at this point.

When *Daphnia magna* becomes heavily infected with spores of the fungus *Monospora bicuspidata,* the infected individual can be grossly recognized by the milk-white color. Infection occurs through ingestion of spores; in the gut, the spore capsule is lost, and the needlelike spore penetrates the intestinal wall to lie outside the alimentary canal. Immediately upon their emergence into the body cavity, the spores are attacked by leukocytes. The leukocytes surround the spore, often fusing together into a "plasmodium," and digest it. Interestingly, the spores are sometimes attacked before they have completed penetration of the gut, resulting in the destruction of the emerged portion, while the part remaining in the gut wall is unaffected. However, Metchnikoff pointed out that the development of a hard, chitinous cuticle as an exterior covering and as a lining for the intestine has been accompanied by a diminution of phagocytic capability in many crustaceans. The added protection of the impervious body covering eliminates much of the need for internal protective functions, and some copepods, for example, apparently lack amoebocytes.

In the literature, there are numerous reports of diseases, teratologies of many origins, and autotomy and regeneration in crustaceans, but there is a surprising paucity of studies of the reaction to injury and wound repair in the group from the pioneer studies of Metchnikoff and other early workers to the present. Bruntz (1905) discussed the process of phagocytosis in crustaceans, noting that small phagocytes are present in the blood and that larger, fixed macrophages also occur. The macrophages form phagocytic organs, most commonly in the dorsal regions of the head, in the laterodorsal parts of the thorax and appendages, and, to a lesser extent, in the dorsal part of the abdomen. Subsequently, Bruntz (1906a,b,c) described lymphogenous organs in a stomatopod (*Squilla mantis*), schizopods (*Mysis chamaelo* and *M. vulgaris*), arthrostracids, and decapods. He noted that the lymphogenous organs are similar in these forms, but are lacking in amphipods and isopods in which amoebocytes multiply normally through mitosis of circulating cells and "accidentally" through amitosis.

Considerable information on the initial reaction to trauma and the wound-repair phases preceding regeneration has resulted from studies of the regenerative phenomenon and are pertinent in this section. Needham (1965), in discussing regeneration in arthropods, pointed out some of the disadvantages of the arthropodan exoskeleton, noting that wounds occurring at sites other than at the autotomy plane would tend to be large and to close slowly. Chemical and, though he did not so state, amoebocytic defenses are probably inferior to those in animals more naturally vulnerable, and the open circulatory system exacerbates hemorrhage. There is strong evidence for the existence of a wound stimulus of a humoral type, even though there appears to be no "damage-hormone" that stimulates wound repair and regeneration in many other groups. Needham's observations on *Acellus* thoracic limb regeneration show that blood corpuscles accumulate quickly within the stump of an autotomized leg; the infiltration is maximal within 5 minutes and lasts for approximately 24 hours. The infiltrating cells are heavily laden with fuchsinophil material which must be imported since there is so little tissue damage. Within 5 hours, the epidermis has begun to close the wound, moving across proximally to the autotomy membrane. There is, as in other arthropods, no mitotic activity involved. Closure is not completed for several days; the limb stump remains adherent to the seal, perhaps as an anchor. The epidermal cells exhibit typical signs of activation, with enlargement (particularly of the nucleus and nucleolus) and increased chromophil contents. By 24 hours, the cells begin to divide, and an epidermal "shell" elongates rapidly after the second day.

The accumulated internal blood corpuscles disappear shortly after 24 hours, leaving the cavity of the stump empty except for a few pale, fusiform cells which are thought to be "mother cells" for future muscles. The epidermis closes, the proliferation of epidermal cells ceases, and regeneration of the limb in an incubation chamber is subsequently completed.

Dent and Fitzpatrick (1963) reported similar results in studying wound healing in crayfish of the genus *Cambarus*. They made wounds approximately 1 mm × 2.5 mm in the dorsal region of the carapace, plugged them with Gelfoam, then covered them with collodion. Histological observations of animals sacrificed at various intervals up to 180 days revealed that a blood clot formed immediately in and around the Gelfoam. The blood clot was invaded, after approximately 2 months, by epidermal cells forming a stratum that laid down, by approximately 80 days, an external covering of protochitin. Within a few days the protochitin was converted to chitin. Even though irregularities in the healed wounds interfered with molting, five experimental animals completed the molting process.

The formation of a blood clot at the site of injury, particularly when the integrity of the exoskeleton is broken, appears to be a common phenomenon in crustaceans and has attracted considerable interest. Hardy (1892) suggested the occurrence of "fibrin-ferment" within the amoebocytes that acts on a fibrinogenlike substance in the plasma and causes the plasma to coagulate. He described "explosive corpuscles" that burst, releasing the cell contents and leaving only a transformed nucleus. Haliburton (1885) described two phases of crustacean blood coagulation: cellular clumping and coagulation of the plasma. He assumed that the blood cells provided "fibrin-ferment," a procoagulent substance coagulating a fibrinogenlike material in the plasma. Gruzewska (1932) found that the rate of coagulation in lobsters and crabs was somewhat dependent on the individual animal tested and believed that retardation of coagulation might be due to transient impoverishment of the fibrinogen and fibrin-ferment in the blood, but he noted that lobster blood coagulates rapidly even in the absence of amoebocytes. Glavind (1948) also commented on the rapidity of coagulation in lobsters and showed that the clot consisted of an accumulation of blood cells. Although heparin will eliminate or retard clotting, much greater amounts are necessary than are required to prevent clotting of vertebrate blood.

Some light has been shed on the gross process of wound healing in crustaceans by a recent study (Fontaine, 1971) of the gross response of the brown shrimp (*Penaeus aztecus*) to tagging with the Petersen disc tag. Insertion of the pin (Fig. II.70) through the abdomen between the first and second segments creates a large open wound resulting in high mortalities, presumably from secondary infection.

Brown shrimp, tagged and held in the laboratory, develop exoskeletal intrusions at the portals of entry and exit of the pin. These intrusions surround the pin, are distinct by the fifth day postwounding, continue to grow inward to cover approximately half the pin by the tenth day (Fig. II.71), and fuse to form a complete tube by the twenty-ninth day (Fig. II.72). A group of 142 brown shrimp recaptured in a mark-recapture study were examined to ascertain whether surviving tagged shrimp responded similarly to the wound in nature. The involuted exoskeletal tube was present in 117 of the recaptured shrimp, but although well developed, fusion into a complete tube did not occur.

Unfortunately, the histological aspects of this wound repair process have not yet been studied. However, the long-term survival, up to 18 months, of shrimp tagged by this technique demonstrates that the "open encapsulation" formed around the pin by exoskeletal intrusions is an effective wound repair process.

MYRIAPODOUS ARTHROPODA

Four groups of mandibulate arthropods—the centipedes, millipedes, pauropods, and symphylans—share the characteristics of a body consisting of a head and an elongated trunk with many leg-bearing segments. This common feature resulted for many years in the four groups being placed in a single class, the Myriapoda. Since they differ greatly in other characteristics, they are now broken into separate classes but are often grouped together for convenience (Barnes, 1963). Since their reaction to injected materials and, presumably, to injury is similar and since they

Fig. II.70. Brown shrimp, *Penaeus aztecus,* with Petersen disc tag in place. Details of the pin and tag are demonstrated in the assembled and unassembled pins and tags beneath the shrimp. (From Fontaine, 1971.)

Fig. II.71. Exoskeleton of segment through which pin was inserted 10 days previously. Note exoskeletal intrusions formed around pin. (From Fontaine, 1971.)

Fig. II.72. Exoskeletal intrusion 29 days postinsertion. Note that the intrusion has fused to form a complete tube around the pin. (From Fontaine, 1971.)

have been studied concurrently, it seems appropriate to consider the group together.

Palm (1953) investigated the elimination of vital dyes in a number of myriapodous arthropods. The dyes, primarily trypan blue, were injected by means of fine glass capillary cannulae fitted to an AGLA micrometer syringe by a short length of small-diameter rubber tubing. The amount of dye injected depended on the size of the animal. The injections, according to Palm, are easily made after slight narcosis with chloroform, but certain technical difficulties do occur: the symphylan *Scutigerella* is small and delicate and often fails to survive the narcosis or the injections; the iulids, *Polydesmus* and *Craspedosoma*, are difficult to pierce because of the hard cuticle, and serious damage occurs in their handling; there is some question as to whether the dyes are taken up by tissues or whether the tissues are simply vitally stained. The organs studied, both fresh and by tissue section, were those generally considered excretory in nature in these animals, including the malpighian tubes, nephridia, nephrocytes, and the phagocytes and hemocytes.

Palm's experiments demonstrated that the malpighian tubes in the myriapods have little function in the removal of injected material. Since the amounts of materials taken up by the malpighian tubes were so small as to be insignificant, Palm was tempted to doubt the excretory nature of the tubes, but noted that they might well function in excretion of normal waste products of the body.

The head region of myriapodous arthropods contains numerous glands, one pair of which, in Palm's opinion, are homologous to the nephridia of other arthropods. Although they have been given various names, Palm suggests that they simply be called nephridia. The nephridia consist of three distinct parts—the sacculus, labyrinth, and utriculus—the latter being absent in Diplopoda and Symphyla. The sacculus blood lacunae contain blood cells that readily take up dyes, but do not excrete them into the labyrinth of the gland. Excretion does occur with injection of certain dyes: lithion and ammonium carminate, indigo-carmine, and sometimes neutral red. In diplopods, trypan blue is absorbed and excreted into the labyrinth, but this does not occur in the chilopod *Lithobius*.

Certain cells in the myriapods have a conspicuous ability to absorb injected carmine particles and were called "cellules a carminate" by early workers (Kowalewsky, 1894; Duboscq, 1896; Bruntz, 1905, 1906a–e). Most later authors have adopted the name nephrocytes for these cells. The location of the nephrocytes varies among the myriapods; they may occur as lateral bands along the malpighian tubules, around the salivary glands, along the sides of the ventral or supraneural blood vessel, at the sides of the sacculus of the nephridium, or associated with the fat body.

The nephrocytes are large cells or syncytial groups of cells with pale-staining nuclei and slightly basophilic cytoplasm containing numerous small, dark-violet pigment granules and varying numbers of larger, yellow-brown pigment granules. The groups of cells (the syncytia) are held together by fine strands of violet-pigmented connective tissue. In animals injected with ammonium carminate, lithion carminate, or trypan blue, the nephrocytes become bright red or blue, and microscopic examination reveals that the dyes are stored in the form of granules of varying size, depending on the amount injected. Sometimes the cells become swollen and degenerated after massive or repeated injections. The nephrocytes may disintegrate, and the granules may be phagocytized by blood cells. Other dyes may also be taken up by the nephrocytes and, though Palm states that they are not phagocytic, india ink is occasionally absorbed, though not to the extent that it is phagocytized by blood cells.

The reaction of the nephrocytes in myriapods is virtually identical to that occurring in the pericardial cells of insects, with a permanent storage in both of trypan blue and carmine granules. Unquestionably the nephrocytes serve a protective function by absorbing foreign substances and storing them in an insoluble form. Palm estimates that 90% of the trypan blue injected into *Lithobius* is stored in the nephrocytes; within 4 hours of injection of 10 ml of a 2% solution of trypan blue, the blood is completely cleared of the dye.

The myriapods are also provided with cellular elements that function in the defense mechanism by ingesting solid particles from the blood. Cuénot (1891) found phagocytes in the blood and at the sides of the dorsal blood vessel of the chilopods in arrangements that resemble the pericardial cells in insects, but the latter cells (with few exceptions) are not phagocytic. The phagocytes in the chilopod *Scolopendra* studied by Cuénot occur in scattered groups embedded in the fat body and connected with the alary muscles of the heart. Kowalewsky (1894) injected bacteria, vertebrate blood, milk, iron saccharate, and various dyes into *Scolopendra* and demonstrated phagocytosis in the blood cells and in "lymphatic glands"; these "glands" correspond to the above-mentioned cell groups and are arranged in two longitudinal rows more ventrolaterally situated than Cuénot described them. The groups of cells, which Duboscq (1896) termed "Kowalewsky's corpuscles," readily phagocytize the foreign materials listed above and are considered to be the site of origin of blood cells.

Bruntz (1906b,c) studied phagocytosis in *Glomeris, Iulus,* and *Polydesmus* by injection of india ink. Carbon particles were ingested by blood cells and by special phagocytes variously arranged in the three genera.

In *Glomeris,* the phagocytes are located along the ventral septum above the perineural sinus in two bands. In *Iulus,* the phagocytes form most of the wall of the perineural sinus; in *Polydesmus* they occur at the sides of the intestine, are associated with the fat body, and are arranged in groups resembling the corpuscles of Kowalewsky in *Scolopendra.*

The phagocytes of *Lithobius, Pachymerium,* and *Geophilus* apparently are not arranged in special structures, but are scattered singly or in small groups at the sides of the fat-body lobules, gonads, and accessory glands or are virtually anywhere in the connective tissue. Palm believes that these fixed phagocytes are genetically related to blood cells, perhaps circulating in the blood and then attaching to the connective tissue and remaining as "fixed hemocytes."

Phylum Echinodermata

There is considerable information available on the reaction to injury and wound repair in some of the echinoderm groups; much of it is reported as the early phases of regeneration, and there are a number of studies describing the types and functions of circulatory cells occurring in the perivisceral fluid of members of the phylum. Phagocytosis of foreign material by circulating cells, usually called coelomocytes, appears to be universal in echinoderms, but the role of the various types of cells varies considerably. Encapsulation of foreign bodies too large to be phagocytized, a common feature in invertebrates, may not occur in the echinoderms; but all groups are capable of removing foreign materials, healing wounds, and repairing tissue rapidly and efficiently. Boolootian and Giese (1958) attempted to clarify the classification and functions of the coelomic corpuscles of echinoderms, utilizing modern techniques and equipment such as phase microscopy. Their study was comparative, using 15 species representing all classes of echinoderms. Their investigation showed 13 types of fairly distinct cells. Some of these, such as the bladder amoebocytes and the filiform amoebocytes (which are actually different phases of the same cell), are common to most members of the phylum; others are common to two or three classes, and still others are unique to one class.

Class Asteroidea

The earliest report on reaction to injury in starfish was by Metchnikoff (1893); he noted that introduction of small, sharp objects into the body of bipinnaria larvae of asteroids resulted in an accumulation of mesodermic phagocytes around the invading object and the formation of a plasmodium completely surrounding the foreign body by fusion of the

infiltrating cells. Metchnikoff stated that no proliferation of the phagocytes occurred and that edema was not part of the reaction. The injection
of vertebrate blood cells, as noted by Metchnikoff, results in the same
response, with intracellular digestion of the erythrocytes. Bacteria and
fat globules of milk are attacked in the same fashion. Subsequently,
Durham (1888) showed that the starfish amoebocytes migrate from the
coelom to the injured area, phagocytize small foreign particles, then carry
them to the outside via the dermal papulae.

Bang and Lemma (1962) noted that trauma, including that induced
by collecting with the typical starfish mop, initiates an agglutination of
amoebocytes in the circulating fluid that is observable in the dermal
papulae with the aid of a dissecting microscope. Transient flocculation of
amoebocytes can be induced experimentally by injecting various substances into the coelom; seawater alone produces a brief, loose agglutination, and heat-produced starfish toxin and bacterial suspensions cause
rapid clumping (a tight association of cells in a compact, persistent ball)
at the injection site that spreads throughout the arm within one-half hour
and persists for several hours. The most marked and persistent effect
found by these authors was obtained by injection of a dried extract of
amoebocytes from healthy starfish. Edema of amoebocytes may be observed on the surface of the body in obvious dilation of the papulae; it
begins at the injection site and spreads slowly throughout the arm. The
starfish remain edematous for several hours to several days, particularly
when Versene injection precedes the injection of amoebocyte extract; the
reaction persists long after clumping has disappeared. Dye injections
accompanying injection of amoebocyte extracts do not indicate any
localization of capillary permeability (increased capillary permeability is
characteristic of vertebrate inflammation).

The coelomic fluid of starfish is normally free of bacteria (Bang and
Chaet, 1959), but traumatized animals commonly contain varying numbers of bacteria that disappear from the coelomic fluid within a few days.
However, bacteria-injected starfish consistently lose weight in the interim,
with the greatest weight loss occurring in the specimens containing the
greatest numbers of bacteria.

Ghiradella (1965) pointed out the advantages of using asteroids as
experimental animals in tissue-transplant experiments and noted that
starfish are hardy; they repair injuries readily and regenerate lost parts
rapidly. In a study of visceral regeneration of a starfish, *Henricia leviuscula*, Anderson (1962) briefly described the gross features of surface-
wound healing. It was necessary for him to make a median longitudinal
incision through the aboral body wall of one ray to remove the paired
pyloric caeca and the associated Tiedemann's bodies. The unsutured
edges of the incision gradually approach each other, beginning proximally,

and the closure is completed with a thin, delicate, white web of connective tissue within 1 week. Healing is well advanced by the end of the second week, with the wound site recognizable only by a persistent furrow in the body wall that is somewhat lighter in color than surrounding normal tissue and by the absence of papulae and spines. Although these differences become less marked with time, the furrow is still persistent 4 weeks postinjury. During the 2 weeks necessary for healing of the surface wound, the stumps of the transected pyloric ducts round up and heal over with no indications that new growth has been initiated. Regeneration is thus delayed until the healing process is completed.

Subsequently, Anderson (1965a) described visceral regeneration in three species of starfish from another family, Asteriidae, in which more massive damage was done to the internal organs, including the cutting or tearing of the mesenteries attaching the caeca to the body wall and, in some instances, removing all the pyloric caeca. The operative incisions and damaged internal tissues healed well and, since no exceptions were noted, presumably in much the same fashion as in *Henricia*.

In still another study, Anderson (1965b) described the wound healing preceding regeneration after surgical removal of the cardiac stomach in *Asterias*. Anderson's technique leaves disconnected all the former lines and points of the cardiac stomach, leaves the perivisceral coelomic cavity in open communication with the external environment through the mouth, and, of course, the coelomic fluid with most of its coelomocytes is lost. After recovery from anesthesia ($MgCl_2$), the starfish draws the roof and floor of the disc tightly together; this is undoubtedly a variant of the process typical of asteroids after autotomy of a ray, whereby contraction of the body wall musculature draws the edges of the resulting wound together and seals the opening with a coelomocyte clot. The starfish presents a markedly flattened appearance, clings tightly to the substrate, and remains immobile for a day or two. "The initial sealing-off of the wound, contributing to the adhesion of cut surfaces and restoration of the integrity of the gut wall and body cavity, presumably involves the formation of a typical asteroid coelomocyte clot" in the manner described by Boolootian and Giese (1959). Since most of the coelomocytes are lost in the operation, Anderson postulated that amoebocytes from nearby tissues must contribute, along with residual coelomocytes, to the healing of the incision. Sections of specimens fixed after 48 hours postinjury show that a similar sealing-off process has restored the integrity of the coelomic cavity. A thick mass of connective tissue, which Anderson calls the scar area, draws together and fuses the widely separated attachments of the stomach. Phagocytic amoebocytes infiltrate the area in large numbers and remove the tissue debris.

Class Echinoidea

As shown by Kindred (1921), when carmine and india ink suspended in seawater are injected into the perivisceral fluid of *Arbacia* through perforation of the peristomial membrane, a rapid phagocytic response is initiated. Examination of perivisceral fluid 30 minutes after injection reveals considerable amounts of the foreign material still free in the fluid, but all leukocytes (bladder amoebocytes) contain phagocytized granules. At the end of an hour the perivisceral fluid is comparatively free of granules, and the leukocytes are heavily loaded with them. Kindred (1924) notes that only the bladder amoebocytes (for which he preferred the term leukocytes) enter into phagocytic activity. Kindred considered the filiform amoebocytes to be passive phases of the bladder amoebocytes, since he observed the latter changing into the filiform phase. He assumed that foreign materials phagocytized by bladder amoebocytes are eliminated by diapedesis through the external surface, as observed in asteroids and holothurians; though it has not yet been demonstrated in echinoids.

Kindred (1921) also studied the clotting mechanism of the perivisceral fluid of *Arbacia*, noting that a drop of perivisceral fluid placed on a cover slip immediately shows signs of agglutination. In his observation, actual clotting took place only as the fluid evaporated and as the cells came into contact with the glass. The leukocytes (bladder amoebocytes) elongated, the numerous projections disappeared, and filamentous projections left, in his opinion, from the membrane extended across the field and came into contact with those of adjacent leukocytes to establish a network to which the other perivisceral cells adhered. He was unable to demonstrate a true plasmodium, since the cells retained their individuality and did not enter into cytoplasmic fusion.

Leukocytes form a clot along the margin of a surface wound in *Arbacia;* the clot gradually extends across and eventually closes the opening. In subsequent investigations, Kindred (1924) injected carmine particles suspended in seawater into the perivisceral cavity of several specimens of *Strongylocentrotus drobachiensis*, left them in a live box for 24 hours to allow the phagocytosis of carmine particles by leukocytes, then removed a portion of the body wall 1 cm square from the aboral surface of each specimen. The resected specimens were replaced in the live box and observations were made over a 4-week period of the changes occurring in the wound area.

The first evidence of wound healing is the formation of a membrane that gradually closes the wound. The membrane grows mesially from the margins of the wound, so that the opening in the body wall slowly diminishes in diameter. In seven of the eight wounded specimens, the wound was closed in 2 weeks. The closing membrane is delicate at the time of formation but gradually becomes firm and tough by the third week. Skeletal material begins encroaching on the membrane where it is contiguous with the original cut surface, showing a partly regenerated test by the fourth week.

The membrane, crimson in color, appears to consist of a syncitium of carmine-containing leukocytes. There are several regions of differentiation; the center of the membrane is thick and reddish in color due to the presence of carmine particles in the cells of the syncytial reticulum of the membrane. The color fades out laterally and also in the region proximal to the stereom, where the syncytium is trabeculated similarly to that of the prestereomal area of the uninjured animal. Since the leukocytes are the only phagocytic cells in the perivisceral fluid, it seems obvious (from the carmine grains in the cells of the membrane) that they at least collaborate with the prestereomal cells in the formation of the membrane and, probably, contribute to the skeletal material in formation of the stereom.

Histological evidence indicates that the repairing membrane is formed by the multiplication of the prestereomal connective tissue cells and of the infiltrating leukocytes that anastamose to form the syncytium that makes up the bulk of the membrane. This region corresponds to the prestereomal area of the normal animal.

Thus, the leukocytes are the principal cells of wound repair in the echinoids. They are phagocytic, thrombocytic, and possibly scleroblastic; they ingest foreign particles and tissue debris; they fuse to form a membrane to close the wound and they possibly contribute to the formation of new skeletal material.

CLASS OPHIUROIDEA

Information on the reactions to injury in the brittle stars is restricted to the description of the cellular components of the perivisceral fluid, largely as a side issue in studies of other echinoderms (Kindred, 1924; Boolootian and Giese, 1958) and to the limited discussion of phagocytes and phagocytosis. Apparently nothing is known of the detailed reaction to injury and the wound repair process in this entire class.

CLASS HOLOTHUROIDEA

The coelomocytes of holothurians have been studied by a number of investigators (Boolootian and Giese, 1958; Endean, 1958; Kindred, 1924; Hetzel, 1963, 1965). Hetzel (1965) injected a suspension of carbon (lampblack) in filtered seawater into the perivisceral coelom of *Cucumaria miniata, Euperitaca qiunquesemita, Psolus chitinoides,* and *Leptosynapta clarki.* One-milliliter samples of coelomic fluid were withdrawn for examination after 30 minutes, 1, 2, 3, and 8 hours, and daily for 30 days. One *Cucumaria* was sacrificed every other day for 1 month and at irregular intervals for 6 months, while one *Psolus* was sacrificed each week for 4 weeks. The number of amoebocytes phagocytizing carbon gradually increased until a peak was reached at 2 weeks. During their phagocytic activity they were incapable of contributing to clot formation. Surprisingly, the hemocytes demonstrated a weak phagocytic activity, the first record of such a function. Carbon was also found in brown bodies, supporting the hypothesis that these cells are excretory in function. Concentrations of carbon were found in various sites in sacrificed animals for as long as 6 months in *Cucumaria.* The elimination of carbon from the ciliated urns of *Leptosynapta,* where it was quickly concentrated, was much more rapid than in other species and occurred by diapedesis through the body wall. The carbon was eliminated from the sea cucumbers to a slight degree in the feces; after the second day, brown bodies composed partly of carbon were found in the aquaria, and masses of amoebocytes laden with carbon were eliminated from the bases of the tube feet of *Cucumaria.* Several specimens of *Cucumaria* were also covered with carbon-laden mucus.

Field observations demonstrate considerable variation in the response of individual *Stichopus badionotus* to cutaneous wounds, some quickly close cutaneous wounds, while others rapidly decompose (Cowden, 1968). Preliminary investigations indicated that wounds penetrating the body wall invariably lead to rapid decomposition, while extensive surface wounds that do not perforate the body wall heal without complication.

In a laboratory study, Cowden made wedge-shaped incisions approximately 4 mm wide and 6–8 mm deep along 6–8 cm of the dorsal surface of more than sixty individuals that were then returned to seawater aquaria. Specimens were sacrificed at appropriate intervals from 1 to 18 days and cross sections of the body wall containing the lesion were prepared for histological and histochemical study.

Grossly, the edges of the incisions spread apart for the first 6–8 hours, then become reapproximated. The wounds are easily reopened by the application of gentle pressure during the next 2 days; thereafter, considerable pressure is required to reopen them. The line of closure was identifiable as a thin fissure throughout the 18-day experiment.

Histologically, the edges of the wound are slightly retracted at 24 hours postwounding, with the epidermis and the underlying dermal layer,

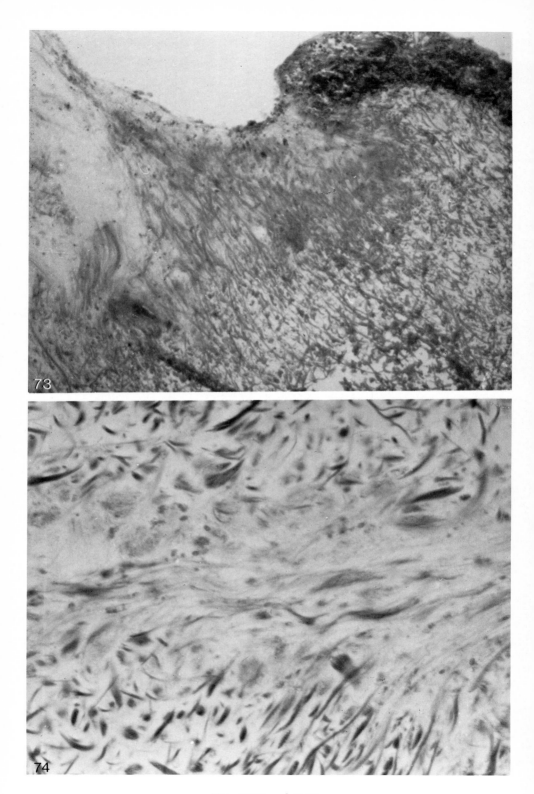

Figs. II.73 and II.74

containing free pigment granules, turned inward (Fig. II.73). The wound channel separating the edges of the incision is filled with coagulum containing some infiltrated, free connective tissue cells. A zone of degeneration develops by this time in the deeper layers of collagen bundles, with the collagen bundles becoming unraveled or broomed, particularly at the margins of the wounds.

By the second day, a column of fibroblasts moves along the surface of the wound channel, and the fibroblasts are oriented so that their long axes are perpendicular to the surface epidermis (and parallel with the wound channel). Fine connective tissue fibers are deposited in the center of the wound channel where the fibroblasts occur.

Considerable fibrosis is apparent in the wound channel by the third day. The fine, newly formed fibers are oriented in the same direction as the long axes of the fibroblasts producing them (Fig. II.74). By this time, the epithelial basement membrane is reformed and continuity of the epidermal epithelium is reestablished. The coagulum in the wound channel does not stain (Fig. II.75), although collagen bundles outside the wound channel stain by the various methods for demonstrating acid mucopolysaccharides. However, some delicate staining occurs in the area within the wound channel where new connective tissues are formed.

Considerable amounts of new collagen are elaborated and organized into thicker bundles by the fourth day after wounding (Fig. II.76). These bundles radiate in all directions from the dermal apex of the wound channel, rather than being oriented parallel to it. Fibrogenesis is essentially completed by the tenth day (Fig. II.77), but the extracellular pigment granules in the superficial, loose connective tissue layer of the dermis do not reunite and are responsible for the gross appearance of the wound fissure. Some passive movement of this material 18 days after wounding indicates that the restoration of the layer of extracellular pigment granules, with concomitant disappearance of a recognizable wound fissure, is the eventual successful completion of the wound repair process in *Stichopus*.

Significantly, no mitotic figures are apparent among the fibroblasts during wound repair, and mitotic proliferation first occurs in the epidermal epithelium on the eighth day postwounding. Thus, proliferation does not play a role in wound closure.

Fig. II.73. Wound healing in the sea cucumber, *Stichopus badionotus*. Shoulder of wound at 24 hours showing retraction of epidermis and superficial dermal layer, some unraveling of cut collagen bundles and the wound coagulum. Masson's trichrome. 56×. (From Cowden, 1968. Courtesy of R. R. Cowden.)

Fig. II.74. Wound healing in sea cucumber. Fibroblasts and fine, new collagen in 3-day wound. Stained with Lillie's picronaphthol blue black-azofuchsin-brilliant purpurin (PNAB). 1053×. (From Cowden, 1968. Courtesy of R. R. Cowden.)

Figs. II.75–II.77

Wound healing in S. *badionotus* differs, as Cowden pointed out, in two important aspects from the process in vertebrate and in most coelomate invertebrates, where initial trauma or irritation initiates a cellular exudative response that is followed by a considerably delayed onset of fibrogenesis. There is no marked infiltration of any cellular elements, particularly coelomocytes, in the wound coagulum, and the rapid onset of fibrogenesis is achieved by migration of fibroblasts from the superficial dermal layer of connective tissue. The latter phenomenon occurs without proliferation and only sufficient numbers migrate to accomplish fibrogenesis. Thus, *Stichopus* has a specialized exudative response to inflammation that resembles that of the freshwater mussel (Pauley and Heaton, 1969) and of those coelenterates lacking wandering amoebocytes.

Cowden's finding that the "morula cells," large spherule amoebocytes, do not contribute to the deposition of collagen in the wound repair process in S. *badionotus* contrasts with Rollefson's (1965) conclusions about their role in wound healing in *Stichopus tremulus*. Rollefson (1965) found morula cells to be abundant in a 20-hour-old wound; they aggregated at the exposed surface and apparently formed a protective barrier between the wound and the exterior. The morula cells infiltrate at the wound surface by 3 days postwounding and histochemical observations indicate that the collagen fibers deposited to repair the wound are derived from the morula cells. Rollefson not only stated that the morula cells play a significant role in wound healing in S. *tremulus*, but also agreed with previous workers that, while sufficient evidence is not yet available that morula cells of *Stichopus* and mast cells of vertebrates are homologous, they are similar in staining reactions and behavior.

Since, as Cowden further noted, S. *badionotus* lives in an environment in which the chances of its sustaining superficial trauma are high, the ability to repair surface wounds rapidly is of great survival value. Further, the catastrophic results from wounds perforating the body wall necessitate the rapid repair of superficial wounds to prevent their extension to vulnerable tissues. Cowden felt that the modifications of cutaneous wound healing in *Stichopus* were for the purpose of speeding up

---

Fig. II.75. Wound healing in sea cucumber. Three-day wound. Note staining of intact collagen fibers but only delicate staining of newly formed fibers in wound. Dilute toluidin blue. 99×. (From Cowden, 1968. Courtesy of R. R. Cowden.)

Fig. II.76. Wound healing in sea cucumber. Four-day wound showing collagen fibers radiating from apex of wound channel. Masson's. 99×. (From Cowden, 1968. Courtesy of R. R. Cowden.)

Fig. II.77. Wound healing in sea cucumber. Ten-day wound. Note fibrogenesis is complete but dermal pigment layer is not closed. Masson's. 99×. (From Cowden, 1968. Courtesy of R. R. Cowden.)

the rate of repair. The source of the substances responsible for dissolution of the organism when the body wall is perforated remains unknown, as does the chemical nature of the coagulum.

CLASS CRINOIDEA

There appear to be no reports in the literature on the reaction to injury and wound repair in crinoids other than a few references to naturally occurring abnormal conditions assumed to be the result of disease or wounds in fossils (Etheridge, 1880) or in living (Anonymous, 1876; Arendt, 1961) crinoids.

Phylum Chordata—Subphylum Urochordata

Little is known of the processes of reaction to injury and wound repair in the tunicates. This is surprising because it is here that closest comparison of these phenomena in the vertebrates and the invertebrates could be made. They are easily obtainable, live well in aquaria, are convenient animals to work with, and their morphology and embryology have been studied in great detail relative to their phylogenetic relationships with the vertebrates.

Cantacuzène (1919) studied the phagocytic response of *Ascidia mentula* to injection of bacteria from a mollusk, *Aplysia punctata;* he demonstrated that the bacteria increased in number in the blood through the first 4 days postinoculation and then decreased, and became rare by the seventh day. Accompanying the decrease in bacteria was an increase in phagocytes and phagocytosis. A typical acute inflammatory response is initiated in *A. mentula* by the inoculation of *Bacterium tumifaciens;* an abscess is formed and surrounded by infiltrating mesenchyme cells of fusiform and stellate shape, small chromophilic leukocytes, and vacuole cells (Thomas, 1931b). When aqueous solutions of arsenic, indole or coal tar and ground *Maia squinado* eggs are injected into the same species, however, an inflammatory response reportedly does not occur (Thomas, 1931a).

Most of the other published material relative to reaction to injury in tunicates is devoted primarily to the description of the various types of blood cells. Fulton (1920) described ten types of blood cells in *Ascidia atra* and gave the approximate proportions of each type. He also discussed phagocytosis and clotting at some length, noting that internal coagulation occurs if the animal is handled, but it is of short duration. He stated that primary coagulation results from cell agglutination, but that secondary coagulation occurs when the cells undergo cytolysis. However, Andrew (1962) stated that clotting in the tunicates consists of agglutination of cells, making no reference to secondary clotting involving cytolysis, and he noted that there appears to be no clotting mechanism involving the plasma. In the same paper, Andrew summarized the findings of previous workers and gave a general description and classification of the various cell types, but he had little to say about their functions. The primary division was into colored and uncolored cells; the former category included blue, orange, green, and brown or reddish-brown cells. The uncolored cells included small amoeboid leukocytes (lymphocytes), large amoeboid leukocytes (macrophages and reticulated leukocytes), and quiescent, vacuolated, colorless cells.

Apparently only the leukocytes are phagocytic and actively enter into the defense mechanism of tunicates. Andrew noted that the small amoeboid leukocytes resemble the lymphocytes of vertebrates in their nucleocytoplasmic ratio and general structure and stated that they may appropriately be termed lymphocytes. The larger, actively amoeboid leukocytes, which he called macrophages, vary in appearance according to their phagocytic activity. Andrew also described actively undulatory beaded appendages on the vacuolated, colorless cells; the appendages continue their motion when detached. The beads on the appendages bear a superficial resemblance to vertebrate blood platelets and, like platelets, are living cell fragments that are markedly adhesive.

In addition to the wandering blood cells among the various tissues, there are in the tunicates fixed tissue counterparts of the blood cells. These nonmotile cells correspond to, and are almost certainly derived from, the various blood cells found in the circulatory system and wandering in the tissues. Several pigment cells of the tests of various tunicates are clearly fixed-tissue counterparts of colored blood cell types, and the macrophages are represented as large phagocytic cells (resembling vertebrate fibroblasts in appearance) in the test.

Although ascidians undoubtedly repair wounds to the test and, presumably, to other parts of the body, there appears to have been no investigation of the mechanism by which this process is accomplished. Since the tunic is produced by diapedesis of cellular elements from the body with subsequent proliferation, it may tentatively be assumed that wounds are repaired quickly and by a similar process.

References

Allgen, C. A. (1959). Free-living marine nematodes. II. On destruction and regeneration in free-living marine nematodes. *Further Zool. Results Swedish Antarctic Expedition, 1901–1903* **2**, No. 2, 268–271.

Anderson, J. M. (1962). Studies on visceral regeneration in sea stars. I. Regeneration of pyloric caeca in *Henricia leviuscula* (Stimpson). *Biol. Bull.* **122**, 321–342.

Anderson, J. M. (1965a). Studies on visceral regeneration in sea stars. II. Regeneration of pyloric caeca in Asteriidae, with notes on the source of cells in regenerating organs. *Biol. Bull.* **128**, 1–23.

Anderson, J. M. (1965b). Studies on visceral regeneration in sea stars. III. Regeneration of the cardiac stomach in *Asterias forbesi* (Desor). *Biol. Bull.* **129**, 454–470.

Andrew, W. (1962). Cells of the blood and coelomic fluids of tunicates and echinoderms. *Amer. Zool.* **2**, 285–297.

Anonymous. (1876). Notes and observations on injured or diseased crinoids. *Proc. Natur. Hist. Soc. Glasgow* **3**, 91–95.

Arendt, J. A. (1961). On the injuries of crinoids made by schizoproboscina. *Paleontol. Zh.* No. 2, pp. 101–106 (in Russian).

Balbiani, E. G. (1888). Recherches experimentales sur la merotomie des infusoires cilies. *Rec. Zool. Suisse* **5** (pagination unknown).

Bang, F. B. (1956). A bacterial disease of *Limulus polyphemus*. *Bull. Johns Hopkins Hosp.* **98**, 325–351.

Bang, F. B. (1961). Reaction to injury in the oyster *Crassostrea virginica*. *Biol. Bull.* **121**, 57–68.

Bang, F. B., and Bang, B. C. (1962). Studies on sipunculid blood. Immunological properties of coelomic fluid and morphology of "urn cells." *Cah. Biol. Mar.* **3**, 363–374.

Bang, F. B., and Chaet, A. B. (1959). The effect of starfish toxin on amebocytes. *Biol. Bull.* **117**, 403–404.

Bang, F. B., and Lemma, A. (1962). Bacterial infection and reaction to injury in some echinoderms. *J. Insect Pathol.* **4**, 401–414.

Barnes, R. D. (1963). "Invertebrate Zoology." Saunders, Philadelphia, Pennsylvania.

Beaver, P. (1937). Experiments on regeneration in the trematode, *Echinostoma revolutum*. *J. Parasitol.* **23**, 423–424.

Boolootian, R. A., and Giese, A. C. (1958). Coelomic corpuscles of echinoderms. *Biol. Bull.* **115**, 53–63.

Boolootian, R. A., and Giese, A. C. (1959). Clotting of echinoderm coelomic fluid. *J. Exp. Zool.* **140**, 207–229.

Bruntz, L. (1905). Etude physiologique sur les Phyllopodes Branchiopodes. Phagocytose et excrétion. *Arch. Zool. Exp. Gen.* [4] **4**, 183–198.

Bruntz, L. (1906a). Un organe globuligène chez les Stomatopodes. *C. R. Soc. Biol.* **60**, 428–430.

Bruntz, L. (1906b). Sur l'existence d'un organe globuligène chez les Schizopodes. *C. R. Soc. Biol.* **60**, 832–833.

Bruntz, L. (1906c). Les globules sangiuns des Crustacés Arthostracés. Leur origine. *C. R. Soc. Biol.* **60**, 835–836.

Bruntz, L. (1906d). La phagocytose chez les Diplopodes (Globules sanguins et organes phagocytaires). *Arch. Zool. Exp. Gen.* [4] **5**, 491–504.

Bruntz, L. (1906e). L'organe phagocytaire des *Polydesmes*. *C. R. Soc. Biol.* **61**, 252–253.

Cameron, G. R. (1932). Inflammation in earthworms. *J. Pathol. Bacteriol.* **35**, 933–972.

Cantacuzène, J. (1919). Etude de'une infection expérimentale chez *Ascidia mentula*. *C. R. Soc. Biol.* **82**, 1019–1022.

Cantacuzène, J. (1922a). Sur le rôle agglutinant des urnes chez *Sipunculus nudus* (1). *C. R. Soc. Biol.* **87,** 259–262.

Cantacuzène, J. (1922b). Réactions d'immunité chez *Sipunculus nudus* vacciné contre une bactérie. *C. R. Soc. Biol.* **87,** 264–267.

Cantacuzène, J. (1922c). Sur le sort ultérieur des urnes chez *Sipunculus nudus* au cours de l'infection et de l'immunisation. *C. R. Soc. Biol.* **87,** 283–285.

Cantacuzène, J. (1928). Recherches sur les réactions d'immunité chez les Invertébrés. *Arch. Roum. Pathol. Exp. Microbiol.* **1,** 7–80.

Cheney, D. P. (1969). The morphology, morphogenesis and reactive response of ³H-thymidine labeled leucocytes in the Manila clam, *Tapes semidecussata* (Reeve). Ph.D. Thesis, University of Washington, Seattle.

Cheney, D. P., and Sparks, A. K. (1969). The behavior of ³H-thymidine labeled leucocytes of the bivalve mollusc, *Tapes semidecussata. Second Annu. Meet. Soc. Invertebr. Pathol. 1969.*

Cheng, T. C., and Streisfeld, S. D. (1963). Innate phagocytosis in the trematodes *Megalodiscus temperatus* and *Haematoloechus* sp. *J. Morphol.* **113,** 375–380.

Cheng, T. C., Shuster, C. N., Jr., and Anderson, A. H. (1966a). A comparative study of the susceptibility and response of 8 species of marine pelecypods to the trematode *Himasthla quissetensis. Trans. Amer. Microsc. Soc.* **85,** 284–295.

Cheng, T. C., Shuster, C. N., Jr., and Anderson, A. H. (1966b). Effects of plasma and tissue extracts of marine pelecypods on the cercariae of *Himasthla quissetensis. Exp. Parasitol.* **19,** 9–14.

Cheng, T. C., Rifkin, E., and Yee, H. W. F. (1968a). Studies on the internal defense mechanism of sponges. II. Phagocytosis and elimination of India ink and carmine particles by certain parenchymal cells of *Terpios zeteki. J. Invertebr. Pathol.* **11,** 302–309.

Cheng, T. C., Yee, H. W. F., and Rifkin, E. (1968b). Studies on the internal defense mechanism of sponges. I. The cell types occurring in the mesoglea of *Terpios zeteki* (de Laubenfels) (Porifera: Demospongiae). *Pac. Sci.* **22,** 395–401.

Cheng, T. C., Yee, H. W. F., Rifkin, E., and Kramer, M. D. (1968c). Studies on the internal defense mechanisms of sponges. III. Cellular reaction in *Terpios zeteki* to implanted heterologous biological materials. *J. Invertebr. Pathol.* **12,** 29–35.

Cowden, R. R. (1968). Cytological and histochemical observations on connective tissue cells and cutaneous wound healing in the sea cucumber *Stichopus badionotus. J. Invertebr. Pathol.* **10,** 151–159.

Cuénot, L. (1891). Etudes sur le sang et les glandes lymphatiques dans la série animale (2 e partie: Invertébrés). *Arch. Zool. Exp. Gen.* [2] **9,** 13–90, 365–475, and 593–670.

Cuénot, L. (1913). Excrétion et phagocytose chez les Sipunculiens. *C. R. Soc. Biol.* **74,** 159–161.

Day, M. F., and Bennetts, M. J. (1953). Healing of gut wounds in the mosquito *Aedes aegypti* (L.) and the leafhopper *Orosius argentatus* (Ev.). *Aust. J. Biol. Sci.* **6,** 580–585.

Dent, J. N., and Fitzpatrick, J. F., Jr. (1963). Wound healing in crawfish. *ASB Bull.* **10,** 26–27.

Des Voigne, D. M., and Sparks, A. K. (1968). The process of wound healing in the Pacific oyster, *Crassostrea gigas. J. Invertebr. Pathol.* **12,** 53–65.

Drew, G. H. (1910). Some points in the physiology of lamellibranch blood corpuscles. *Quart. J. Microsc. Sci.* **54,** 605–623.

Drew, G. H., and de Morgan, W. (1910). The origin and formation of fibrous tissue produced as a reaction to injury in *Pecten maximus*, as a type of the Lamellibranchiata. *Quart. J. Microsc. Sci.* **55**, 595–610.

Duboscq, O. (1896). La terminaison des vaisseaux et les corpuscles de Kowalewsky chez les Scolopendrides. *Zool. Anz.* **19**, 391–397.

Durham, H. E. (1888). The emigration of amoeboid corpuscles in the starfish. *Proc. Roy. Soc., Ser. B* **43**, 327–330.

Endean, R. (1958). The coelomocytes of *Holothuria leucospilata*. *Quart. J. Microsc. Sci.* **99**, 47–60.

Etheridge, R. J. (1880). Observations on the swollen condition of carboniferous crinoid stems. *Proc. Natur. Hist. Soc. Glasgow* **4**, 19–36.

Florey, H. (1958). "General Pathology." Saunders, Philadelphia, Pennsylvania.

Fontaine, C. T. (1971). Exoskeletal intrusions—a wound repair process in penaeid shrimp. *J. Invertebr. Pathol.* **18**, 301–303.

Fried, B., and Penner, L. R. (1964). Healing of wounds following partial or complete amputation of body parts in a trematode, *Philopthalmus hegeneri*. *J. Parasitol.* **50**, 266.

Fulton, J. F. (1920). The blood of *Ascidia atra* Lesueur; with special reference to pigmentation and phagocytosis. *Acta Zool.* **1**, 381–431.

Galtsoff, P. S. (1925). Regeneration after dissociation (an experimental study of sponges). II. Histogenesis of *Microciona prolifera*. Verr. *J. Exp. Zool.* **42**, 223–255.

George, W. C., and Ferguson, J. H. (1950). The blood of gastropod molluscs. *J. Morphol.* **86**, 315–327.

Ghiradella, H. T. (1965). The reaction of two starfishes, *Patiria miniata* and *Asterias forbesi*, to foreign tissue in the coelom. *Biol. Bull.* **128**, 77–89.

Glavind, J. (1948). "Studies on Coagulation of Crustacean Blood," pp. 7–137. Nyt Nordisk Forlag, Arnold Busck, Copenhagen.

Gruzewska, Z. (1932). Sur la coagulation du sang chez les Crustacés. *C. R. Soc. Biol.* **110**, 920–922.

Haliburton, W. D. (1885). On the blood of decapod Crustacea. *J. Physiol. (London)* **6**, 300–335.

Hardy, W. B. (1892). The blood corpuscles of the Crustacea, together with a suggestion as to the origin of the crustacean fibrin-ferment. *J. Physiol. (London)* **13**, 165–190.

Haughton, I. (1934). Amoebocytes and allied cells in Invertebrata. *J. Roy. Microsc. Soc.* **54**, 246–262.

Hetzel, H. R. (1963). Studies on holothurian-coelomocytes. I. A survey of coelomocyte types. *Biol. Bull.* **125**, 289–301.

Hetzel, H. R. (1965). Studies on holothurian coelomocytes. II. The origin of coelomocytes and the formation of brown bodies. *Biol. Bull.* **128**, 102–111.

Hyman, L. H. (1940). "The Invertebrates: Protozoa through Ctenophora," Vol. I. McGraw-Hill, New York.

Hyman, L. H. (1951). "The Invertebrates: Acanthocephala, Aschelminthes, and Entoprocta," Vol. III. McGraw-Hill, New York.

Ishiwaka, C. (1890). Trembley's Umkehrungsversuche an *Hydra* nach neuen Versuchen erklärt. *Z. Wiss. Zool.* **49** (pagination unknown) (from Hyman, 1940).

Iwati, S. (1934). Some experimental studies on the regeneration of the plerocercoid of Manson's tapeworm, *Diphyllobothrium erinacei* (Rudolphi) with special reference to its relationship with *Sparganum proliferum* Ijima. *Jap. J. Zool.* **6**, 139–158.

Jacquemain, R., Jullien, A., and Noel, R. (1947). Sur l'action de certains corps cancérigènes chez les Céphalopodes. *C. R. Acad. Sci.* **225**, 441–443.

Janisch, R. (1964). Sub-microscopic structure of injured surface of *Paramecium caudatum*. *Nature* (*London*) **204**, 200–201.

Janisch, R. (1966). The mechanism of the regeneration of surface structures of *Paramecium caudatum*. *Folia Biol.* (*Prague*) **12**, 65–73.

Johnson, P. T. (1968). "An Annotated Bibliography of Pathology in Invertebrates other than Insects." Burgess, Minneapolis, Minnesota.

Jullien, A. (1928). Sur la transformation des cellules sanguines de la seiche au cours des réactions inflammatoires aseptiques. *C. R. Acad. Sci.* **186**, 526–529.

Jullien, A. (1940). Sur les réactions des Mollusques Céphalopodes aux injections de goudron. *C. R. Acad. Sci.* **210**, 608–610.

Kaestner, A. (1967). Class Hirudinea. *In* "Invertebrate Zoology Vol. I," pp. 541–566. Interscience Publ., John Wiley and Sons, New York.

Kedrovsky, B. (1925). Reaktive Veräenderungen in den Geweben der Teichmuschel (*Anodonta* sp.) bei Einfuschrung von sterilem Celloidin. *Virchows Arch. Pathol. Anat. Physiol.* **257**, 815–845.

Kepner, W. A., and Reynolds, B. D. (1923). Reactions of cell-bodies and pseudopodial fragments of *Difflugia*. *Biol. Bull.* **44**, 22–47.

Kindred, J. E. (1921). Phagocytosis and clotting in the perivisceral fluid of *Arbacia*. *Biol. Bull.* **41**, 144–152.

Kindred, J. E. (1924). The cellular elements in the perivisceral fluid of echinoderms. *Biol. Bull.* **46**, 228–251.

Kowalewsky, A. (1894). Etude expérimentale sur les glandes lymphatiques des invertébrés. *Bull. Acad. Imp. Sci. St.-Petersbourg* [4] **36**, 273–295.

Kudo, R. R. (1954). "Protozoology," 4th ed. Thomas, Springfield, Illinois.

Kuhl, W. (1954). Zeitrafferfilm-Untersuchungen über die Wirkung von Zentrifugierung und Pressung auf die Cytoplasmastrukturen, den Plasmogamieablauf und die Zellrestitution bei *Actinosphaerium eichorni* Erhrbg. *Protoplasma* **43**, 3–62.

Labbé, A. (1929). Réactions experimentales des Mollusques à l'introduction de stylets de cellodine. *C. R. Soc. Biol.* **100**, 166–168.

LeGore, R. S. (1970). A histological examination of wound repair and reaction to large foreign bodies in the rhynchobdellid leech, *Piscicola salmositica*. M.S. Thesis, University of Washington, Seattle.

LeGore, R. S., and Sparks, A. K. (1971). Repair of body wall incisions in the rhynchobdellid leech, *Piscicola salmositica*. *J. Invertebr. Pathol.* **18**, 40–45.

Loeb, L. (1902). On the blood lymph cells and inflammatory processes of *Limulus*. *J. Med. Res.* **7**, 145–158.

Mackin, J. G. (1961). Oyster leucocytes in infectious disease. *Amer. Zool.* **1**, 371.

Mann, K. H. (1962). Leeches (Hirudinea). *Int. Ser. Monogr. Pure Appl. Biol., Div.: Zool.* **2**, 1–224.

Manter, H. W. (1933). The genus *Helicometra* and related trematodes from Tortugas, Florida. *Carnegie Inst. Wash. Publ.* **435**, 167–182.

Manter, H. W. (1943). One species of trematode, *Neorenifer grandispinus* (Caballero) attacked by another, *Mesocercaria marcianae* (LaRue). *J. Parasitol.* **29**, 387–392.

Messing, S. (1903). Ueber Entzundung bei den Neidere wirbellosen Tieren. *Zentralbl. Allg. Pathol. Pathol. Anat.* **14**, 915–920.

Metchnikoff, E. (1893). "Lectures on the Comparative Pathology of Inflammation" (Transl. from the French by F. A. and E. H. Starling). Kegan Paul, Trench, Trübner and Co., Ltd., London.

Myers, R. J. (1935). Behavior and morphological changes in the leech, *Placobdella parasitica*, during hypodermic insemination. *J. Morphol.* **57**, 617–648.

Mikhailova, I. G., and Prazdnikov, E. V. (1961). Two questions on the morphological reactivity of mantle tissues in *Mytilus edulis* L. *Tr. Murmansk. Morsk. Biol. Inst.* **3**, 125–130.

Mikhailova, I. G., and Prazdnikov, E. V. (1962). Inflammatory reactions in the Barents Sea mussel (*Mytilus edulis* L.). *Tr. Murmansk. Morsk. Biol. Inst.* **4**, 208–220.

Mix, M. C., and Sparks, A. K. (1971). Repair of digestive tubule tissue of the Pacific oyster, *Crassostrea gigas*, damaged by ionizing radiation. *J. Invertebr. Pathol.* **17**, 172–177.

Needham, A. E. (1965). Regeneration in the Arthropoda and its endocrine control. "Proceedings of the Regeneration in Animals." North-Holland Publ., Amsterdam.

Orton, J. H. (1922). The blood cells of the oyster. *Nature (London)* **109**, 612–613.

Pai, S. (1934). Regenerationsversuche an rotatorien. *Sci. Rep., Univ. Chekiang* **1**, No. 2.

Palm, N. B. (1953). The elimination of injected vital dyes from the blood of Myriapods. *Ark. Zool.* [2] **6**, 219–246.

Pauley, G. B., and Heaton, L. H. (1969). Experimental wound repair in the freshwater mussel *Anodonta oregonensis*. *J. Invertebr. Pathol.* **13**, 241–249.

Pauley, G. B., and Sparks, A. K. (1965). Preliminary observations on the acute inflammatory reaction in the Pacific oyster, *Crassostrea gigas* (Thunberg). *J. Invertebr. Pathol.* **7**, 248–256.

Pauley, G. B., and Sparks, A. K. (1966). The acute inflammatory reaction in two different tissues of the Pacific oyster, *Crassostrea gigas*. *J. Fish. Res. Bd. Can.* **23**, 1913–1921.

Pauley, G. B., and Sparks, A. K. (1967). Observations on experimental wound repair in the adductor muscle and the Leydig cells of the oyster *Crassostrea gigas*. *J. Invertebr. Pathol.* **9**, 298–309.

Perkins, F. O., and Menzel, R. W. (1964). Maintenance of oyster cells *in vitro*. *Nature (London)* **204**, 1106–1107.

Prazdnikov, E. V., and Mikhailova, I. G. (1962). A note on the problem of the character of the early inflammatory reaction in some coelenterates (*Staurophora mertensii* Brandt, 1935, *Aurelia aurita* L., *Beroe cucumis* Fabr.). *Tr. Murmansk. Morsk. Biol. Inst.* **4**, 22–228.

Reynolds, B. D. (1924). Interaction of protoplasmic masses in relation to the study of heredity and environment in *Arcella polypora*. *Biol. Bull.* **46**, 106–142.

Rollefson, I. (1965). Studies on the mast cell-like morula cells of the holothurian *Stichopus tremulus* (Gun.). *Arbok Univ. Bergen, Mat.-Natur. Ser.* **8**, 3–12.

Ruddell, C. L. (1969). A cytological and histochemical study of wound repair in the Pacific oyster, *Crassostrea gigas*. Ph.D. Thesis, University of Washington, Seattle.

Schlumberger, H. G. (1952). A comparative study of the reaction to injury; the cellular response to methylcholanthrene and to talc in the body cavity of the cockroach (*Periplaneta americana*). *AMA Arch. Pathol.* **54**, 98–113.

Sinitsin, D. F. (1932). Studien über die Phylogen ie der Trematoden. VII. Regeneration in the digenetic trematodes. *Biol. Zentralbl.* **52**, 117–120.

Sparks, A. K., and Pauley, G. B. (1964). Studies of the normal post-mortem changes in the oyster *Crassostrea gigas* (Thunberg). *J. Insect Pathol.* **6**, 78–101.

Stauber, L. A. (1950). The fate of India ink injected intracardially into the oyster, *Ostrea virginica* Gmelin. *Biol. Bull.* **98**, 227–241.

Stauber, L. A. (1961). Immunity in invertebrates, with special reference to the oyster. *Proc. Nat. Shellfish. Ass.* **50**, 7–20.

Stein, J. E., and Mackin, J. G. (1955). A study of the nature of pigment cells of oysters and the relation of their numbers to the fungus disease caused by *Dermocystidium marinum. Tex. J. Sci.* **7**, 422–429.

Takatsuki, S. (1934). On the nature and function of the amebocytes of *Ostrea edulis. Quart. J. Microsc. Sci.* **76**, 379–431.

Tartar, V. (1970). Transplantation in protozoa. *Transplant. Proc.* **2**, 183–190.

Thakur, A. S., and Cheng, T. C. (1968). The formation, structure, and histochemistry of the metacercarial cyst of *Philophthalmus gralli* Mathis and Léger. *Parasitology* **58**, 605–618.

Thomas, J.-A. (1931a). Réactions de deux Invertébrés. *Ascidia mentula* Müll. et *Nereis diversicolor* O.F.M., à l'inoculation de substances à propriétés cancérigènes. *C. R. Soc. Biol.* **108**, 667–669.

Thomas, J.-A. (1931b). Sur les réactions de la tunique d'*Ascide mentula* Müll. à l'inoculation de *Bacterium tumefaciens* Sm. *C. R. Soc. Biol.* **108**, 694–696.

Tokin, B. P., and Yericheva, F. N. (1959). Phagocytic characteristics of cells of *Hydra oligacis* Poll. *Sci. Rep. Higher Sch., Biol. Ser.* No. 2 (in Russian).

Tokin, B. P., and Yericheva, F. N. (1961). Phagocytal reaction in the course of regeneration and somatic embryogenesis in lower coelenterates. *Tr. Murmansk. Morsk. Biol. Inst.* **3**, No. 7.

Triplett, E. L., Cushing, J. E., and Durall, G. L. (1958). Observations on some immune reactions of the sipunculid worm *Dendrostomum zostericolum. Amer. Natur.* **92**, 287–293.

Tripp, M. R. (1958). Disposal by the oyster of intracardially injected red blood cells of vertebrates. *Proc. Nat. Shellfish. Ass.* **58**, 143–147.

Tripp, M. R. (1961). The fate of foreign materials experimentally introduced in the snail *Australorbis glabratus. J. Parasitol.* **47**, 745–749.

Tripp, M. R., Bisignani, L. A., and Kenny, M. T. (1966). Oyster amebocytes *in vitro. J. Invertebr. Pathol.* **8**, 137–140.

Wagge, L. E. (1955). Amoebocytes. *Int. Rev. Cytol.* **4**, 31–78.

Yonge, C. M. (1926). Structure and physiology of the organs of feeding and digestion in *Ostrea edulis. J. Mar. Biol. Ass.* **12**, 295–386.

Zawarzin, A. (1927). Über die reactiven Veranderungen des Epithels bei der Einfhrung eines fremd Korpers in den Mantel von *Anadonta. Z. Mikrosk.-Anat. Forsch.* **11**, 215–282.

CHAPTER III

Physical Injuries

The numerous physical agents that may cause injury to invertebrates can be broadly classified into several major categories: trauma, pressure changes, and temperature changes.

Trauma

Trauma is one of the most common sources of injury to invertebrates; it occurs when a mass strikes the body with sufficient force to cause physical damage to tissues. In the case of vertebrates (especially birds and mammals) and insects, injury occurs as readily when the body hits a stationary mass or when both body and mass are moving at the moment of impact. It is unlikely, however, that any invertebrates (other than insects) under natural conditions move with sufficient acceleration to be traumatized by contacting a stationary mass.

The injuries resulting from trauma might be classified in several ways; e.g., an injury can be crushing or penetrating, depending on whether the contacting mass does or does not penetrate the outer surface (epithelium, integument, epidermis, etc.) into the underlying tissue. The severity of the injury varies from the equivalent of a human contusion or bruise to complete disruption of the body architecture in crushing injuries, and from the most limited local damage to amputation or bisection of the body in penetrating injuries.

Secondary effects may develop from localized trauma, such as hemorrhage, loss of body fluids, and infection by microorganisms that enter at the site of injury. Vertebrates are often subject to shock after physical injuries; the shock is either the primary neurogenic type (fainting) or the more serious hemorrhagic and secondary shock. Robbins (1957) suggests that primary neurogenic shock should not be considered as shock, and, since the remaining two types have the same clinical manifestations, adaptive physiological responses, resultant tissue damage, and loss of effective circulating blood volume, he recommends that hemorrhagic and secondary shock be considered the same.

Hemorrhage in vertebrates may be clinically significant in two ways: the actual loss of blood from the organism resulting in anoxia to vital tissues and the pressure effect created by hemorrhage into the brain or pericardial sac. Acute loss of 10–20% of the blood volume and even greater losses over a longer period may have no clinical significance in man, while relatively small amounts of hemorrhage into the pericardial sac or brain may cause death (Robbins, 1957). Although loss of the circulating fluid of invertebrates, which can be considered as hemorrhage, undoubtedly results in anoxia and desiccation, it is far less critical than in the vertebrates, and the pathological effects of hemorrhage or its equivalent in invertebrates has been virtually ignored.

Separation of the two forms of vertebrate shock (hemorrhagic and secondary) has been made on the basis of the cause of loss of effective blood volume. Effective blood volume is lost by hemorrhage in hemorrhagic shock, while in secondary shock due to crushing injuries, sepsis or putrefaction, excessive exposure to heat or cold, etc., the cause for reduction in circulating blood is not hemorrhage and is often obscure (Robbins, 1957). Pathological changes from shock occur primarily in the kidneys, lungs (congestion and edema), heart (fatty degeneration resulting from systemic anoxia), liver (fatty degeneration), and adrenal glands.

Numerous theories have been proposed to explain the loss of effective blood volume in the absence of hemorrhage. The three most widely accepted are increased permeability of the capillary walls; plasma loss; generalized capillary atony and dilation. Another theory proposes that under stress a toxic humoral agent, once thought to be histamine, is absorbed into the blood stream where it exerts a systemic effect. More recently, another humoral theory has been advanced, based on the discovery that two vasotropic principles (vasoexcitor material and vasodepressor material) are produced whenever tissue anoxia develops. Robbins presents what he terms as the best working hypothesis for the genesis of shock in the absence of gross hemorrhage; "neuro-vascular collapse acting alone or synergistically with loss of whole plasma, or plasma water, probably initiates reduction in circulatory volume and drop in blood pressure. This vascular collapse then activates a chain of circulatory dearrangements, through mechanisms as yet unknown, that progress to peripheral pooling of blood and profound loss of effective circulatory volume."

Although nonhemorrhagic or secondary shock has not been studied in the invertebrates and may not even exist, several kinds of physiological responses to mechanical stimulation have been demonstrated in various insects. These have been concisely reviewed by Day and Oster (1963). Cockroaches have been the subject of many of the experiments; they produce, during struggles to free themselves, a substance in the blood that causes paralysis of other cockroaches into which it is injected. The substance can also be produced experimentally by extreme mechanical or electrical stimulation. Cockroaches stimulated in this manner contained abnormally small amounts of hemolymph; the hindgut contained abnormally large amounts of water, suggesting to the authors that the water-absorbing mechanism was not working properly. It appears to me, however, that fluid was *removed* from the circulatory system. If this is the case, then the cockroach under stress has undergone a physiological condition that is obviously analogous to secondary shock in vertebrates. The paralysis-inducing substance lends credence to the humoral theories for secondary shock in humans.

The acetylcholine (ACH) content of nerve cords of cockroaches that are paralyzed by mechanical stimulation is approximately twice that of normal roaches from the same colony. The ACH content of the nerve cord is also elevated by desiccation, but only to about one-third the level induced by mechanical stimulation.

A characteristic response to stress in vertebrates is the release of adrenaline into the blood stream. Day and Oster noted that a comparable reaction may occur in insects. They cited work by other authors showing that the corpora cardiaca of insects contains substances with at least some of the pharmacological properties of adrenaline; forced hyperactivity or electrical stimulation results in a marked reduction in the amount of histologically recognizable secretory material in the glands, accompanied by a loss of potency of extracts, and extracts of the corpus cardiacum of cockroaches inhibit contractions of rat uterus in a manner similar to that caused by adrenaline.

Day and Oster lament the lack of a systematic assemblage of the normal histology of insects and the neglect of histopathology studies. They do point out, however, that insects are generally able to withstand remarkable amounts of trauma, and this has made them suitable subjects for experimentation in physical injuries.

The comments of Day and Oster are even more appropriate for the rest of the invertebrates. The normal histology of many groups is inadequately known or scattered through the literature, even though Hyman and other workers have done remarkable jobs of assembling such information. The situation in regard to the histopathology of physical injuries is even less encouraging. There have been few studies of the effects of physical agents on invertebrates, and even these investigations have generally ignored the histopathological effects.

Changes in Atmospheric Pressure

There have been several studies on the effects of blasting on various invertebrates in relation to seismic operations in petroleum exploration, ordnance research, or channel blasting in marine engineering. In the first of these investigations, Gowanloch and McDougall (1945) placed shrimp (*Penaeus setiferus*), crabs (*Callinectes sapidus*), and oysters (*Crassostrea virginica*) in slatted wooden cages at middepth at intervals from 50 to 400 feet from a point where one 200-pound and two 800-pound dynamite charges were fired. The experimental animals were examined just before and immediately after each firing and at 24 and 48 hours. No conclusive data were obtained for oysters, but shrimp were uninjured by even the heavier charges at 50 feet.

A much more elaborate investigation was undertaken by the Chesapeake Biological Laboratory (Anonymous, 1948) in collaboration with the U. S. Naval Mine Warfare Station and the Naval Ordnance Laboratory. Not only were these experiments more sophisticated, but the collaboration of the physical scientists and engineers in the naval organizations was extremely valuable. For example, an apparent discrepancy between the seemingly small effects of the 800-pound charges of Gowanloch and McDougall's experiments and the more lethal effects of higher charges in the Chesapeake Bay studies was readily explained by the Naval Ordnance Laboratory as being due to the slatted wooden frame of the Louisiana experimental cages. They produced a decrement in the shock and pressure reaching the experimental animals. To avoid the error, the Chesapeake studies first used 4-foot cubes of ½-inch wire supported by 2 × 4 wooden frames. Again, the Naval Ordnance Laboratory questioned the use of wood-framed cages and showed that the wooden frames partially absorb the shock wave; the amount of error depends on the position of the test animal in relation to the wooden frame. Subsequently, cages were modified to 2-foot cubes constructed of ½-inch mesh wire supported by ¼-inch iron frames. Copper-ball crusher gages were used to approximate the maximum pressure created by the explosive.

The underwater detonation of an explosive generates a shock wave that reaches its maximum pressure almost instantly, then decays exponentially. This peak pressure or shock wave is a function of the size and type of the charge and the distance from it; the wave is greater at any given point with larger charges and decreases at increasing distances from a given charge. The patterns of the waves initiated by underwater explosions are quite complicated because of a number of factors, and the effects may be greatly modified by the geometry of the system, wave velocities, and character of the bottom. It was assumed in this work that peak pressure was the principal cause of damage to organisms, even though it was noted that momentum or energy were other possible causes.

Effects of Underwater Explosions on Mollusks

As noted previously, Gowanloch and McDougall (1945) did not obtain any conclusive data on the effects of explosions on oysters. The results obtained by the Chesapeake Biological Laboratory (Anonymous, 1948) were not as clear-cut as had been anticipated and were not considered conclusive. They noted that immediate mortality from the explosions was apparently not influenced by the effects of handling and holding the oysters. Only 2 of 106 (slightly under 2%) of the oysters exposed in suspended cages within a 100-foot radius of 27- to 31-pound charges were found to be gaping immediately after the explosion. With charges of 300 pounds, 4 of 184, or less than 2.2%, were immediately killed within 200 feet of the charge. Monitoring the experimental oysters for an additional 2 weeks revealed that all subsequent mortalities occurred within the 200-foot radius and attained a level of 5.4%, slightly more than double the immediate death rate.

Additional experimentation on the effect of underwater explosions on oysters was conducted during 1949–1950 in the Barataria Bay Region of Louisiana (Sieling, 1951). The study was initiated by the Texas A & M Research Foundation (Project 9) as a result of assertations by commercial oystermen that the explosives utilized in seismic operations kill oysters immediately, that gases bubbling out of the shot holes are lethal to oysters, and that the bottom is softened or shifted by the explosion causing the oysters to be buried and subsequently die.

The experimental design followed the usual operating procedures for seismic exploration for oil in the shallow water and marshlands along the coast of the Gulf of Mexico. A hole is drilled by a powerful jet of water through a nozzle into the mud; a string of 3-inch diameter threaded pipe is placed in the hole; extra sections of pipe are screwed on as the drilling continues to the desired depth, often over 100 feet. The explosive, usually nitramon, is packed in 1-pound cans that also have screw threads on each end so that a long stick of explosive may be lowered to the desired depth within the pipe and detonated by electrical detonators. Thus, the explosions actually take place at a considerable depth underground and result in a small geyser of mud and water rising from the top of the pipe to a height of 50–100 feet. According to Sieling, a person standing in a boat near the explosion, or shot, feels a sharp slap on the bottom of his feet.

The details of the elaborately designed experiment basically consisted of detonation of a 50-pound charge at 70 feet and of a 20-pound charge at 30 feet. The depths at which the charges were fired were those required by Louisiana law for the size of charge used, and the sizes and number were typical of seismic operations. Oysters were placed in trays on the bottom at distances of 20, 60, 130, and 250 feet from shot points. Oysters were examined immediately after the shots, crusher-type pressure gauges located at 20 and 60 feet from the shot point were picked up, and the pressure was computed.

The results of the experiments showed conclusively that no oysters were killed outright, even at 20 feet, and there was no significant difference between mortalities of oysters exposed to the explosions at 20-, 60-, 130-, and 250-foot distances or between any of these groups and controls over a 4- or 8-month period. There was no indication that toxic substances were liberated in the gases from the shot hole or that oysters sank into or were covered by silt as a result of the explosion.

As Sieling, who had worked on the Chesapeake Bay study, pointed out, the seismic explosion with its buried charge is quite different from those explosions in which the charges were either suspended in water or placed on the bottom. Undoubtedly, the primary shock wave is directed into the air through the protruding pipe. This was substantiated by the fact that the soft copper balls in the pressure gauges showed no measurable deformation, indicating a very weak pressure wave, while all explosions in the Chesapeake Biological Laboratory study caused some degree of deformation of the copper balls.

Tollefson and Marriage (1949) took advantage of an opportunity to observe the effects of blasting on certain beach-dwelling invertebrates when the Port Commission of Bay Ocean, Oregon, made two test blasts in Tillamook Bay to determine the feasibility of using explosives to cut a channel for small-boat moorage. Although no commercially valuable shellfish were in the vicinity (except for oysters some 500 feet from the blast), the tests were run to determine what effect such blasting would have if done in clam- or crab-producing areas.

Four cases of 50% dynamite were buried 3 feet below the surface and were fired as a single shot along with a smaller charge along a 95-foot line in a sandy mud bottom of +1.0- to +3.0-foot tide level. Specimens of clams (cockles, *Cardium corbis*) and oysters (*Crassostrea gigas*) were placed at varying intervals perpendicular to the line. In general, the blasts created a ditch 10–15 feet wide at the top and 4–6 feet deep. Immediately after the blast, water began to "boil" very strongly on the test site, and a great deal of mud was scattered about. Although cockles and oysters placed at the blast site were blown 150 feet away and destroyed and buried cockles 10 feet away were killed and their shells smashed, there was no appreciable damage to surface oysters or cockles located 10 feet or further from the blast line or to subsurface cockles located 15 feet or further from the center. Tollefson and Marriage also observed that a small gastropod, *Thais* sp., 15 mm in height, was found unharmed within 25 feet of the center, as was a mud clam (*Macoma* sp.), 38 mm in length.

It appears fairly obvious that mollusks, particularly oysters and hard clams, are highly resistant to damage from underwater explosion shock waves. There are many complicating features, such as interference by reflected shock waves from the surface and bottom, texture of the bottom, protection or "shadows" from bottom objects, roughness of water surface, and drag of the wave along the bottom (as noted by the Chesapeake Biological Laboratory report) that must be considered in evaluating the effects of subsurface blasts on oysters. These factors are further complicated by the shape and position of the oyster and whether the shock wave approaches towards the hinge, bill, or side. Nevertheless, it is unfortunate that exact data are not available as to the mechanism of injury to mollusks by explosives and the amount required to cause injury or death.

Effects of Underwater Explosions on Crustaceans

All of the typical bottom-dwelling crustaceans, including crabs, shrimp, and lobsters, are as difficult to study as oysters. Additionally, crabs may injure one another unless their chelipeds are "gloved" with rubber tubing, and particular attention must be paid to the postmolting stage, since the exoskeleton varies from extremely soft to a brittle, "paper" stage to a hard, resistant stage. As pointed out in the Chesapeake Biological study, susceptibility to the pressures initiated by an explosion probably varies according to the postmolting stage. The report also warns that, although a large sample is always needed, overcrowding in a test cage must be avoided to prevent shielding of some of the animals by those animals between them and the shock wave.

Crabs were affected by explosions, although not as seriously as fish. The gross effects were loss of part or all of the carapace, cracking of the carapace, heart rupture, broken spines, and frequent (apparently autotomous) loss of one or both chelipeds. The report noted that many of the crabs killed by the explosions exhibited no grossly discernable damages. It is unfortunate that no microscopic examination of the tissues for histopathological effects was made.

Some data were obtained on both the lethal range and the peak pressures causing mortalities from 30-pound charges. Approximately 90% of the blue crabs were killed at 25 feet, with peak pressures exceeding 800–900 pounds/square inch; very few (7%) died at 150 feet from the blast where the pressure reached about 270 pounds/square inch. Intermediate distances gave erratic results, ranging from 38% to 55% mortality, but it appears that the major drop in lethality with the 30-pound charge occurred between 125 and 150 feet from the source.

Tollefson and Marriage also found that Dungeness crabs (*Cancer magister*) within 25 feet of a dynamite blast are killed, often with the carapace broken or cracked, but the crabs located more than 30 feet from the charge are apparently unharmed. Some additional information on the effects of blasting on crustaceans was provided by these authors finding numerous ghost shrimp (*Upogebia pugettensis* and *Callianassa* sp.) within the 25-foot zone. A total mortality of 51.3% of the *Upogebia* and 77.8% of the *Callianassa* occurred, with 18.4 and 33.3%, respectively, broken up or obviously mutilated by the blast. There was some indication that the larger individuals were more susceptible to blast injury than smaller forms.

Some miscellaneous invertebrates were found within 25 feet of the charge, including three large sand worms, two of which were torn in half and the third suffered rupture of the body cavity. Several nemertine worms were found within the 25-foot zone; all were fragmented and killed.

From the rather fragmentary information available, it appears that many invertebrates are remarkably resistant to damage from shock waves, and that those forms with resistant coverings, such as calcareous shells or chitinous carapaces, are more resistant than the soft-bodied, unprotected forms.

GRAVITY

There is little information available on the effects of increased gravitational forces on invertebrates, though a number of authors have shown that insects are unharmed when exposed to gravitational fields hundreds of times those (approximately 20 g) causing structural damage in man (Day and Oster, 1963). Abbott and Dawson (1966) studied the responses of selected anomuran and brachyuran crabs to high-speed centrifugation. Their original purpose was to attempt to remove hermit crabs from their "host" shells without harming them. Presuming that mild centrifugation might disturb the crab enough to cause it to leave its shell, preliminary trials were run using a hand centrifuge. Since these were completely unsuccessful, the investigators utilized an electric high-speed centrifuge to produce a force of 8.5 g for 5 minutes (200 rpm) on four hermit crabs (*Pagurus floridanus*). The crabs appeared completely normal after this treatment; they exhibited typical responses to various stimuli. Further tests on these animals with maximum acceleration to and deceleration from speeds as high as 2000 rpm (800 g) apparently did not disturb their normal behavioral patterns.

Additional tests were made on another hermit crab, *Clibanarius vittatus*, which was more readily available. Crabs of several sizes in *Thais* and *Polinices* shells and crabs removed from shells were exposed for varying periods to centrifugal forces in the range of 800–2700 g. Maximum acceleration and deceleration within the capability of the centrifuge and electric brake were used, with acceleration times ranging from 15 to 26 seconds and deceleration times from 13 to 25 seconds.

Only one of twelve shelled crabs died within 24 hours after exposure to maximum forces up to 2700 g with immediate deceleration. Maintaining the exposure at maximum force (2700 g) for short periods (3 minutes) did not increase the mortality, but two of four crabs exposed to only 800 g for 30 minutes died within 24 hours. Tolerance to the extremely high centrifugal forces is an innate characteristic of the anomurans; their survival was affected little or not at all by removal from their host shell prior to centrifugation.

Similar experiments on two genera of brachyurans, *Sesarma reticulum* and *Uca* sp., demonstrated that brachyurans, or at least these two species, are much less resistant than anomurans to high centrifugal forces. Half the *Sesarma* and most of the *Uca* were dead within 3 hours after exposure to acceleration up to 2700 g followed by rapid deceleration. Most individuals of both species survived accelerations of 800 g followed by immediate deceleration, while one of two of each species was killed by prolonged exposure (30 minutes) to 800 g.

Although the numbers of individuals tested were few and the scope of the experiments was not exhaustive, this work does provide a clear indication that crabs can tolerate high centrifugal forces and that there is considerable variation between the anomurans and brachyurans in the level tolerated. The authors questioned some of the data published on the capacity of grasshopper nymphs to withstand high centrifugal forces. Eberly *et al.* (1963) reported 50% survival (after 8 days) of *Melanoplus* nymphs subjected to 9 g and 50% survival (after 20 minutes) when subjected to 10,000 g. Abbott and Dawson were unable to obtain specimens of *Melanoplus*, but found that centrifugation at 800 g for 10 minutes eviscerated and killed the one lubber grasshopper (*Romalea* sp.) nymph tested, and centrifugation at 2700 g reduced their other specimen to mush.

In view of the sketchy data indicating high resistance to centrifugal forces in arthropods, further work should be undertaken to study the effects of centrifugation on other invertebrates. The mechanism of damage by centrifugal force should, of course, be studied, with particular attention given to physiological, biochemical, and histopathological effects.

Changes in Environmental Pressure

Somewhat surprisingly, it has not been possible to locate a single reference on the pathological effects of pressure changes on any invertebrate other than insects. The resistance to damages resulting from rapid changes in atmospheric pressure in insects is well known (Day and Oster, 1963). Individual cockroaches, for example, have been repeatedly taken to a simulated altitude of 50,000 feet and rapidly returned to normal pressures without apparent distress.

Many invertebrate groups have representative species living under extreme pressure differences, and many benthic marine forms such as the king crab (*Paralithodes*) and tanner crabs (*Chionocetes tanneri*) migrate over a wide depth range and are, therefore, subjected to great pressure changes. In my experience in collecting benthic invertebrates for a number of years, no invertebrate brought aboard ship from any depth down to 1150 fathoms exhibited markedly obvious distress or gross pathological effects.

Despite the apparent independence of most invertebrates from deleterious effects of pressure changes, it seems obvious that research should be undertaken under controlled conditions in the laboratory to determine the effects, if any, of pressure changes on representatives of all the phyla of invertebrates.

Temperature

According to Gunter (1957), temperature is the most important single factor governing the occurrence of life, since all organisms are affected directly when their specific optimum temperature range is exceeded, and all organisms are affected indirectly by effects of temperature on the physical and biological environment. As Gunter pointed out, temperature largely determines the distribution of life on earth. This may be a direct effect—with animals dying when their temperature range is exceeded— or indirect—for example, when the temperature range for reproduction is more restricted than the range for survival. An even more subtle control may be operative in the effect of temperature on food production, particularly during the early stages of life when rapid growth through larval stages requires an abundance of suitable food. Only the direct pathological effects of temperature extremes will be considered.

Heat

If the temperature of any organism is sufficiently elevated, combustion of the protoplasm occurs, leaving only ash. Long before this point, death occurs, and damage to the organism is initiated at far lower temperatures. The precise mechanism by which hyperthermia causes cell injury or death is not yet completely understood. There are several possibilities which probably act jointly: (1) denaturation or coagulation of cell proteins, (2) inactivation or destruction of essential enzyme systems, (3) acceleration of metabolism accompanied by accumulation of toxic metabolites, (4) relative anoxia resulting from vascular inadequacy to compensate for the hypermetabolism.

Gunter (1957) discussed several hypotheses to explain the mechanism of heat death. He noted that when cells are exposed to increased heat the viscosity of the protoplasm is increased. This reaction is reversible up to a point if the stimulus is removed; beyond the critical point, however, it becomes irreversible and death ensues. Usually, coagulation of the protoplasm occurs at the time of heat death and appears to be the immediate cause. However, many organisms die before their protoplasm has coagulated. Also, many animals are killed by heat before their enzymes are destroyed.

There is, as Gunter noted, evidence of a relationship between water content and susceptibility to heat death; increased water content enhances susceptibility. This probably accounts for the fact that marine organisms survive heat better in high salinities or hypertonic solutions than in low salinities or hypotonic solutions. Day and Oster (1963) cited a review by Edwards (1953) of theories advanced to explain heat and cold injury to insects; he found no theory completely satisfactory, but he noted that it is likely that death from heat or cold is often attributable to changes in hydration of the tissues.

In the homeothermic animals, hyperthermia may be of either exogenous or endogenous origin. Endogenous hyperthermia, usually called "fever," results from an imbalance in the amount of heat produced metabolically and dissipated by the body. This type of hyperthermia, which can result in death without producing specific morphological alterations, does not occur in the invertebrates because they are poikilothermic.

Heat of exogenous origin is absorbed at the body surface or internally, in the case of those animals with an internal respiratory system. The effect of heat may be local if it is from a point source, or systemic if all or most of the body surface is subjected to the temperature elevation or if heated air is inspired.

Surface absorption of excess heat results in tissue damage, "burns," the extent and depth of which depend on the level of the temperature and the duration of exposure. The nature of the heat also influences the tissue response; dry heat results in desiccation and charring, while wet heat causes an opaque coagulation. Obviously, the most intense effects of local absorption of excess heat would, in those animals covered by an epidermis or surface epithelium, be in the outermost layer of the body, with a progressive decline in the temperature gradient and a corresponding decline in the severity of tissue damage in the deeper tissues. However, this may not be the case in animals covered by chitinous exoskeletons, tests, or acellular cuticles.

The pathological effects of heat on invertebrates includes a number of somewhat related facets: the immediate effects on the tissues, the immediate effect on survival, the effect on gonadal development and reproduction, and the effects on metabolic activities. Although there is little information available on the tissue responses of invertebrates to excess heat, there are much more data on the effect of elevated temperatures on survival. Gunter (1957) noted that marine invertebrates and fish die from heat at lower temperatures than do terrestrial poikilotherms; the thermal death point usually ranges 30°–35°C in marine and 42°–45°C in terrestrial forms. However, intertidal marine invertebrates have a tolerance for high temperatures about equal to that of terrestrial forms, and there are some geographic differences in lethal high temperature levels. Gunter cited reports noting that animals died from heat at 30°C in the North Sea, while some survive temperatures of 42°C; he pointed out that *Limulus* has a thermal death point of 46.3°C in Florida and

41°C at Woods Hole, Massachusetts. Gunter also made a general comment to the effect that tropical animals normally live at temperatures much nearer their thermal death point than do polar forms.

Phylum Protozoa

Most protozoans respond to a temperature gradient by selecting an optimum temperature through trial and error. The optimum temperature for *Paramecium* is in the range of 24°–28°C. The optimum is somewhat dependent on previous conditioning and can also be altered by other environmental conditions (Hyman, 1940). It appears that the entire surface is equally thermosensitive, and local applications of temperatures above the optimum elicit avoidance reactions. A common protective reaction to heat, as well as to other unfavorable conditions, in many freshwater protozoans is encystment. In the encysted state, protozoans are capable of withstanding wide temperature ranges; encysted *Colpoda*, for example, can survive a temperature of 106°C for 1 hour if thoroughly dry.

Numerous workers have shown that Protozoa occur naturally in hot springs, even in temperatures as high as 56°C. It has also been shown that various species can, over a period of several years, be acclimated to temperatures as high as 70°C. In nature, however, the thermal death point of most free-living protozoans appears to lie between 36° and 40°C, with an optimum temperature between 16° and 25°C (Kudo, 1947). The only published information on the pathological effect of excess heat appears to be, as Kudo stated, that excessive heat induces destructive chemical changes in the protoplasm.

Phylum Platyhelminthes

Hyman (1951) stated that the majority of turbellarians are undoubtedly eurythermous—withstanding wide ranges of temperature—but some species appear to be relatively stenothermous. As in many invertebrates, reaction to temperature is considerably influenced by prior conditioning.

Planarians react to changes in environmental temperature of as little as 2 or 3 degrees. Evidently the entire body is temperature sensitive. Most planarians possess a high capacity for temperature acclimation. Thermal death points of *Dugesia dorotocephala* and *Phagocata gracilis* are approximately 42° and 30°C, respectively.

Watson and Price (1960) found that exposure of infected cyclopids to temperatures of 30°–35°C, even for short periods of time, caused the death and rapid autolysis of procercoids of the pseudophyllidean tapeworm genus *Triaenophorous*. Later, Watson (1963) demonstrated that the hatching of embryonated *Triaenophorous* eggs was retarded above 30°C, and there was increased mortality of the larvae in comparison to embryonated eggs held at 24°C.

Phylum Annelida

Cameron (1932) described the effects of cauterization of earthworms by a heated needle applied to the external surface. Such treatment results in complete destruction of the epithelial covering and of portions of the underlying subcutaneous connective tissue and muscular layers. Muscular contraction of surrounding uninjured tissue causes the necrotic area to protrude. It seems obvious that both the lateral dimension and the depth to which there is the complete tissue destruction depends on the temperature level and duration of exposure. Although Cameron did not give details on either, it appears that the application of a red-hot needle to the epidermis of earthworms causes tissue changes analogous to a third degree burn in mammals.

Repair of such burns is described in detail in Chapter II as the typical reparative process in response to injuries of the body wall of earthworms. However, the salient points should be reiterated here. There is an almost immediate increase in the number of coelomic corpuscles and an infiltration of these cells to the area of the coelom just beneath the injury. At about this time, the parietal epithelial cells undergo marked proliferation, but the reaction is still purely coelomic with no invasion of the necrotic area.

Infiltration of the necrotic tissue by migrating coelomic corpuscles is initiated at approximately 6 hours postinjury, and within 24 hours there is a marked thickening of the body wall accompanied by an apparent decrease in the size of the coelom resulting from the formation of a mass of cellular tissue that forms a plug between the necrotic superficial tissue and the coelom. The plug, though bounded on each side by well-differentiated normal tissue, consists of undifferentiated elongated cells and coelomic corpuscles; its great cellularity contrasts sharply with the adjacent, undamaged body wall. The plug gradually becomes organized and invaded by muscle fibers, connective tissue fibers, and blood vessels.

Desquamation of the necrotic area occurs in about 2 days, followed by ingrowth of adjacent epithelium to completely cover the denuded surface with a thinner than normal epithelium within 5 days. There is almost complete repair of the damaged tissue within 10 days; the epithelial covering and body wall are relatively normal, but some reparative processes still continue.

LeGore (1970) investigated the tissue damage and response to cauterization with a heated dissecting needle of the leech, *Piscicola salmositica*.

The extent of the lesion produced by cauterization is relatively uniform, provided that the needle is heated and applied uniformly; in LeGore's experiments it resulted in local destruction of the cuticle, epithelium, fat cells, dermal connective tissue, and most of the circular muscle layer, leaving a thin layer of ash over the wound (Fig. III.1).

Immediately after the injury, epithelial cells adjacent to the site of needle application become tall columnar rather than the normal low columnar or cuboidal epithelium, probably as a result of constriction of the affected and adjacent segments. Staining intensity of these cells with hematoxylin and eosine is normal, but the cytoplasm appears coagulated. With Mallory's triple stain, the cytoplasm exhibits an abnormal purplish-pink cast. Epithelial abnormality decreases with distance from the burn, with normal epithelial cells reappearing 25–30 cells from the lesion.

Circular muscle beneath the lesion appears normal when stained with hematoxylin and eosine, but stains brick red with Mallory's triple stain. Circular muscle adjacent to the lesion appears normal, while longitudinal muscle is obviously affected for a distance of 1–5 muscle bundles on each side of the lesion, staining a deep blue-black with Mallory's and exhibiting increased eosinophilia with hematoxylin and eosine.

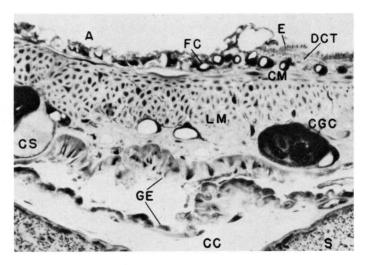

Fig. III.1. Burn in leech, *Piscicola salmositica*, at 0 hour postinjury. Note the tall columnar epithelium (E), the darkly staining medullae of longitudinal muscles (LM), and the ash (A) on the surface of the wound. GE, gut epithelial cells; CM, circular muscle; CGC, cocoon gland cell; CC, coelomic channel; CS, collecting sac; DCT, dermal connective tissue; FC, fat cell; S, stomach. Hematoxylin and eosine. Approximately 140×. (From LeGore, 1970.)

Connective tissue stains normally, but the dermal connective tissue adjacent to the wound swells to approximately twice its normal size, and the cytoplasm of the fat cells in the dermal connective tissue exhibits the same purplish-pink staining reaction noted in the necrotic epithelium. Cocoon gland cells in the burn appear normal.

By 4 hours postinjury, many of the tissues beneath and adjacent to the burn begin to disintegrate, including the cocoon gland cells and portions of the deep connective tissue. Muscle bundles fragment, and epithelium, dermal connective tissue, and fat cells adjacent to the burn begin to slough into the wound. At this time, oppleocytes appear among the muscle bundles along the edge of the lesion. These cells range from round to slightly oval, have a diameter of 5–6 μm, and contain a round, centrally located nucleus approximately 2 μm in diameter. The nucleus is densely basophilic, but the cytoplasm is virtually achromatic with hematoxylin and eosine.

After 8 hours, the oppleocytes congregate at the edge of the wound and form a loosely organized sheet that begins to extend beneath the sloughing epithelium, connective tissue, and fat cells (Fig. III.2). Meanwhile, the cocoon gland cells begin to lose their basophilia and fragment. Muscle cells, though intact, become disorganized because of the disintegration of intermuscular connective tissue.

At 24 hours postinjury, the mass of necrotic tissue becomes elevated to form a conspicuous weal (Fig. III.3). Oppleocytes continue to congregate at the edge of the wound (Fig. III.4) and 12 hours later begin to organize into parenchymatous connective tissue that covers the exposed tissue as the edges of the necrotic mass break away from the uninjured tissue (Fig. III.5). As the oppleocytes form connective tissue, the outer layer of cells becomes modified in staining reaction, with the cytoplasm staining golden-yellow and the nuclei red with Mallory's.

The new connective tissue continues to infiltrate toward the center of the lesion filling all the space between the necrotic tissue and the uninjured tissue (Fig. III.6). By 96 hours postinjury, the layer of new connective tissue, 20–30 cells thick, covers the surface of the wound. Most of the necrotic tissue has been sloughed, but fragments are incorporated into the new connective tissue in capsulelike structures.

From 24 hours postinjury onward, many of the leeches in LeGore's experiment died, and the experiment was terminated after 5 days because no leeches survived to be examined. Most of the leeches that died during the experiment were badly decomposed when discovered, but all contained fish blood in their stomachs, indicating that death was not related to starvation. Among the few fixed soon after death, response to the burn varied from no discernable cellular reaction to the characteristic

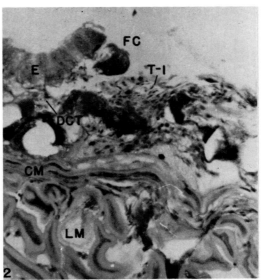

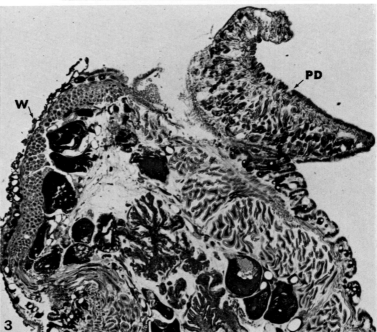

Fig. III.2. Burn in leech at 8 hours postinjury. Note loose sheet of oppleocytes (T-1) beneath sloughing epithelium (E), dermal connective tissue (DCT), and fat cells (FC). CM, circular muscles; LM, longitudinal muscles. Hematoxylin and eosine. Approximately 350×. (From LeGore, 1970.)

Fig. III.3. Burn at 24 hours postinjury. Note the large weal-like swelling (W) of necrotic tissue. PD, posterior disc. Hematoxylin and eosine. Approximately 100×. (From LeGore, 1970.)

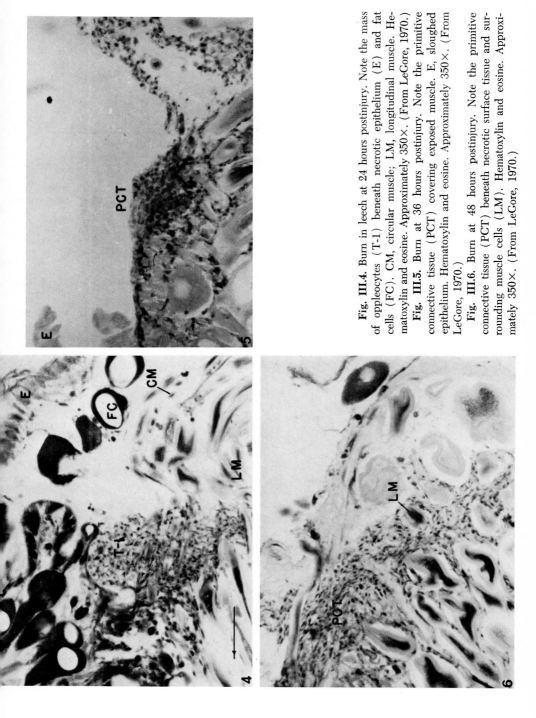

Fig. III.4. Burn in leech at 24 hours postinjury. Note the mass of oppleocytes (T-1) beneath necrotic epithelium (E) and fat cells (FC). **CM**, circular muscle; LM, longitudinal muscle. Hematoxylin and eosine. Approximately 350×. (From LeGore, 1970.)

Fig. III.5. Burn at 36 hours postinjury. Note the primitive connective tissue (PCT) covering exposed muscle. E, sloughed epithelium. Hematoxylin and eosine. Approximately 350×. (From LeGore, 1970.)

Fig. III.6. Burn at 48 hours postinjury. Note the primitive connective tissue (PCT) beneath necrotic surface tissue and surrounding muscle cells (LM). Hematoxylin and eosine. Approximately 350×. (From LeGore, 1970.)

response described above. Although leeches respond to focal burns in a manner indicating they can successfully eliminate necrotized tissue and restore the integrity of the body, it is not yet certain that they can repair the wound and survive such injury.

Phylum Mollusca

Gunter (1957) presented a table of temperatures lethal to a large number of species of normally submerged marine mollusks from several areas of the world and from several depth zones. Dickie and Medcof (1963) discussed the role of temperature in the relatively common mass mortalities of sea scallops (*Placopecten magellanicus*) in eastern Canada. They pointed out the weakness in the method of determination of lethal temperatures in which test animals are subjected to increases of 1°C every 5 minutes until they die; they noted that lethal temperatures so determined are unlikely to occur in the natural environment. In the case of the scallop, for example, the lethal temperature obtained by this method is 30.6°C, which is far above any water temperature likely to be experienced by scallops in the Gulf of St. Lawrence.

The authors cited the work of Fry (1947) and his associates showing that sudden, even frequent, and great temperature changes are not lethal so long as these changes are within the "zone of thermal tolerance" characteristic of the species. Above this, however, is a series of threshold temperatures for each species, the "incipient upper lethal temperatures" related ("corresponding") to the temperatures at which they have been acclimated. Thus, animals with the same recent thermal history behave similarly and show a characteristic median survival time when exposed to temperatures above their upper lethal. This work indicates that lethal temperatures determined by the gradual heating method have little value other than as a measure of relative sensitivity of different species to high temperatures and have little meaning in interpreting field observations. Dickie and Medcof found that the considerably lower incipient upper lethal temperatures determined by Fry's method do have meaning when applied to field conditions in scallop mortalities. Thus, scallop mortalities attributable to direct effects of high temperature may be expected, and seem to occur, when sustained environmental temperatures rise beyond the zone of thermal tolerance, which is lower than the lethal temperature reported by Stevenson.

Data on the seasonal, upper temperature tolerance limits of scallops in the Gulf of St. Lawrence are summarized in Fig. III.7 (from Dickie and Medcof, 1963). The incipient upper lethal temperatures for scallops seasonally acclimated to various environmental water temperatures are shown. For example, if scallops living at −2°C, the approximate freezing

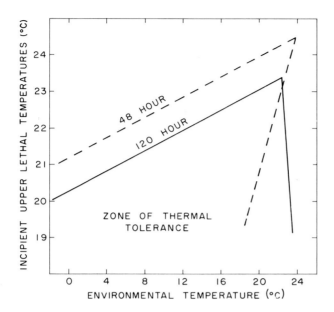

Fig. III.7. Relationship between upper incipient lethal temperatures (temperatures causing 50% mortality: upper line in 48 hours, lower line in 120 hours) and temperatures to which scallops are acclimated during seasonal changes in environment. (Redrawn from Dickie and Medcof, 1963.)

point of seawater, are suddenly exposed to a temperature of 20°C, there will be a 50% mortality by 120 hours. Dickie (1958) ascertained that the seasonal upper lethal temperature rises 0.14°C for each 1.0°C rise in the acclimation temperature. Therefore, scallops acclimated to 5°C have an upper lethal temperature of 21°C, those living at 10°C have an upper lethal limit of 21.7°C, and those living at 15°C have an upper lethal temperature of 22.4°C. The 50% mortality figure occurs in a shorter time if the temperatures to which the scallops are suddenly subjected exceed those mentioned above (see Fig. III.7), thus, death occurs at 48 hours at 21°C for the −2°C conditioned animals as compared to 120 hours at 20°C.

Since the young of most invertebrates (starfish, sea urchins, clams, etc.) are more resistant to high temperatures than adults, it is likely that very young scallops are more resistant to high temperature than old ones. Nevertheless, the authors stated that direct temperature kills of most scallops can be expected when temperatures of 20°–23.5°C are reached. The occurrence and severity of the mortality will depend on previous conditioning and length of exposure to the increased temperature.

Dickie and Medcof also noted that any substantial but nonlethal temperature change within the zone of temperature tolerance resulted in exhibitions of shock and debility by scallops, including copious mucous secretions, inability to close valves, and inability to escape predators.

A localized heavy mortality of the bean clam, *Donax gouldi,* caused by heat and freshwater from hot water springs uncovered by ebbing spring tides, was described by Johnson (1966). Although *D. gouldi,* like many species of the genus, migrates up and down the beach with the tide, the beach at which the mortality occurred is flat and is uncovered so rapidly during ebbing of spring tides that the clams are stranded.

The hot springs area was described as consisting of rock and sand areas emitting hot freshwater and extensive but spottily distributed areas of hot sand where freshwater was not detected. All the hot sand areas were littered with dead clams, and windrows of dead clams occurred in the rocky freshwater springs. Histological examination of the dead clams was not possible because the temperatures were sufficiently high for the protein to be denatured. The spring-water temperature exceeded 50°C and exceeded 70°C at 2-cm depth in the hot sand areas. However, the thermal areas were sharply delimited from normal ones; live clams were found in an area of cold sand (16.5°C) within 15 m of a hot spring.

Phylum Arthropoda

Little work has been done on the pathological effects of high temperature on arthropods other than insects. McLeese (1956) studied the tolerance of the American lobster to the interrelated factors of temperature, salinity, and oxygen. Sprague (1963) investigated the resistance of four species of freshwater crustaceans to high temperatures using, as Dickie and Medcof (1963) did, the methods developed by Fry and his co-workers. The crustaceans studied included an isopod, *Asellus intermedius,* and three amphipods, *Hyallela azteca, Gammarus fasciatus,* and *G. pseudolimnaeus.*

Resistance to high temperatures, as in many invertebrates, decreases with increasing size in *Asellus intermedius* and *Gammarus fasciatus,* but no size differential in heat susceptibility was found in *Hyallela azteca.* Female *Gammarus* are more resistant to heat than males; there is no sex-related difference in resistance in *Asellus. Hyallela* exhibited variable relations, with the females more resistant under some acclimation conditions and the males more resistant under other conditions of acclimation.

These crustaceans can increase their heat resistance at rates of 2.5°–5°C per day under laboratory conditions and at temperatures above 14°C. The crayfish is capable of much more rapid increase in heat resistance, 1 day or less when the environmental temperature is raised from 4° to

24°C. The lobster, on the other hand, requires a minimum of 24 days to acclimate from 14.5° to 23°C (McLeese, 1956).

Raising the acclimation from 10° to 20°C results in a higher lethal temperature level (50% mortality in 24 hours) for *G. fasciatus* (1.9°C), *A. intermedius* (1.3°C), and *G. pseudolimnaeus* (0.5°C), but apparently has no effect on the lethal temperatures for *H. azteca.* The ultimate 24-hour lethal temperatures for these species are estimated to be 34.6°C for *A. intermedius* and *G. fasciatus,* 33.2°C for *H. azteca,* and 29.6°C for *G. pseudolimnaeus.*

Apparently no one has studied the actual mechanisms by which heat death in the arthropods (other than insects) occurs or the histopathological effects, if any, from increased environmental heat.

COLD

Invertebrates are much more variable in their resistance to cold than to heat, but, as Gunter (1957) pointed out, death from cold is as universal in nature as death from heat. Although the precise mechanisms by which chilling or freezing of tissues induces injury are not known, a number of possibilities have been suggested. Robbins (1957) lists three phenomena that occur in man and notes they may all act conjointly. (1) Cell metabolism may be slowed or blocked by lowering the temperature to the point where inhibition or suppression of vital metabolic activities results in cell damage or death. (2) Crystallization of the water contained in cells may cause mechanical damage or toxicity by increasing the concentration of electrolytes (from the withdrawal of the water). (3) Vascular damage may occur with attendant exudation of serum and plasma. Severe injury from cold induces intravascular clotting and results in anoxic changes even to the point of infarction necrosis of affected tissues.

Although the mechanisms listed above apply specifically to the effects of cold on man, the cellular mechanisms are probably common to all animals, despite great variation in the exact temperatures at which cell metabolism is affected and in the temperatures at which formation of ice crystals in the cells occurs. Obviously, vascular damage would be limited to the vertebrates and, perhaps, to those invertebrates with well-developed circulatory systems.

The detailed discussions of the effects of cold on insects by Salt (1958, 1961a,b) were summarized by Day and Oster (1963). They remark that the problems are better defined than those associated with heat injury, but note that the mechanism of injury by freezing is also far from being understood. At least four theories of the lethal effect of freezing in insects have been proposed, but Salt considered none of them to be completely satisfactory. The theories have been designated as (1) the bound water theory; (2) the electrolyte concentration theory; (3) the mechanical theory, and (4) the site-of-freezing theory. The last theory comes the nearest to being acceptable to most insect pathologists; it holds that the freezing of extracellular water is innocuous, but the freezing of intracellular water causes irreparable cell damage. The precise nature of this damage, however, is not known. Some insect cells can be frozen and subsequently thawed without apparent injury. Since water in biological systems is associated with colloids and ions that lower its freezing point, supercooling generally occurs in insect tissues. Frequently, insects undercool to between $-20°$ and $-30°C$ before they freeze, although ice crystals form in many insects held in the undercooled state. There is a great deal of variation in the temperature at which freezing occurs in insects, and it is altered by prior treatment of the insect. Insects have large concentrations of glycerol in their hemolymph, and it has been suggested that it may have a protective role in cold resistance. A correlation between the presence of glycerol and cold hardiness has been demonstrated, as has the increased cold hardiness resulting from increased quantity of glycerol in the tissues, but its mode of action in the cell has not been determined.

It should be noted that insects can be damaged or even killed by temperatures above freezing. Such damages, probably caused by metabolic disturbances, may exert effects on longevity, oviposition, and the viability of eggs of surviving adults.

Even though it is true that organisms in which gelation of protoplasmic colloids takes place before ice crystals are formed can stand much lower temperatures, Gunter (1957) noted that these conditions are impossible for most organisms to attain in nature. In most instances when gelation of protoplasm occurs, it is preceded by dehydration, which enhances its occurrence. Thus, dehydration increases the resistance of cells or organisms to cold, as it was previously noted to do for heat injury.

Where death is due to the formation of ice crystals in the protoplasm, high salt content tends to increase the resistance (Gunter, 1957). Gunter pointed out that most marine organisms die at a few degrees below 0°C, but many die at even higher temperatures. For example, mass mortalities of invertebrates (and fish) occur along the Texas Coast when water temperatures reach a minimum of 4°C (Gunter and Hildebrand, 1951). Gunter believed that some derangement of metabolic activity causes death in these instances; the cause may be as simple as cessation of breathing movements or heartbeat and consequent anoxia.

Of course, death from cold may result indirectly from inactivation, but it may occur directly from some other cause, such as predation. Gunter (1941) observed that shrimp, *Penaeus,* and pistol-shrimp, *Crangon,* were "numbed" when washed up on the beach, and though they were revived by the sun, most were consumed by gulls prior to their recovery. He also noted that large numbers of gastropods, pelecypods, and echinoids were stranded on the beach by the surf; they revived, but were unable to return to the water, and thus eventually died of desiccation and anoxia.

Virtually all discussions of the effect of cold on invertebrates in the literature ignores the histopathological effects. Although temperatures that are lower than tolerated levels control the distribution of untold species of invertebrates and cause mass mortalities of others in warmer areas that are subjected to irregular cold waves, discussion of the temperature minima tolerated by various invertebrates does not lie within the scope of this book.

It should be noted that, at least in oysters, mechanical damage from ice crystals occurs when the frozen organism is moved about. If frozen but left undisturbed until they thaw, the majority will survive. However, if shaken or subjected to other rough handling, heavy mortalities ensue. Loosanoff (1946) reported that in experiments, 100% of shaken frozen oysters died, while almost all unshaken frozen oysters were alive after thawing and were apparently healthy after 10 weeks. Loosanoff surmised that the mortality among the shaken oysters was probably due to damage caused by the rearrangement of ice crystals within the cells. It is also probable that movement causes damage or destruction of cell membranes and vital cell constituents by the ice crystals.

The response of *Lumbricus terrestris* to locally applied freezing has been described by Firminger *et al.* (1969). Mature earthworms, approximately 10 cm long, were stretched longitudinally between two layers of Parafilm, and a 5 ml beaker partly filled with finely crushed solid CO_2 was applied to the dorsal surface of the Parafilm-covered worms for 20 seconds. The injured worms were maintained for several months, with specimens sacrificed for histological study at appropriate intervals. All worms were examined daily for grossly observable pathological effects of the cold-induced injury.

Immediately after freezing, the area contacted by the beaker presents an opaque yellowish-white appearance (Fig. III.8) which quickly fades as the area thaws and the injury site is no longer evident. Copious amounts of mucus are secreted immediately upon thawing. After a few days, the traumatized area is recognizable as duller, drier, and a paler tan than adjacent uninjured tissue. Segments involved in the injury appear slightly shortened with a slight bilateral protrusion that does not change caliber as do the adjacent segments during movement. The

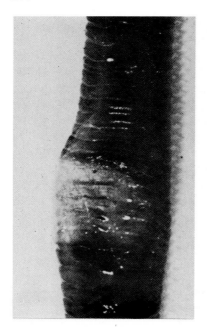

Fig. III.8. Cold injury of earthworm (*Lumbricus terrestris*). Gross appearance of injury immediately after removal of solid CO_2. Note light area of injury. (From Firminger *et al.*, 1969.)

traumatized area may be slightly indurated (hardened) and can be felt on palpation as a small firm plaque or button. Eight to 10 days postinjury, the wound consists of a light-tan "button" of tissue clearly visible beneath the smooth, glistening surface of the worm and bulging laterally on each side. Some loss of function results from the focal cold trauma, since the authors (Firminger *et al.*, 1969) noted that the segments involved were "slightly shortened and did not stretch like the normal segments with the progressive undulating movements that characterize the normal locomotion of the earthworm." The dorsal surface appears to heal completely with a slight umbilication in some animals, denoting the center of the underlying plaque.

Histologically, in the least severely damaged worms, the cuticle is lifted off the underlying tissue within the first day after freezing. There is vacuolar swelling at the base of the epithelial cells, accompanied by linear distortion of nuclei from below or from the sides by the pressure of the vacuoles. Epithelial cells are occasionally ruptured, resulting (when widespread) in the formation of an amorphous, granular mass that contains randomly scattered, small round nuclei.

Intermittently, the epithelium may be sloughed from the underlying basement membrane, but the basement membrane remains intact in the milder wounds. Beneath the basement membrane, the cytoplasm of the sarcolemma cells becomes much more conspicuous, particularly in the more internal layers of circular muscle, and forms basophilic rings around the eosinophilic sarcoplasm.

Aggregations of coelomocytes appear in the coelomic cavity and infiltrate between the bundles of the deeper longitudinal muscle (Fig. III.9) and, occasionally, along the septa separating the segments. Degenerating eleocytes, though not common, may sometimes be present in the underlying coelomic cavity.

With more severe cold injury, the basement membrane is ruptured (Fig. III.10), and the granular degeneration mentioned previously extends into the circular muscle layer within the first or second day postinjury. Repair of the overlying and adjacent epithelium apparently originates from the basal cells. The replacement epithelial cells are somewhat flattened and squamous in appearance and are less organized than the normal columnar epithelium cells. Although the reepithelization appears clearly proliferative in nature, mitotic figures do not occur in

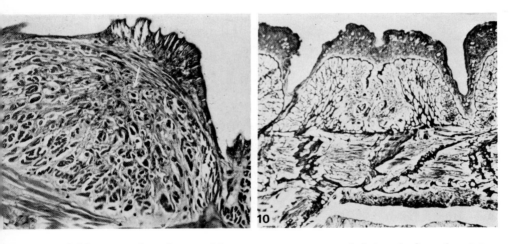

Fig. III.9. Cold injury of earthworm. Microscopic appearance of lesion 1 day after injury. The surface epithelium is distorted but uninterrupted, and a thin cuticle covers most of the surface. Note the layers of flat, spindle-shaped epithelium lying beneath the columnar cells and extending downward into the circular muscle through the ruptured basement membrane (arrow indicates end of basement membrane). Note further the shrunken sarcoplasm of muscle cells and swollen, prominent, intervening interstitial collagen. Hematoxylin and eosine. 189×. (From Firminger *et al.*, 1969.)

Fig. III.10. Cold injury of earthworm. Lesion 1 day after injury. Note the down-turned ends of the ruptured basement membrane and the mass of epithelial cells that have poured through and displaced the reticular network that characteristically outlines the circular muscle cells. The longitudinal muscle beneath the circular muscle is intact, and immediately beneath it is a collection of leukocytes and eleocytes in the coelomic cavity. Reticulum stain. 123×. (From Firminger *et al.*, 1969.)

tissue sections. Continuing from the third to the sixth day postinjury, epithelial cells "pour" through the break in the basement membrane, "filling the defect in the underlying circular muscle and, at times, deeper longitudinal muscle; the edges of the basement membrane" are "turned in, and the reticular network, which normally surrounds every muscle fiber," is "swept aside or lost" (Firminger *et al.*, 1969). The traumatized area becomes filled with solid masses of spindle-shaped epithelial cells, haphazardly arranged and frequently extending between adjacent muscle fibers. As the authors point out, the cellular architecture greatly resembles infiltration by tumor cells. It is indeed probable that such a diagnosis would be made if the nature of the stimulus inspiring this proliferative, invasive cellular response were not known. Occasional fragments of reticulum or trapped muscle fibers may be seen along the edge of the otherwise solid mass of epithelial cells.

A surface layer of epithelium, consisting of a single layer of fairly well-oriented cells covered with cuticle, typically covers the cell mass by the seventh day. However, mucous cells occur only at the edges of the traumatized area and are not yet present in the regenerated epithelium.

In particularly deep wounds, epithelial cell invasion extends into the coelom and may involve the wall of the gut. In such cases, aggregations of degenerating, necrotic eleocytes may occur adjacent to the gut. In one specimen examined 22 days after freezing, the mass of epithelial cells had extended into the coelom and apparently spread within the coelomic cavity. Despite the inherent difficulty of determining in a single specimen whether the cells were infiltrating inward from the epithelium or outward from the body cavity, or both, sequential sampling strongly indicates the origin of the cells in question to be inward growth of cells of epithelial origin. The alternative possibility, that the cell mass is of coelomocytic origin similar to the repair of surface burns in earthworms, can probably be eliminated since the authors showed that the coelomocyte response to freezing injury is early and transitory. No coelomocyte response persisted through 22 days postinjury, and coelomocytes are rare in the chloragogen tissue and in the coelomic cavity.

One month after freezing, the epithelium in the traumatized area is still thickened, but the surface is covered with normal cuticle. Many surface epithelial cells contain PAS-positive, diastase-resistant particles, but typical goblet cells are not present. The characteristics of the lesion are maintained both grossly and microscopically for at least several months duration (for as long as the experimental animals were maintained).

Firminger and his co-workers believe that epithelial proliferation constitutes the mechanism of wound healing in the earthworm, and they are of the opinion that coelomocyte proliferation is an early phenomenon, to ensure phagocytosis of necrotic tissue, which subsequently recedes. They suggest that the tumorlike masses found by Hancock and described as "granular cell myoblastoma" may represent epithelial cell proliferation in healing wounds. They did not discuss Cameron's work that indicated that proliferating coelomocytes contribute greatly to the healing of surface burns by plugging the wound channel as well as by phagocytizing necrotic tissue debris. I accept the probability of epithelial proliferation in cold trauma of earthworms, but unpublished work in our laboratory strongly supports Cameron's findings of the importance of the coelomocytes in response to surface burns. Such a contrast in response to burning and freezing is enigmatic indeed, but it must be assumed that it exists until careful comparative experimentation is undertaken.

As Firminger *et al.* point out, the epithelial hyperplasia in response to cold injury has certain characteristics of a neoplasm, including the formation of a relatively disorganized mass of cells infiltrating adjacent normal tissue and persisting long after removal of the initiating stimulus. It is highly probable that an invertebrate pathologist examining the lesion, without prior knowledge of the initiating stimulus and the progression of the lesion, would diagnose it as a tumor. Neoplastic characteristics missing are altered nucleocytoplasmic relationships, hyperchromaticity, variations in cell size, and frequent and often bizarre mitotic figures.

References

Abbott, W., and Dawson, C. E. (1966). Responses of some anomuran and brachyuran crustaceans to high speed centrifugation. *Nature* (*London*) **211**, 1320–1321.

Anonymous. (1948). "Effects of Underwater Explosions on Oysters, Crabs and Fish—A Preliminary Report," Publ. No. 70. Chesapeake Biol. Lab., Solomons Island, Maryland.

Cameron, G. R. (1932). Inflammation in earthworms. *J. Pathol. Bacteriol.* **35**, 933–972.

Day, M. F., and Oster, I. I. (1963). Physical injuries. *Insect Pathol.* **1**, 29–63.

Dickie, L. M. (1958). Effects of high temperature on the survival of the giant scallop. *J. Fish. Res. Bd. Can.* **15**, 1189–1211.

Dickie, L. M., and Medcof, J. C. (1963). Causes of mass mortalities of scallops

(*Placopecten magellanicus*) in the southwestern Gulf of St. Lawrence. *J. Fish. Res. Bd. Can.* **20**, 451–482.

Eberly, L., Cogswell, S., Jr., and Wunder, C. C. (1963). Growth and survival of grasshoppers during continual exposure to high gravity. *Amer. Zool.* **3**, 533.

Edwards, G. A. (1953). Respiratory metabolism. *In* "Insect Physiology" (K. D. Roeder, ed.), Chapter 5. Wiley, New York.

Firminger, H. I., Antoine, S., and Adams, E. (1969). Epithelioma-like lesion in *Lumbricus terrestris* after cold injury. Neoplasms and related disorders of invertebrate and lower vertebrate animals. *Nat. Cancer Inst., Monogr.* **31**, 645–653.

Fry, F. E. J. (1947). Effects of the environment on animal activity. *Univ. Toronto Stud., Biol. Ser.* **55**, Publ. Ont. Fish. Res. Lab., No. 68.

Gowanloch, J. N., and McDougall, J. E. (1945). Effects from the detonation of explosives on certain marine life. *Oil* **4**, 13–16.

Gunter, G. (1941). Death of fishes due to cold on the Texas Coast, January, 1940. *Ecology* **22**, 203–208.

Gunter, G. (1957). Temperature. Chapter 8. Treatise on Marine Ecology and Paleoecology. Ecology. *Geol. Soc. Amer., Mem.* **67**, 159–184.

Gunter, G., and Hildebrand, H. H. (1951). Destruction of fishes and other organisms in the south Texas Coast by the cold wave of January 28–February 3, 1951. *Ecology* **32**, 731–736.

Hyman, L. H. (1940). "The Invertebrates: Protozoa through Ctenophora," Vol. I. McGraw-Hill, New York.

Hyman, L. H. (1951). "The Invertebrates: Acanthocephala, Aschelminthes and Entoprocta. The Pseudocoelomate Bilateria," Vol. III. McGraw-Hill, New York.

Johnson, P. T. (1966). Mass mortality in a bivalve mollusc. *Limnol. Oceanogr.* **11**, 429–431.

Kudo, R. R. (1947). "Protozoology," 3rd ed. Thomas, Springfield, Illinois.

LeGore, R. S. (1970). A histological examination of wound repair and reaction to large foreign bodies in the rhynchobdellid leech, *Piscicola salmositica*. M.S. Thesis, University of Washington, Seattle.

Loosanoff, V. L. (1946). Survival and mortality of frozen oysters (*O. virginica*). *Anat. Rec.* **96**, 90.

McLeese, D. W. (1956). Effects of temperature, salinity and oxygen on survival of the American lobster. *J. Fish. Res. Bd. Can.* **20**, 387–415.

Robbins, S. L. (1957). "Textbook of Pathology with Clinical Application." Saunders, Philadelphia, Pennsylvania.

Salt, R. W. (1958). Cold-hardiness of insects. *Proc. Int. Congr. Entomol., 10th, 1956* Vol. 2, pp. 73–77.

Salt, R. W. (1961a). Resistance of poikilothermic animals to cold. *Brit. Med. Bull.* **17**, 5–8.

Salt, R. W. (1961b). Principles of insect cold-hardiness. *Annu. Rev. Entomol.* **6**, 55–74.

Sieling, F. W. (1951). "Experiments on the Effects of Seismographic Exploration on Oysters in the Barataria Bay Region," Proj. No. 9, unpublished mimeo rept. Texas A&M Res. Found., College Station, Texas.

Sprague, J. B. (1963). Resistance of four freshwater crustaceans to lethal high temperature and low oxygen. *J. Fish. Res. Bd. Can.* **20**, 387–415.

Tollefson, R., and Marriage, L. D. (1949). Observations on the effect of intertidal blasting on clams, oysters and other shore inhabitants. *Oreg. Fish. Comm. Res. Briefs* **2**, 19–23.

Watson, N. H. F. (1963). A note on the upper lethal temperature of eggs of two species of *Triaenophorus*. *J. Fish. Res. Bd. Can.* **20**, 841–844.

Watson, N. H. F., and Price, J. L. (1960). Experimental infections of cyclopid copepods with *Triaenophorus crassus* Forel and *T. nodulsoa* (Pallas). *Can. J. Zool.* **38**, 345–356.

CHAPTER IV

Chemical Injuries

Virtually any chemical substance can injure an animal if taken into the body in sufficient quantity or concentration. Some chemicals, however, are harmful even in minute quantities and are classified as poisons. In mammals, the mode of action and pattern of tissue vulnerability of most poisons has been established. These parameters are well defined for each poison, and fairly standard and distinctive patterns of tissue damage, varying only with the toxic level of the agent, are evoked.

Because of the tremendous economic and medical importance of many deleterious insects, there has been a phenomenal growth in insect toxicology. The insect toxicologists use the same techniques but work toward a completely different goal than human toxicologists. The increasing sophistication of the insect toxicologist has been matched by that of the insect pathologist working in the area of chemical injuries.

With a few exceptions, which will subsequently be discussed in some detail, systematic and extensive information on the toxic properties of various agents and tissue vulnerability of other invertebrates does not exist. This is neither surprising nor particularly discouraging in view of the vast effort devoted to the determination of these data in mammals and insects. Despite obvious differences in susceptibility to specific toxins within the animal kingdom, there are certain generalizations applicable to all chemical injuries.

General Principles of Chemical Injury

The potential effect of a chemical agent upon the body of any animal depends on three variables: (1) the vulnerability of individual tissues, (2) the mode of action of the agent, and (3) the concentration of the agent. Tissues and cells within the same animal vary greatly in their susceptibility to chemical injury, and, obviously, there is an even greater variation among the divergent invertebrate groups. The mode of action of the chemical determines the location of the major damage. Some toxins exert their principal effects locally at the portal of entry, thus causing damage to the external surface of the body (if present in the environmental media), the proximal portion of the gastrointestinal tract (if ingested), or the pulmonary tissue of air breathers. Other toxic substances cause no deleterious effects at the portal of entry, but systemically affect tissues or organs in which they are concentrated or stored. Still other substances exert their maximum effects at the site of excretion or portal of exit. However, toxic chemicals frequently have combined sites of damage: e.g., the portal of entry and portal of exit (site of excretion) or site of storage and site of excretion. Since the sites of damage are generally the places of maximum concentration of the poison, the effect of chemical concentrations on the severity of injury is emphasized.

Virtually all knowledge of pathological changes induced in insects by chemicals has resulted from investigation of insecticide agents. Brown (1963) notes that insect toxicology is becoming more and more relevant to insect pathology. The nature of the poisoning process and the cause of death becomes intelligible to the insect pathologist when recognizable changes occur in the cells and tissues, while the insect toxicologist gains understanding of the mode of action of a chemical by the discovery of enzyme inhibition or of characteristic changes in the electrophysiology of nerves. The pathological effects of the many chemicals used as insecticides are discussed by Brown; he describes the symptoms, effect on respiration and heart, changes in nerve electrophysiology, inhibition of enzymes, and recognizable histological changes caused by each chemical.

It is not appropriate at the present time to attempt a classification system for chemicals toxic to all invertebrates. Nevertheless, there are a number of substances that appear to be damaging to all animals even in small quantities and, thus, warrant consideration here. Although the physiological and histopathological effects of these substances have not been worked out in detail for the various invertebrate groups, information on their effects on vertebrates and insects is probably indicative of their general protoplasmic effects. Thus, phenol, which is corrosive to vertebrate tissue, would also be corrosive to invertebrate tissue.

A second, and possibly the most important, group of toxins affecting invertebrates are the pesticides and herbicides. These may, and frequently do, include substances from the category listed above. An invertebrate may be the specific target for a pesticide, the accidental victim of an attempt to control other organisms, or may be subjected to an environment, such as an estuary, in which pesticides from distant sources are concentrated. Some of the pesticides and heavy metals may be concentrated and stored in invertebrates and thus may exert toxic effects on organisms higher in the food chain.

Universal Poisons—Organic

Ethyl and Methyl Alcohol

In mammals, methyl alcohol undergoes oxidation after absorption and forms formaldehyde and formic acid, both of which are more toxic than the methyl alcohol; these products cause severe central nervous system damage. Patchy edema and hemorrhage occur in the stomach where maximum exposure occurs. Since methyl and ethyl alcohol are used both alone and in combination with other substances as fixatives, they are clearly of universal toxicity. Their effects on invertebrates, other than causing necrosis and fixation of cells where absorbed, remain largely undescribed. It seems likely that additional study will demonstrate chronic toxic effects involving nervous tissue.

Metaldehyde

Metaldehyde, a white, solid polymer of acetaldehyde, is the primary constituent of many of the commercial slug baits. According to Cragg and Vincent (1952), it may act as either a contact or internal poison. Its greatest effect is probably as an internal poison, since slugs are attracted to and avidly consume commercial slug bait containing metaldehyde. Within 3 minutes postingestion, the animal becomes markedly distressed (S. G. Martin, personal communication). Cragg and Vincent described the characteristic gross symptoms of metaldehyde poisoning in slugs as immobilization broken by outbursts of uncoordinated muscular activity and sliming, resulting in severe water loss. Oral doses of 0.6 mg of solid metaldehyde are lethal to slugs of 400–500 mg body weight (Figs. IV.1–IV.4).

Cragg and Vincent reported no grossly recognizable gut lesions, but they apparently made no histological observations. S. G. Martin (personal communication) investigated the gross and histological effects of

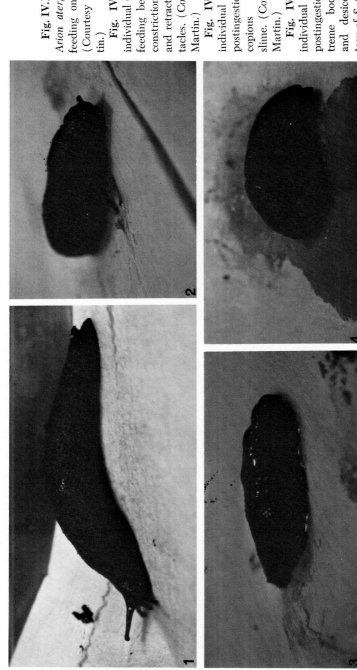

Fig. IV.1. Garden slug, *Arion ater*, just prior to feeding on metaldehyde. (Courtesy of S. G. Martin.)

Fig. IV.2. The same individual 6 minutes after feeding began. Note the constriction of the body and retraction of the tentacles. (Courtesy of S. G. Martin.)

Fig. IV.3. The same individual 17 minutes postingestion. Note the copious secretion of slime. (Courtesy of S. G. Martin.)

Fig. IV.4. The same individual 55 minutes postingestion. Note extreme body contraction and desiccation. (Courtesy of S. G. Martin.)

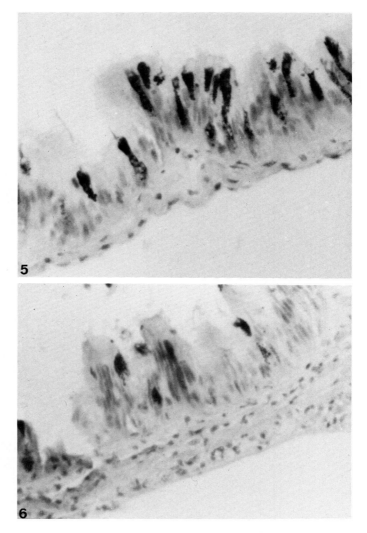

Fig. IV.5. Normal intestine of *Arion ater*, secretory area. Hematoxylin and eosine. (Courtesy of S. G. Martin.)

Fig. IV.6. Secretory area of *A. ater* intestine 2 hours postingestion of metaldehyde. Note necrosis of epithelial cells and loss of goblet cell secretion. Hematoxylin and eosine. (Courtesy of S. G. Martin.)

metaldehyde poisoning on a common garden slug (*Arion ater*) of the Pacific Northwest. Although his study was not exhaustive, it did provide some information on the sites of damage and tissue destruction of metaldehyde.

Initial tissue damage occurs in the gastrointestinal tract. The oral mucosa and the esophagus, which are covered by a protective cuticle, are not affected. The epithelial cells of the stomach present apical cytoplasmic protrusions, clumped cilia, faded cytoplasm, and pycnotic nuclei. The anterior loop of the intestine (Figs. IV.5–IV.7) presents the same appearance as the stomach, but the posterior loop is apparently unaffected. The basal ends of the digestive tubules (lime cell areas) are vacuolated or foamy (Figs. IV.8–IV.11). Adjacent blood vessels appear necrotic with an apparent loss of epithelial linings.

Epithelial cells in the tubular portion of the kidney exhibit decreased affinity for eosinophilic stains, and many cells appear to be lysed, with some containing pycnotic nuclei. Several abnormalities are prominent in the saccular portion of the kidney: the epithelium is necrotic and the cells are distorted; nuclei are pycnotic and karyolytic, and cytoplasm is faded; empty or "ghost" epithelial cells are frequent, and enlarged, vacuolated cells are conspicuous. Complete disruption of the normal architecture of the saccular epithelium occurs in some areas.

The remainder of the histological appearance of the slug is unremarkable or can be attributed to the fluid loss through desiccation, physical trauma in the violent reaction after poisoning, or the onset of autolysis. For example, the reproductive and nervous systems are normal; the lung appears traumatized with collapsed trabeculae, necrotic epithe-

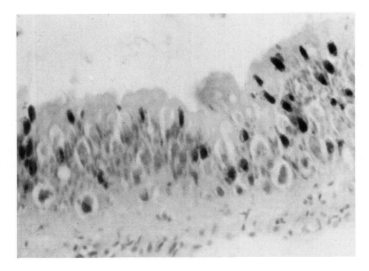

Fig. IV.7. Secretory area of *A. ater* intestine 8 hours postingestion of metaldehyde. Note increased necrosis and vacuolation. Hematoxylin and eosine. (Courtesy of S. G. Martin.)

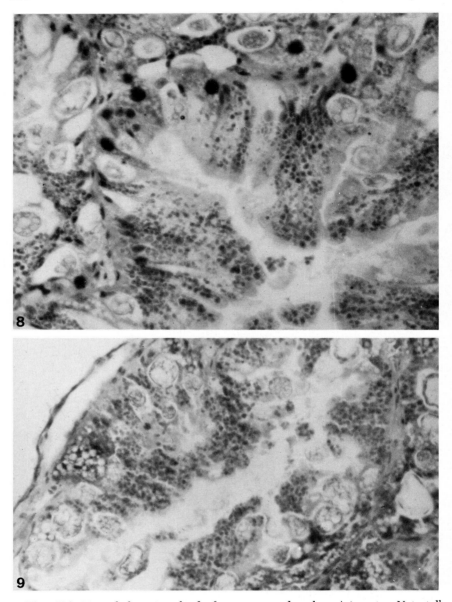

Fig. IV.8. Normal digestive gland of common garden slug, *Arion ater.* Note tall digestive cells filled with secretory granules and less numerous, shorter lime cells with dark, rounded nuclei. Hematoxylin and eosine. (Courtesy of S. G. Martin.)

Fig. IV.9. Digestive gland of *A. ater* 2 hours postingestion of metaldehyde. Note increased secretion due to gland cell hyperactivity; contraction and partial dissolution of tall digestive cells, and beginning necrosis of shorter lime cells. Hematoxylin and eosine. (Courtesy of S. G. Martin.)

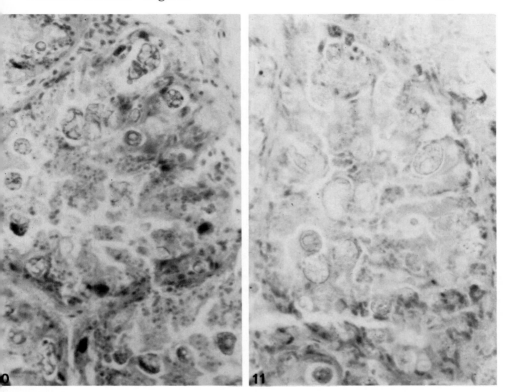

Fig. IV.10. Digestive gland of *A. ater* 4 hours postingestion of metaldehyde. Note extensive ·rosis of all cellular components with only cell fragments remaining; only nuclei of lime cells ·ain intact. Hematoxylin and eosine. (Courtesy of S. G. Martin.)

Fig. IV.11. Digestive gland of *A. ater* 8 hours postingestion of metaldehyde. Note complete ·rosis and large secretion vacuoles. Hematoxylin and eosine. (Courtesy of S. G. Martin.)

lial cells, and ruptured blood vessels; the external epithelium is normal, except that the lime and protein gland cells are shrunken and empty because of the excessive discharge of mucus.

Maximum tissue damage occurs at the site of absorption of the metaldehyde (stomach and anterior intestine) and the site of excretion (the saccular kidney). Interestingly, ingestion of metaldehyde by man results in severe intestinal irritation and kidney and liver damage. Other histological abnormalities in the slug are almost certainly related to mucous discharge, desiccation, trauma, and autolyses.

FORMALDEHYDE

Formaldehyde gas is intensely irritating to mucous membranes. It is highly soluble in water, and it is called "formalin" in this form. It is universally toxic to animal tissues, and because of its rapid penetration and characteristic fixation of tissue, it is widely used as a histological fixative and preservative.

PHENOLICS

Phenol and cresol are both corrosive compounds that invoke immediate necrosis of epithelium on contact. In mammals, external applications result in extensive chemical burns that subsequently ulcerate. Ingestion by mammals causes rapid coagulation necrosis of the upper alimentary tract followed by sloughing necrosis. Absorption may also lead to systemic effects including central nervous system depression, vascular collapse, and necrosis of renal and hepatic parenchymal cells.

Although the effects of these substances on invertebrates have not been investigated, it is almost certain that the immediate corrosive effects would be similar, if not identical, on the outer surface of any invertebrate with an external epithelial covering and in the digestive epithelium of those forms with a digestive system. The systemic effects in the various invertebrate groups remain unelucidated.

STRONG ACIDS AND STRONG ALKALIES

Corrosive strong acids and alkalies cause local destruction of tissues, particularly epithelium. Although data are not available on the effects of these substances on invertebrates, their tissues undoubtedly respond similarly to mammalian tissues. Sulfuric acid causes hardening of epithelial tissue with a bright-red color or a black, tarry appearance due to carbonization. Nitric acid induces extensive ulceration and sloughing of epithelium. Hydrochloric acid produces a reddened or blackened, shriveled epithelium. Corrosive alkalies, such as lye, cause swelling, softening, and ulceration of epithelial tissues. If the injury is not fatal, healing is accompanied by marked fibrosis. Since these substances are highly soluble in water and dissociate to release ions, their effect on aquatic invertebrates is restricted to changes in pH rather than to direct corrosive damage.

Universal Poisons—Inorganic

CYANIDE

Hydrocyanic acid and cyanides are respiratory-enzyme poisons; they inhibit the cytochrome–cytochrome oxidase system of electron transport.

Therefore, any aerobic animal is susceptible to poisoning by the cyanides, either by inhalation of hydrocyanic gas in air breathers or by ingestion of cyanide compounds that are rapidly absorbed and converted to hydrogen cyanide.

As pointed out by Brown (1963), respiration of developing insect embryos is sensitive to inhibition by cyanide, but some diapausing insect embryos that characteristically have low respiration rates are insensitive to cyanide. Adult insects placed in HCN vapor are quickly narcotized after a brief excitatory period. If removed immediately, most insects recover. Some insects recover from a fumigation period of several hours and a paralysis of several days. This is apparently caused by a great decrease in respiratory movements resulting from the initial narcosis; this "protective stupefaction" causes a decreased intake of the poison (Brown, 1963).

Injection of cyanide salts into cockroaches induces rapid narcosis and quick death. Lower concentrations result in recovery from narcosis followed by the classical symptoms of excitation, tremors, and paralysis leading to death.

The effect of cyanide on invertebrates other than insects has not been described, but it can be assumed that its effects on air breathers will be quite similar to those in insects and air-breathing vertebrates and that ingestion of cyanide salts will cause necrosis of the gastrointestinal tract.

Mercuric Salts

Although metallic mercury is highly insoluble and therefore nontoxic, the mercuric salts are violent poisons with a protein-precipitating action resulting in coagulation necrosis of tissues. The actual poisoning is due to absorption of the highly soluble toxic mercuric ion.

The protein-precipitating and lethal effects of the mercury ion on invertebrates are demonstrated by the common use of mercuric chloride in fixatives for invertebrate tissue. Brown (1963) briefly summarized the effects of mercuric chloride on insects by noting that it precipitates the hypodermal cells on contact, that oral poisoning in cockroaches drastically reduces the number of cells in the hemolymph, and that it greatly inhibits dehydrogenase and catalase activity.

Some invertebrates, particularly the mollusks, are capable of concentrating mercuric compounds without fatal damage to themselves. Oysters (*Crassostrea gigas*) were collected from Minimata Bay, Japan, where a severe human neurological disorder and a number of deaths from mercury poisoning resulted from ingestion of seafood contaminated by the mercury-containing effluent of a chemical plant. The oysters were found to contain appreciable quantities of mercury. Although complete, systematic information is not available, it can be assumed that the mercuric salts are generally toxic and that the maximum pathological effects from its absorption will be exerted at the portal of entry and pathways of excretion.

LEAD

Soluble salts of inorganic lead are strong protoplasmic poisons that may be stored by progressive accumulation in an animal's body until they reach toxic levels and then cause tissue damage. Therefore, lead poisoning may be acute, from the sudden absorption of large quantities, or chronic, with insidious accumulation of small doses. Although the two forms of lead poisoning differ in their rate of development, the morphological changes are identical. In man, the major lesions caused by lead are in the hematopoietic, gastrointestinal, and nervous systems.

Galtsoff (1964) presented data from other sources indicating that uptake and accumulation of lead occurs in oysters. Shuster and Garb (1967) noted increased mortality in oysters (*Crassostrea virginica*) exposed to lead concentrations of 0.1 and 0.2 ppm for 10 weeks (as compared to controls, 10% greater average cumulative mortality in 0.1 ppm and 15% in 0.2 ppm), but no significant increase in mortality occurred in 0.025 and 0.05 ppm concentrations of lead over the same period. Grossly, the lead-challenged oysters had conspicuously edematous mantles, though whether this was directly attributable to the lead is not known. Histological examination of the 0.2 ppm-challenged oysters (the only ones examined to date) demonstrated lead chromate crystals and granular aggregates (possibly lead proteinate) most commonly in the basal portion of the gut epithelium and less commonly throughout the connective tissue of the visceral mass, gills, and mantle, and even less in the reproductive tissues.

ARSENIC

Arsenic is a strong protoplasmic poison by Robbins' (1957) definition. Brown (1963) classifies it as a necrotic tissue poison. In humans, acute arsenic poisoning is of fairly frequent occurrence, both because of its role as a favorite for homicidal purposes and its accidental ingestion through wide use as a rat poison and insecticide. Since it is commonly used in small amounts in numerous household preparations and accumulates in the human body, it is an extremely dangerous poison to man. Because of its widespread use as a pesticide, arsenic poisoning is of far greater importance to the invertebrates.

With ingestion of massive dosages of arsenic, death in mammals occurs rapidly without marked morphological lesions and appears to be due to depression of the central nervous system. Caterpillars (*Euxoa segetum*) first respond to ingested arsenic by convulsion of the anterior sphincter resulting in regurgitation; this is a defense mechanism for species that are refractive to control with arsenicals (Brown, 1963). Cockroach nymphs orally poisoned by arsensious oxide become progressively feebler in their movements until a motionless state is reached. The following progression of symptoms occurs in cockroaches after ingestion of arsenates: (a) decrease in activity, (b) loss of equilibrium, (c) loss of recovery reflexes, (d) general asthenia, and (e) motion only on stimulation (Brown, 1963). The general effect in insects is alimentary hypersecretion with resultant watery feces.

Although insects display considerable variation in their susceptibility to arsenic and even greater variation in the reactions to different arsenical compounds, there is a fairly consistent pattern of damage; the most consistent pattern is necrosis and subsequent desquamation of gastrointestinal epithelium. There is remarkable similarity of symptoms of arsenic poisoning in man and insects. The acute or most violent syndrome in man is characterized by rapid onset of vascular collapse and nervous system depression, followed by coma and death within a few hours. In less severe cases, there is often vomiting, followed by severe, persistent, watery diarrhea. The major histopathological changes in human chronic poisoning are also comparable to those in insects, involving the gastrointestinal tract, nervous system, and skin. The intestinal tract is characterized by congestion, edema, and small, superficial ulcerations of the stomach and small intestine.

Arsenicals also cause changes in insect hemocytes; affected cells respond by mitosis, vacuolization, chromatolysis, and eventual cytolysis (Brown, 1963). This is again comparable to the violent onset of vascular collapse, thrombosis, and widespread petechial hemorrhages (apparently due to necrosis of capillary walls) characteristic of the most violent form of arsenic poisoning in man.

There is virtually nothing in the literature on the toxic properties of arsenic in invertebrates other than insects, despite the fact that calcium arsenate is a common constituent of poison baits for slugs, sowbugs, and pillbugs. Galtsoff (1964) noted that oysters in the northeastern United States had been shown by Hunter and Harrison (1928) to contain up to 3.0 mg/kg dry weight of arsenic (As_2O_3). There is no question that arsenic in its many forms is a poisonous substance of great significance to invertebrates. Its effects on invertebrates, as well as the ability to accumulate it without lethal effects, should be investigated.

COPPER

Although copper poisoning is of minimal concern in vertebrates, it is of major importance to the invertebrates, particularly those living in an aquatic environment. Despite general knowledge of the widespread toxic effect of the copper ion (commonly in the form of copper sulfate as a pesticide), there is surprisingly little information available as to its pathological effect on invertebrates. Copper sulfate has been used to control oyster pests and predators. Immersion in a concentration of 1:2000 CuSO₄ for 30 minutes is fatal to *Urosalpinx* embryos and adult *Crepidula fornicata* (Hancock, 1959). Longer immersion at lower concentrations also kills adult *Urosalpinx*. Adult oysters are relatively resistant to copper sulfate, but young oysters, spat, and developing larvae are highly susceptible to its toxic effects.

Adult mollusks, especially oysters, are not only resistant to the toxic effects of copper, but are also capable of accumulating and storing it in their bodies at surprisingly high levels. Marks (1938) investigated the copper content and tolerance of ten species of mollusks; he noted that the upper limit of copper tolerance for most of the species investigated (which did not include any oysters) lies in a range of 0.10–0.20 mg of added copper per kilogram of seawater, or about 100–200 times that normally present. However, the clam, *Paphia staminea*, can tolerate much higher concentrations. Adult oysters undoubtedly can also tolerate higher concentrations since they frequently accumulate copper in sufficient amounts to cause a green discoloration of the gills and mantle and an unpleasant coppery taste. Ryder (1882), Boyce and Herdman (1897), and Herdman and Boyce (1899) demonstrated that the copper was accumulated in leukocytes that aggregate in cysts under the epithelium of the mantle and near the surface of the gills.

Because of the proximity of valuable oyster beds to refineries that sometimes discharge toxic quantities of copper, Fujiya (1960) studied the effects of copper on the survival of oysters and the histopathology of copper poisoning. Microscopically, Fujiya described necrosis and desquamation of the stomach and noted regressive changes in digestive diverticula of oysters maintained in 0.1–0.5 ppm copper for 2 weeks. Histochemical studies showed that RNA and polysaccharides decreased in the digestive diverticula during exposure.

Glude (1957), in search of a means of controlling the damage to oysters by the predaceous snail *Urosalpinx cinerea*, experimented with the use of barriers made of various metals. Although *U. cinerea* readily crawled across strips of zinc and iron, those snails placed in an aquarium and surrounded by a horizontal 3-inch strip of 0.024-inch thick copper sheet not only did not cross the copper, but all retracted the foot and remained motionless as long as the copper plate was in position. Subsequent laboratory experiments verified these observations, but showed that the oyster drill (*U. cinerea*) and the mud snail (*Nassa obsoleta*) had to contact the copper or approach it very closely to be affected. When the copper concentration was allowed to increase by ionization in standing, aerated seawater, all the drills died within 10 days. Currents up to almost ½ knot have no effect on the response, but small amounts of sand or silt covering the barrier effectively offset the toxic effects of the copper. The laboratory investigations were essentially corroborated in the field with *Nassa obsoleta*, which is more active at the lower temperatures of Maine waters where the research was carried out.

Copper, along with a number of other metal ions, has been shown to cause what Harry and Aldrich (1963) call the distress syndrome in freshwater pulmonate snails. When subjected to higher concentrations, the snail retracts into the shell and remains there until it dies or until the concentration is lowered sufficiently for it to resume normal activities. There is an intermediate range of concentrations (between the concentrations that are too low to affect the snails and high enough to initiate retraction) in which the distressed snail is extended but unable to attach its foot and carry on locomotion, feed, or breathe atmospheric air (by renewing its pulmonary air bubble at the surface of the water). The typically distressed snail lies on the substrate with the cephalopedal mass extended. It frequently attempts to attach the foot to the substrate but is unable to do so. After several hours, the tentacles become swollen at their bases and slough large numbers of cells at the distal ends. Movement of the foot becomes more feeble and occurs less frequently, but infrequent spasmodic contractions of the body stalk may occur. Muscular activity may eventually cease; ciliary action of the surface and pulmonary epithelium appear to be unaffected, at least during the first 24 hours of exposure. Although the rate of heartbeat appears to be depressed, the heart continues to beat while the snail is in distress. Finally, sand grains, which are normally retained in the stomach of pulmonate snails, are voided by defecation.

The actual physiological mechanisms and the histopathology involved in the distress syndrome are still unknown, although Harry and Aldrich believe the syndrome to be a neuromuscular phenomenon. Ciliary activity, which is notoriously autonomous in mollusks and difficult to arrest by poisons, is apparently unaffected. They noted the gross resemblance of distress to the response evoked by such "relaxing" agents as menthol and nembutal commonly used by malacologists on gastropods prior to fixation.

Bernard and Lane (1963) discussed the effects of the copper ion on the planktonic larvae of a barnacle (*Balanus amphitrite*), particularly in its relationships to oxygen uptake. They had previously (1961) shown that the copper ion is absorbed by the cyprid larvae, primarily through the surface epithelium of the thorax, and that it is excreted into the lumen of certain regions of the gut by the lining epithelium.

When the copper concentration is raised from 0.005 mg/liter, which is the normal level in seawater, to 0.5 mg/liter, oxygen uptake of barnacle larvae is elevated. However, further increases in the amount of copper ion up to 500 ppm results in a steady decrease in oxygen consumption. The authors noted that the survival rate of test animals was drastically reduced at 1000 mg/liter, and the survival time was markedly reduced between 100 and 200 mg Cu^{++} per liter (Table IV.1). Since the activity of the barnacle larvae is reduced above 50 mg/liter and since no swimming or antennal movements occur in concentrations above 100 ppm, it is apparent that attachment is prevented at a lower concentration than that required to kill the cyprids.

Bernard and Lane suggest that the toxicity of the cupric ion in barnacle cyprids results from several interrelated metabolic responses, some of which are stimulated by copper while others are depressed. Generally, the larvae adapt to each copper concentration by adjusting their metabolism to a level at which no more copper is absorbed through the respiratory surfaces than can be excreted. This results in increased motor activity and oxygen consumption at low copper concentrations, but increased concentrations are accompanied by decreases in motor activity

TABLE IV.1

Effects of Copper on Survival of Barnacle Cyprids[a]

Cu^{++} (mg/liter)	Percent survival	Average time of death (minutes)
0.005	100	—
0.5	80	51
1.0	77	63
5	91	80
10	91	30
20	91	80
50	83	73
100	77	95
200	71	18
500	83	22
1000	30	24

[a] From Bernard and Lane (1963).

and in oxygen uptake which reflect the energy cost of active transport of copper ions across excretory membranes of the intestinal epithelium.

Copper sulfate has also been used, though without great success, against starfish on oyster beds. Laboratory tests showed that concentrations of 0.15 ppm were effective if exposure was prolonged over several days. The characteristic symptoms of copper poisoning in starfish include muscular weakness, with the poisoned animal incapable of righting itself if turned over. There is a marked lack of coordination in the movements of the rows of tube feet, the mouth often relaxes, and the stomach becomes everted. Eventually the arms slough from the body, and the entire animal disintegrates (Galtsoff and Loosanoff, 1939).

Chlorine

In both its elemental gaseous state or as free chloride ions in water, chlorine is a highly poisonous, universal toxin. However, considerable variation in levels of susceptibility exists among the invertebrates, thereby allowing its use under certain conditions as a selective bactericide or pesticide.

Chlorination of domestic water supplies for public health purposes is, of course, common throughout much of the civilized world. Chlorination of swimming pools is also common, particularly in the United States. Chlorination techniques are used widely in industry to prevent fouling by sedentary organisms and bacteria of water that is circulated for cooling, but this is not well known to the laymen. Through all these legitimate uses of chlorine, numerous invertebrates are killed; additional effects on invertebrate populations occur with intentional or accidental discharge of chlorine into the environment.

Two types of fouling occur in cooling water operations: bacterial slime in the condenser tubes and macroscopic sedentary invertebrates in the intake and outfall culverts. Both types are generally considered to be controlled by chlorination, but many industrial operations utilizing seawater for cooling must be periodically shut down so that tons of mussels (*Mytilus edulis*) and other organisms can be removed from the system (Waugh, 1964). This suggests that chlorine may not be as toxic to marine invertebrates as is generally assumed.

There is a limited amount of data on the effect of chlorine on sedentary invertebrates. Waugh (1964) cites a translation of work by Haou and Chian (in 1958) in which 380 ppm residual chlorine for 10 minutes was required to cause detachment of adult *M. edulis*. Five days of continuous treatment with chlorine at a residual of 10 ppm was required to kill a mussel population, and 100% of a mussel population was killed by a residual of 1.0 ppm for 15 days. In general, most investigators believe that the larvae of marine organisms are much more sensitive to low concentrations of chlorine than the adults, though little supporting evidence has been presented.

Waugh (1964), in his investigation of the effects of chlorine on oyster larvae (*Ostrea edulis*) and nauplii of a barnacle (*Elminius modestus*), showed that oyster larvae were unharmed at chlorine concentrations normally used by industry (10 ppm for 10 minutes). Even at 20 ppm there was considerable growth and survival of larvae, but between 50 and 200 ppm virtually all larvae were killed. The barnacle nauplii were much more sensitive; the majority were killed by concentrations of chlorine as low as 2.0 ppm.

Free chlorine has been advocated and sometimes used for the purification of shellfish since 1914. Numerous experiments have been conducted on its use both in Great Britain and the United States. Galtsoff (1946) summarized past findings and reported on the physiological reactions of oysters to chlorination. There is considerable variation in the individual sensitivity of oysters (*C. virginica*) to chlorine, but experimental evidence indicates that initial concentrations of 0.01–0.05 ppm chlorine materially interferes with normal physiological functioning. Repeated exposure results in the development of some tolerance, but Galtsoff maintained that there was no evidence of effective water pumping by oysters in concentrations exceeding 1.0 ppm. Oysters kept in chlorinated water secrete unusually large quantities of mucus, primarily as pseudofeces, probably as a protective covering over the tentacles, mantle, and gills.

Galtsoff and Loosanoff (1939) cited some studies by Palmer on the effect of free chlorine on starfish. Starfish exposed to 10 and 20 ppm were unaffected. Palmer was unable to kill starfish with 20 minutes

TABLE IV.2

EFFECT OF FREE CHLORINE ON STARFISH[a]

Concentration (ppm)	Exposure (minutes)	Temperature (°C)	Remarks
1000	15	23.5	Dead
500	20	24.0	Recovered in 24 hours
250	5	18.5	Recovered
100	25	20	Motionless 4 hours, then recovered
50	60	26	Recovered in 23 hours
20	30	26	Recovered in 4 hours
15	60	25	Recovered in 30 minutes

[a] From Galtsoff and Loosanoff (1939).

exposure at 500 ppm. At 1000 ppm, however, small starfish were killed in 1 minute (Table IV.2).

Pesticides

NARCOTIC GASES

Vapors of diethyl ether or chloroform quickly anesthetize air-breathing vertebrates and insects. One must presume they would have the same effect on any air-breathing invertebrate. Among the narcotic gases there may be additional toxic effects, as have been demonstrated with other gases, on insects. Narcotics such as chloroform, diethyl ether, carbon tetrachloride, carbon disulfide, and ethylene dichloride accumulate in the nervous system (Brown, 1963). It has been further suggested that narcosis involves tissue anoxia, particularly of the nerves. Carbon disulfide vapors cause a loss of blood volume in the American cockroach, and carbon tetrachloride applied to the body louse induces nuclear enlargement and chromatin swelling in the hemocytes. There is, to my knowledge, no information available on the effects of the narcotic gases on invertebrates other than the insects.

CHLORINATED HYDROCARBONS

Benzene Derivatives

Paradichlorobenzene (PDB) has been a useful insecticide for many years. When injected or contact applied to insects, it typically causes hyperactivity, tremors, and eventual paralysis (Brown, 1963). It has also been shown to induce repetitive discharge of action potentials when applied to a crayfish nerve preparation (Welsh and Gordon, 1947). These symptoms in insects resemble those of DDT poisoning and poisoning by other benzene derivatives such as phenol, aniline, and hydroquinone.

Orthodichlorobenzene, a commonly used insecticide causing similar effects, has been used both alone and in combination with other pesticides for control of shellfish pests and predators. This substance, a heavy oil, is usually mixed with dry sand or another inert carrier prior to application (Loosanoff *et al.*, 1960a). The treated sand can be placed around the edges of shellfish beds, preventing their invasion by predators, or spread over infested areas for direct application. Under both laboratory and field conditions, it appears that direct contact with the orthodichlorobenzene is required for toxic effects to be elicited. Oyster drills (*Urosalpinx cinerea*) contacting the treated sand become swollen, disabled, and die (Figs. IV.12 and IV.13). Starfish (*Asterias forbesi*)

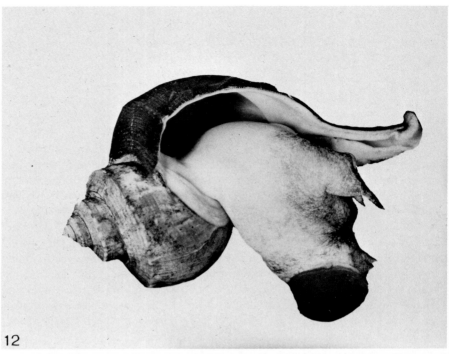

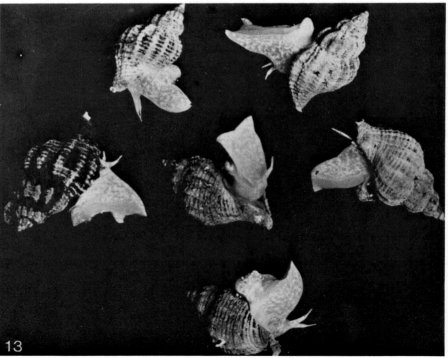

Figs. IV.12 and IV.13

are "burned" and decompose (Fig. IV.14). Oysters are not harmed by 3 hours immersion in 100 ppm orthodichlorobenzene, but almost all are killed by 24 hours immersion in the same solution.

Paradichlorobenzene, trichlorobenzene, and tetrachlorobenzene alone, in various chlorobenzene combinations, and with other pesticides such as Sevin and lindane have been tested for efficiency as barriers against starfish (Fig. IV.15). A combination of ortho- + tri- + tetrachlorobenzene + Sevin + lindane was found to be the most effective.

Insoluble crystals of paradichlorobenzene in concentrations above 10 gm/square foot of bottom affect all gastropods studied (*Busycon, Urosalpinx, Murex, Nassa, Polinices,* and *Thais*). All the heavy chlorinated oils cause marked swelling of the foot of gastropods (Loosanoff *et al.,* 1960b).

Unless treated sand was contacted directly, setting and survival of all the normal components of the fauna associated with oysters was not affected by the barriers in laboratory and field studies. Soft-shell clams (*Mya arenaria*), for example, that undergo metamorphosis and settle in the treated sand suffer high mortality. No significant effects have been shown on oysters kept in a concentration of 1 ppm orthodichlorobenzene for 48 hours. Raising the concentration to 10 ppm causes some oysters to gape, but they quickly recover if returned to uncontaminated seawater (Loosanoff *et al.,* 1960b).

DDT and Related Compounds

Dichlorodiphenyl trichloroethane, commonly known as DDT, is a residual contact insecticide characteristically causing "DDT jitters" (Brown, 1963), a persistent tremulousness of the entire body of poisoned insects. DDT-poisoned insects become unusually active, irritable, and uncoordinated.

This pesticide has been shown to be pathogenic to a wide spectrum of other invertebrates, although it was at first thought to be a useful pesticide for the oyster industry. Experiments (Loosanoff, 1947) demonstrated that settling of barnacles on oyster shells and other spat collectors could be prevented if the collectors were treated with a DDT solution. Subsequently, DDT was used extensively in Great Britain to control fouling of cultch and oyster beds by barnacles (principally

---

Fig. IV.12. A conch, *Busycon,* subjected to a combination of Sevin and orthodichlorobenzene mixed with sand. Note the extended, swollen foot. (Courtesy of C. L. MacKenzie, Jr.)

Fig. IV.13. Oyster drills, *Urosalpinx cinerea,* paralyzed by exposure to Sevin and orthodichlorobenzene mixed with sand. Note the swollen extended foot of each snail. (Courtesy of C. L. MacKenzie, Jr.)

Fig. IV.14. Starfish, *Asterias forbesi*, subjected to orthodichlorobenzene. Note the gross lesions on aboral surface. (Courtesy of C. L. MacKenzie, Jr.)

Fig. IV.15. Starfish, *Asterias forbesi*, reacting to contact with a barrier consisting of sand impregnated with a mixture of orthodichlorobenzene and Sevin. (Courtesy of C. L. MacKenzie, Jr.)

Elminus modestus introduced from Australia during World War II) (Waugh *et al.*, 1952; Waugh and Ansell, 1956).

It eventually became apparent that organisms other than the target were often affected, and that the kinetics of pesticides such as DDT in the marine environment were not understood (P. A. Butler, 1966). Oysters exposed to 0.001–0.002 ppm DDT at a temperature of approximately 30°C appear normal in growth and behavior, but chemical analysis reveals that they concentrate DDT in their tissues. Such biological magnification may reach 70,000 times the amount present in the environment (P. A. Butler, 1966), depending on water temperature, exposure time, and DDT concentration.

In sufficient concentrations, DDT kills oysters; at lesser concentrations it seriously interferes with normal functions. The first obvious effect of low concentrations of DDT on young oysters is inhibition of growth, with significant effects occurring within 24 hours at concentrations as low as 0.1 ppm (P. A. Butler *et al.*, 1962) (Table IV.3). Suppression of shell growth increases uniformly with each tenfold increase in DDT concentration, within the range of 0.0001–0.1 ppm at a water temperature of approximately 17°C. There is virtually no growth at 0.1 ppm, and the growth at 0.0001 is approximately 20% of that of a control group not exposed to DDT. Growth rates return to normal after 10 days of flushing with DDT-free water, although the DDT residue in the oyster's body is still above 10 ppm (P. A. Butler, 1966).

In view of the amazing capacity of the oyster to accumulate DDT, the sites of storage are of considerable interest, particularly because DDT is stored largely in the adipose tissue of vertebrates, and oysters contain less than 5% fat. Analysis of the organ systems of oysters with a total body burden of approximately 25 ppm DDT showed that about

TABLE IV.3

Common Pesticides that Inhibit the Growth of Young Oysters after 24 Hours Exposure[a]

Chemical and percent active ingredients		Minimum effective concentration (ppm)
Aldrin	93	0.1
Chlordane	100	0.01
o-Dichlorobenzene	100	1.0
DDD	50	1.0
DDT	100	0.1
Dieldrin	100	0.1
Endrin	97	0.1
Hepatochlor	74	0.01
Rotenone	98	0.01
Sevin	95+	1.0
Toxaphene	60	0.1

[a] From P. A. Butler *et al.* (1962).

67% of the DDT occurred in the intestinal tract (including the digestive diverticula) and the gonads. The mantle and gills contained approximately 26% of the total, and the remainder occurred in the adductor muscle and tissue fluids (P. A. Butler, 1966). Subsequent experimentation has revealed that the gonads are a major site of DDT storage, with the gametes themselves sometimes containing 25 ppm. Although adequate tests have not been made on embryos resulting from fertilization of gametes in 20–30 ppm DDT, there is a strong possibility that their ability to achieve maturity would be affected since oyster larval cultures exposed to only 1.0 ppm DDT suffer total mortality within 6 days (Davis, 1961).

Little information on the histopathological effects of DDT on oysters is available. Small numbers of oysters exposed to a mixture of DDT, toxaphene, and parathion (1.0 ppb of each) for long periods were examined by consulting pathologists (G. Pauley and R. Taylor) and their findings reported (Lowe *et al.*, 1971). Gonadal maturation was retarded, compared to untreated control oysters, in the female oysters subjected to this mixture for 36 weeks. In addition to fewer mature ova and more numerous immature ova, there was an abnormal leukocytic infiltration into the gonads and hyperplasia of the germinal epithelium. Leukocytic infiltration of the gonad of male experimental oysters also occurred. Treated oysters exhibited slight edema beneath the gut, sometimes accompanied by leukocytic infiltration, and the epithelial cells of the digestive tubules were reduced in height. Eggs from the treated females fertilized by spermatozoa from treated males developed into apparently normal 24-hour trochophore larvae.

All the oysters subjected to the pesticide mixture became infected with an unidentified mycelial fungus that invaded and caused systemic lysis of tissues. Since none of the untreated control oysters developed the fungal infection, it was assumed that the pesticide mixture caused a breakdown in the oyster's defense mechanism, at least against this particular fungus. The challenged oysters appeared to have almost completely recovered after 12 weeks in pesticide-free water.

Oysters examined after 36 weeks in 1.0 ppb of DDT alone did not differ significantly from untreated controls. There were some indications that parathion- or toxaphene-treated oysters were more pathological than controls, but insufficient numbers were examined for definite conclusions to be made.

DDT, like most insecticides, is highly toxic to crustaceans and other arthropods. At less than 10 ppb, half the adult commercial brown and pink shrimp (*Penaeus aztecus* and *P. duorarum*) are killed or immobilized in 48-hour laboratory tests. Paralyzed individuals may live for days or weeks in the laboratory, but such survival is highly unlikely in nature.

Field studies of DDT used in accordance with recommended practice for aerial mosquito control in tidal salt marshes demonstrated the high sensitivity of crustaceans (Springer and Webster, 1951; Springer, 1961). With application rates as low as 0.3 pound/acre, amphipods and isopods were virtually eliminated from the area, and the populations did not show any signs of recovery after 2 years. Grass shrimp (*Palaemonetes* sp.) were somewhat less susceptible. Blue crab populations suffered 10–40% mortality with single applications at 0.3 pound/acre and 95–97% mortality with repeated applications (3–10 times per year for several years). Fiddler crabs (*Uca*) are generally more resistant to DDT poisoning than blue crabs, though contradictory findings on the relative susceptibility of fiddler and blue crabs have been reported (George *et al.*, 1957).

Although virtually no information is available on the histopathological effects of DDT on the highly susceptible crustaceans, such effects on insects may be indicative of the manner in which other invertebrates, especially crustaceans, are affected. DDT is well known for its affinity for and ability to penetrate chitin; it is more effective when applied to the cuticular surface of insects than when injected into the internal body tissues. Though the primary effect of DDT is on the nervous system, it causes no discernable histopathological changes in nerve tissue, and frequently it causes no marked changes in any tissue. However, a number of authors (see Brown, 1963) have described definite histological changes of other tissues in both contact poisoned and orally poisoned insects. The cells of the midgut epithelium of bees orally poisoned with DDT exhibit severe vacuolization and proliferation at their tips; this glandular hyperactivity is believed to be induced by stimulation of the stomatogastric nerve branches by the DDT (Salkeld, 1950, 1951). Body lice contact poisoned by DDT are affected similarly with vacuolization of the midgut epithelial cells; the nuclei swell and the cell membranes become indistinct. Similar, but more pronounced, cellular effects can be seen in the labial glands of poisoned body lice; the hemocytes, epidermis, and malpighian tubes are also affected.

Studies of shrimp chronically exposed to DDT have been underway for some time at the Gulf Breeze Laboratory of the United States Environmental Protection Agency (formerly the Pesticide Field Station of the Bureau of Commercial Fisheries, Center for Estuarine and Menhaden Research), Gulf Breeze, Florida. Since penaeid shrimp die within a few weeks if continuously exposed to 0.1 ppb DDT, the question arises as to the possible effects of long-term continuous exposure to concentrations of DDT below 0.1 ppb (Nimmo and Blackman, 1970).

Prolonged exposure (24 and 45 days) to 0.1 ppb DDT results in a gradual decrease in blood protein levels of pink shrimp (*P. duorarum*). Blood protein levels were determined by acetate electrophoresis on blood drawn from each test animal at 2- to 3-day intervals and were compared to levels of similarly handled control animals maintained in DDT-free water. The only fraction affected appears, from optical densitometer measurements, to be hemocyanin. Statistically significant differences (95% confidence level—with Student's "t" test) existed between test and control shrimp blood protein levels at the twenty-fourth day in one experiment and at the thirty-ninth day in a second experiment. Shrimp exposed to 0.05 ppb exhibited no significant change in protein concentration, suggesting that the threshold of toxicity lies between 0.05 and 0.1 ppb of DDT.

Investigations of DDT accumulation in shrimp have revealed that DDT is most heavily concentrated in the hepatopancreas and least heavily in the muscles. Shrimp do not accumulate large concentrations of DDT in comparison to many other animals, but total body residues may have 2000 times as much DDT as the water. Some individuals in a population appear to be more resistant than others as shown by the range of residues in shrimp dying from DDT poisoning.

The one experiment designed to determine possible histopathological changes resulting from chronic exposure to DDT was inconclusive. Juvenile pink shrimp subjected to 0.1 ppb DDT for 35 days exhibited slight necrosis of the hepatopancreas, but, because some of the test animals were severely parasitized, definite conclusions were not possible. When repeated with grass shrimp (family Palaemonidae rather than Penaeidae), no tissue abnormalities were observed.

Polychlorinated Biphenyls

Polychlorinated biphenyls (PCB's) are structurally related to DDT, soluble in lipid, relatively insoluble in water, and persistent in the environment (Duke *et al.*, 1970). They are typically industrial pollutants, being used as plasticizers, flame retardants, insulating and heat-exchange fluids, and many other products (Nimmo *et al.*, 1971), but they are also used as carriers for some insecticides (Duke *et al.*, 1970). PCB's have been reported in fish and wildlife, including invertebrates, from a wide geographic range. Duke *et al.* (1970) detected a PCB (Aroclor 1254) in the biota, sediment, and water of estuarine areas near Pensacola, Florida, and studied its distribution. Additionally, they investigated the toxicity of Aroclor 1254 to oysters, shrimp, and crabs.

Acute toxicity bioassays (96 hours or less) revealed that shell growth of oysters exposed to 100 ppb for 96 hours was completely inhibited, but all test oysters survived and resumed normal growth rates after being returned to uncontaminated water. Shrimp, however, are susceptible to relatively low concentrations of Aroclor 1254, with 80% mortality after 24 hours exposure to 100 ppb and 100% mortality by 48 hours. No mortality occurred in 10.0 and 1.0 ppb concentrations within 48 hours, but the shrimp accumulated the PCB at a rate equivalent to the concentration.

Longer term (20 days) bioassays were conducted on juvenile pink shrimp (*P. duorarum*) and juvenile blue crabs (*Callinectes sapidus*). Seventy-two percent (18 of 25) of the shrimp exposed to 5.0 ppb Aroclor 1254 died during the 20 days; the first death occurred on the tenth day of exposure, and a few died each day until termination of the experiment. Dying shrimp did not exhibit the typical symptoms of insecticide poisoning (extreme irritability followed by a loss of equilibrium), and several died immediately after molting. Residue analysis of a composite sample of the seven surviving shrimp contained 16 ppm of the PCB. Most of the substance is concentrated in the hepatopancreas (Nimmo *et al.*, 1971).

Juvenile blue crabs are not as sensitive as shrimp to Aroclor 1254. Only one death occurred in the test group during the 20-day exposure to 5.0 ppb of the PCB, despite much heavier accumulation of the substance in the tissues (average of 23 ppm Aroclor 1254 after 20 days exposure). Residues persist long after removal to clean water, averaging 22 ppm after 1 week and 11 ppm after 4 weeks.

Mirex (Dodecachlorooctahydro-1,3,4-metheno-2H-cyclobutal[cd]pentalene)

Mirex, used to control fire ants in coastal areas of the southeastern United States, was reported to have relatively low acute toxicity to marine crustaceans. Juvenile blue crabs and pink shrimp show no symptoms of poisoning during a 96-hour exposure to 0.1 ppm technical mirex in flowing seawater. More recent experimentation (Lowe *et al.*, 1970), however, has demonstrated that all individuals of both species become irritated, paralyzed, and die within 18 days after being placed in clean water. About 30% of the shrimp and 20% of the crabs become paralyzed or die within 10 days, and 80% of the shrimp and 60% of the crabs suffer this fate by 15 days. This is the first documented case of delayed toxicity to marine crustaceans.

Mirex bait (the actual formulation used in the field to control fire ants) also causes delayed toxicity to juvenile blue crabs. Individual crabs held in compartments provided with flowing seawater were paralyzed within 3 to 14 days when one particle (1.5 mg average weight) was present in the container. Crabs were observed to pick up and feed on the bait particles. Thus, mirex can act both as a contact and an internal poison to juvenile crabs.

Lindane

Lindane (1,2,3,4,5,6-hexachlorocyclohexane) is a potent insecticide, with both residual contact and vapor toxicity, that acts so rapidly that the symptomatic stages are telescoped (Brown, 1963). It is also a useful, but dangerous, pesticide in certain estuarine situations. Chemical control of the green crab, *Carcinus maenas* (the most important predator of the softshell clam [*Mya arenaria*] north of Cape Cod, Massachusetts), by the use of lindane-soaked fish has been shown to be effective (Hanks, 1963). Concentrations as low as 1 ppm lindane are lethal to green crabs within 48 hours, and spray applications at dilutions of 6.5 ppm protected young rice seedlings from destruction by another crab, *Sesarma africanum* (Jordan, 1955).

Lindane mixed with fine sand and furnace oil is remarkably effective in eliminating the ghost shrimp, *Callianassa* (a serious pest in oyster beds of the Pacific Northwest; they destroy oyster beds by their burrowing activities) (Lindsay, 1963). Heavy mortalities of ghost shrimp occur within ½ hour after application, but shore crabs (*Hemigrapsus*), shrimp (*Crago*), bentnose clams (*Macoma nasuta*), and eastern soft-shell clams (*M. arenaria*) associated with the ghost shrimp are also killed. Japanese littleneck clams (*Venerupis*) and sand worms (*Nereis*) in Lindsay's investigation were "irritated" but not killed.

As would be expected, lindane is much more toxic to crustaceans than to most of the nonarthropod invertebrates. This can be used advantageously as mentioned above, but can also be of serious consequence when valuable, nontarget crustaceans are affected. P. A. Butler and Springer (1963) reported that concentrations of lindane (or hepatochlor or endrin) of 0.3–0.4 ppb killed or immobilized half the adult penaeid shrimp (*Penaeus setiferus* and *P. aztecus*) in 48 hours exposure. The paralyzed individuals may live for extended periods in the laboratory. Chin and Allen (1957) reported that 50 ppb of lindane was lethal to the same two species. Lobsters (*Homarus americanus*) are highly sensitive to lindane (Hanks, 1963), and it has been shown to be more toxic than DDT to blue crabs (*Callinectes sapidus*) and fiddler crabs (*Sesarma*). When used for mosquito control in marsh areas, lindane can cause extensive mortalities of highly susceptible, important commercial species, such as the penaeid shrimp, which use these low-salinity estuarine areas as nursery grounds. Shrimp being held in pens for sale as fish bait along the Texas and Florida coasts were killed by lindane or DDT applied for mosquito control (Chin and Allen, 1957; deSylva, 1954). The symptoms and pathological effects of lindane poisoning have been worked out for a number of species of insects, but no such information is available for other invertebrates.

Other chlorinated hydrocarbon insecticides, including aldrin, dieldrin, endrin, toxaphene, chlordane, and heptachlor markedly affect shell growth of young oysters and initiate irritative shell movements even at levels below 1.0 ppm (P. A. Butler *et al.*, 1962; P. A. Butler, 1964; P. A. Butler and Springer, 1963). Furthermore, most of the chlorinated hydrocarbons, when present at 1.0 ppm for 4 hours, decrease phytoplankton production by 50–90%. Thus, these pesticides could severely affect phytoplankton feeders by starvation (P. A. Butler and Springer, 1963).

Crustaceans, as would be expected, are particularly sensitive to the remaining chlorinated hydrocarbons. Heptachlor and endrin have the same effects at the same concentrations on penaeid shrimp as lindane; they kill or paralyze half the population exposed to 0.3–0.4 ppb for 48 hours. Similar effects are elicited by concentrations of 1–6 ppb of chlordane, toxaphene, and dieldrin. Juvenile blue crabs, *Callinectes sapidus,* an important commercial and sport species, are about 100 times more resistant than penaeid shrimp to the toxic effects of chlorinated hydrocarbons, while the stone crab is intermediate in susceptibility (P. A. Butler and Springer, 1963).

There are numerous records of accidental mortalities to crustaceans resulting from applications of chlorinated hydrocarbon insecticides (other than DDT and lindane) for control of various insects. Some of these were noted by Butler and Springer and include numerous blue and fiddler crabs killed by a treatment of 0.3 pound/acre of dieldrin for greenhead fly control and complete annihilation after 1 pound/acre treatment for sandfly control. Shrimp mortalities resulted from heavy rains washing in heptachlor applied 6 days earlier at 0.25 pound/acre for fire ant control.

Although the chlorinated hydrocarbons are the most toxic pesticides that have been tested on mollusks (P. A. Butler and Springer, 1963) in the laboratory, field applications for mosquito or greenhead fly control at recommended levels have not elicited recognizable toxic effects on estuarine mollusks. In laboratory experiments, mollusks have been continuously exposed to low levels of chlorinated hydrocarbons for long periods of time in efforts to determine the effects of such exposure (P. A. Butler and Springer, 1963). Juvenile clams were challenged with DDT (1 ppb) and dieldrin (5 ppb), while oysters and mussels were subjected to aldrin (2 ppb), malathion (2 ppm), and toxaphene (50 ppb) for periods of 3–6 months. Under these conditions, growth and survival were virtually identical in experimental and control groups.

ORGANOPHOSPHORUS INSECTICIDES

The numerous reports on the effects of organophosphates on various species of insects have been summarized by Brown (1963). Injection of simple organic phosphates into *Periplanata americana* induces, successively, hyperactivity and hyperexcitability, exaggerated tonus, ataxia, convulsions, paralysis, and death. Symptoms begin quickly with the simpler compounds (within 10–30 minutes), and they are characterized at the onset by violent tremors of the body and appendages. With the more complex organophosphates, the onset of symptoms is considerably delayed. Symptoms do not appear until 2–7 hours after injections of parathion into the cockroach (*P. americana*); however, death occurs in both extremes between 24 and 36 hours. Death from malathion is much slower, usually occurring after 5 days. Response varies among different species; in houseflies contact treated with parathion, hyperactivity develops in 30 minutes, paralysis in 3 hours, and death occurs within 24 hours.

The actual mechanism of damage by these compounds in insects is through inhibition of cholinesterase in the nerve cord, though other esterases are inhibited as well. The toxins are ganglionic poisons that cause about 50% inhibition of cholinesterase in the hyperactive stage and 90–95% inhibition in the prostrate stage.

There are no outstanding histopathological characteristics of organophosphate poisoning in insects. Tigrolysis of the Nissl granules in nervous tissue is common, and midgut epithelial cells may be swollen because of hypersecretion, although honeybees orally poisoned with parathion exhibit no histological changes in the midgut (Salkeld, 1951).

Little is known of the effects of organophosphate insecticides in other invertebrates. P. A. Butler (1966) reports that these compounds are much less toxic to oysters than the chlorinated hydrocarbons (Table IV.4). Few organophosphorus compounds have been tested on crustaceans; those studied have shown varying but high levels of toxicity. Baytex (Bayer 29,493), used in mosquito control, was the most toxic pesticide to crabs and shrimp tested (P. A. Butler and Springer, 1963).

Rather intensive field tests of the effects of "Dibrom 14 Concentrate" or Naled, an organophosphorus compound (1,2-dibromo-2,2-dichloroethyldimethyl phosphate) manufactured commercially and widely used in tideland mosquito abatement programs (Bearden, 1967), indicate that

TABLE IV.4

RELATIVE TOXICITY OF COMMON TYPES OF PESTICIDES TO ESTUARINE
FAUNA (HERBICIDES RATED AS UNITY)[a]

Pesticide type	Plankton	Shrimp	Crab	Oyster	Fish
Herbicide	1	0	1	1	1
Insecticide					
Organophosphorus compounds	$\frac{1}{2}$	1000	800	1	2
Polychlorinated hydrocarbon compounds	3	3	100	100	500

[a] From P. A. Butler (1966).

little or no observable effects occur in test animals held in their natural environment when the chemical is applied by thermal fogging or aerial applications in the concentrations normally used for marshland mosquito control. This is rather surprising since preliminary laboratory experiments indicated that Dibrom is toxic to postlarval brown shrimp (*Penaeus aztecus*) at concentrations slightly over 2.0 ppb. This agreed with previous data that 5.5 ppb caused 50% mortality or loss of equilibrium among adult pink shrimp (*Penaeus duorarum*) over a 48-hour period, and juvenile blue crabs suffered 50% mortality or loss of equilibrium when exposed to 0.30 ppm Dibrom for 48 hours. Animals tested and found unharmed in the field tests included postlarval and juvenile brown shrimp (*P. aztecus*), postlarval and subadult white shrimp (*P. setiferus*), adult hardback shrimp (*Palaemonetes pugio*), and juvenile blue crabs.

Bearden noted a number of factors in the natural environment that tend to decrease the concentration to which the animals are actually exposed, including adsorption or absorbtion of Dibrom by suspended particles, adherence of the spray to stems and leaves of marsh grass (*Spartina*), and tidal flow and flushing. Since Dibrom quickly hydrolyzes, is relatively insoluble in water, and is applied in low concentrations, Bearden suggested its use as preferable to the more stable chlorinated hydrocarbons.

CARBAMATE INSECTICIDES

The carbamate insecticides also inhibit insect cholinesterase. Gross symptoms from carbamate poisoning are high frequency tremors, falling, and prostration.

Some of the carbamates, primarily Sevin, have been used rather widely in predator and pest control on oyster beds. Since the Sevin is usually mixed with one or more chlorinated hydrocarbons in the treatment, it is difficult to establish the toxic effects of the carbamate. P. A. Butler *et al.* (1962) noted that 1.0 ppm of Sevin inhibits the growth of juvenile oysters. Solutions of Sevin (0.7–2.1%) in orthodichlorobenzene mixed with sand have been reported to cause swelling of the foot and eventual death of such gastropods as *Polinices, Busycon,* and *Urosalpinx* (Davis *et al.,* 1961). Pelagic crustaceans and jellyfish are killed by the above mixture when it contacts them in its descent during treatment. Starfish contacted by the treated sand particles, which lodge among the spines on the aboral surface and adhere to the surface, are irritated and move about actively, often curling their rays until they roll along the bottom with water currents.

Sevin has been used in combination with orthodichlorobenzene, Polystream, or furnace oil with dry sand as a carrier in ghost shrimp (*Callianassa*) and oyster drill control in the Pacific northwest. Numerous other invertebrates in addition to the ghost shrimp were killed, but the lethality of Sevin cannot be separated from that of the other agents. Many invertebrates were found dying immediately after application of Drillex (an experimental compound no longer manufactured that contained 98% Polystream and 2% Sevin) treated sand, including shrimp (*Crangon* sp. and *Palaemonetes* sp.), mud crabs (*Xanthidae*), and polychaetes (*Scoloplos* sp.) (Shaw and Griffith, 1967). Some oyster mortality occurred several days after application, but ceased after 2 weeks. Significantly higher numbers of oyster spat were collected on Polystream-treated and Drillex-treated oyster shell cultch than on untreated shell. Other workers (Haven *et al.,* 1966) reported that Polystream–Sevin did not control drills on treated plots, and oyster production was not increased by treatment, but the chemically treated sand killed the invertebrates in the area at the time of application.

Wood and Roberts (1966) cast serious doubt on the desirability of using Sevin in estuarine predator control. They exposed adult oyster drills (*Urosalpinx cinerea*) to various combinations of Sevin and Polystream, including each of them alone, to determine the specific effects of the two agents on *Urosalpinx* at the levels recommended for drill control. Animals treated with Sevin alone suffered low mortalities (maximum of 11% over 7 days), while the combination of Sevin and Polystream did not usually kill more drills than Polystream alone (77 and 78% in one experiment, for example), or was only slightly more effective. They also verified previous workers' reports that Sevin causes swelling of the foot of gastropods within 6 or 7 hours that lasts for 2 or 3 days. Since Sevin was not found to add materially to the control of drills but is apparently highly toxic to crustaceans, the possibility of affecting nontarget, commercially valuable crabs or shrimp either directly or by feeding on paralyzed gastropods led Wood and Roberts to recommend that Sevin not be used in Virginia, where the blue crab (*Callinectes sapidus*) supports a large fishery.

Lamprey Larvicide

A chemical known as TFM (3-trifluoromethyl-4-nitrophenol) is used to control the lamprey (*Petromyzon marinus*) in the Great Lakes of North America. The seriousness of the lamprey depredations in the Great Lakes was responsible for an extensive research program in search of suitable control devices, one of which has been the use of chemicals placed in spawning streams to kill the larvae. Since most of the streams inhabited by sea lamprey larvae also have valuable resident fish or serve as spawning and nursery areas for fish of the Great Lakes, it is important to both protect the fish from direct toxic effects of chemicals used in lamprey control and to ensure that destruction of their food chains (in the form of resident invertebrates) does not occur.

Several investigators had conclusively established that TFM is non-toxic to most species of fish at concentrations used to control larval lampreys. More recently some concern has been expressed as to the effect of TFM on aquatic invertebrates. Applegate *et al.* (1958) reported that it was not harmful to selected invertebrates included in simulated stream tests, and no deleterious effects on invertebrates were noted during actual applications in the stream. However, the taxonomic variety of invertebrates used in the simulated stream tests was quite limited, and close observation of invertebrates under field conditions is difficult. Catastrophic mortalities of freshwater stream invertebrates are less conspicuous than a comparable mortality of the vertebrate fauna and certainly elicit less public alarm.

Because of the problems in assessing the rather meager information on TFM toxicity to invertebrates, a relatively comprehensive laboratory study, consisting of bioassays of TFM toxicity to a wide range of invertebrate groups, was undertaken (Smith, 1967). Great variation in susceptibility to poisoning from TFM exists among various invertebrate groups; concentrations at which mortality approached 100% were 3 ppm for *Hydra*, 8 ppm for turbellarians, and 16 ppm for freshwater clams. Leaches of the family Erpobdellidae suffered approximately 95% mortality at 20 ppm, but only about 10% of leaches belonging to the family Glossiphoniidae died at this concentration, and almost none died at 10 ppm and lower. TFM is also virtually nontoxic to crayfish, isopods, and amphipods.

In actual control applications, concentrations are maintained between the minimum sufficient to produce 100% mortality of lamprey larvae (MLC_{100}) and the maximum concentration that kills no more than 25% of the rainbow trout (MAC_{25}). In Smith's study, the MLC_{100} was 4 ppm, and the MAC_{25} was 10 ppm. By Smith's calculations, a maximum period of exposure at 10 ppm should result in 100% mortality of the hydras and turbellarians, 89% of Erpobdellid leaches, and 50% of the freshwater clams. Even though several insect species (blackflies and mayflies) would also suffer mortalities ranging from over 25% to 100% between these levels, Smith correctly pointed out that TFM at the concentrations used for lamprey control does not constitute a hazard to most invertebrates in a stream.

Herbicides

Invertebrates may be subjected to the possible toxic effects of herbicides in any one of several ways. Terrestrial forms are commonly exposed directly by fallout at the time of application or by residues. Aquatic invertebrates usually encounter herbicides entering their environment by surface run-off or drainage, but there is also extensive and increasing use of weedicides for control of aquatic vegetation. Considerable apprehension has been expressed as to the possible deleterious effects of these compounds on species of interest or to forms serving as food for them.

Most of the interest has been directed toward aquatic forms, with the major effort involving determinations of the toxicity of various herbicidal materials to freshwater fish (Bond *et al.*, 1960). Although concern for the possible effects on invertebrate fish-food organisms is often expressed, little research appears to have been done. In estuarine waters, Ukeles (1962) showed that the floating microscopic marine plants are extremely sensitive to the substituted urea herbicides, such as monuron, diuron, and neburon, succumbing at concentrations as low as 0.5 ppb. Other herbicides are, apparently, much less toxic; 2,4-D, for example, has little effect on phytoplankton productivity even at 1.0 ppm.

Methylurea Compounds

The methylurea group of weedicides mentioned above has been tested for toxicity to clam eggs and larvae (Davis, 1961). These compounds vary greatly in their toxicity in an inverse relation to their solubility. Fenuron and monuron, which are the most soluble, are the least toxic, with no effect on the development of clam eggs and larvae at concentrations up to 5 ppm. Diuron, on the other hand, at 1.0 ppm significantly reduces the percentage of eggs developing normally, and no eggs reach the straight-hinge larval stage as normal larvae in concentrations of 50 ppm. Limited tests of neburon, the least soluble of the group, indicated that it was highly toxic since normal development of clam eggs was prevented at 2.4 ppm (the only concentration tested).

The order of toxicity of these compounds on survival and growth of clam larvae is the same as for egg development. Davis found that larvae actually grew significantly better at all concentrations of fenuron used (0.25, 0.50, 1.00, and 5.00 ppm) than in control cultures. At low concentrations, monuron also increased the growth of clam larvae, but at 1.0 ppm and higher there was evidence of toxicity. Dinuron drastically reduced the rate of growth at 5.00 ppm and caused more than 90% mortality in one culture. Neburon caused 100% mortality in concentrations as low as 2.4 ppm.

2,4-D FORMULATIONS

Information on the possible toxic effects of the various 2,4-D compounds on invertebrates is meager and sometimes contradictory. Prior to 1960, its use was essentially restricted to terrestrial and freshwater-aquatic situations, and the literature reflected this with reference primarily to toxicity to birds, mammals, and freshwater fish. Indications in these studies were that some esters or salts of 2,4-D are relatively toxic to animals, while others can be tolerated at high concentrations (up to 100 ppm). It is likely that additives and impurities also account for much of the toxicity.

More recently, 2,4-D has been utilized for control of noxious plants, particularly watermilfoil, in estuarine situations. As in freshwater, there is great variation in toxicity among the different compounds called 2,4-D. Laboratory studies have shown that Dow Silvex (Code M1847) at 1.0 ppm reduces normal development of oyster eggs by about 80% and will eventually kill all the larvae. The butoxyethanol ester (2,4-D BE) has little effect on egg development at 5.0 ppm, while 10.0 ppm reduces normal development by 75% and 1.0 ppm will kill larvae (Davis, personal communication to G. F. Beaven, cited by Rawls, 1965).

Growth of juvenile oysters is not suppressed at 2.0 ppm of 2,4-D acid or dimethylamine salt, but 3.75 ppm of 2,4-D BE caused a 50% decrease in growth rate in a 96-hour exposure (C. K. Butler, 1965).

Juvenile blue crabs are irritated by the 2,4-D dimethylamine salt at 5.0 ppm (P. A. Butler, 1963), and 2,4-D acid is harmful to fiddler crabs, *Uca pugnax* (George, 1960). Adult penaeid shrimp may suffer loss of equilibrium when exposed to 2,4-D dimethylamine salt (10% at 2.0 ppm in laboratory experiments), but are unaffected at comparable concentrations of 2,4-D BE (P. A. Butler, 1964).

Extensive field testing to determine possible toxic effects of a number of 2,4-D formulations to caged blue crabs (*Callinectes sapidus*), eastern oysters (*Crassostrea virginica*), and soft-shell clams (*Mya arenaria*) was undertaken in Chesapeake Bay where Eurasian watermilfoil has become a serious problem (Beaven *et al.*, 1962; Rawls, 1965; Rawls and Beaven, 1963). Of the compounds tested, only 2,4-D acetamide appeared to be dangerously toxic to the animals tested (Rawls, 1965). However, anaerobic conditions caused by large mats of decomposing milfoil remains on the bottom after treatment can cause severe losses to sedentary invertebrates or even mobile forms in confined areas. It is possible that death is caused by the hydrogen sulfide produced by sulfur bacteria, rather than by oxygen deficiency.

There is, unfortunately, no information available as to how the herbicides cause damage to invertebrates. Also, the potential chronic effects on invertebrates have not been studied. It should probably be reemphasized that the herbicides are relatively nontoxic; P. A. Butler (1966) listed the herbicides as unity in comparing the toxicity of various pesticides, some of which were 1000 times as toxic to certain invertebrates as the herbicides.

The Soft Detergents

Natural soap products have been largely replaced by synthetic detergents and other synthetic surfactants since World War II. Up to mid-1965 most of the detergents manufactured were of the alkylbenzene sulfate (ABS) types, which are degraded very slowly by bacterial action and, therefore, persist for long periods in receiving waters. Subsequent to June 30, 1965, industry-wide conversion to the manufacture of biodegradable linear alkylate sulfonate (LAS) detergents has been accomplished.

Because of the large volumes of detergents discharged into rivers and because of their persistence, particularly of the ABS type, these substances may occur in relatively high concentrations in estuaries where commercially valuable species of shellfish are found. Hidu (1965) investigated the effect of eight detergents on the fertilized eggs and developing larvae of marine bivalves, the hard clam (*Mercenaria mercenaria*) and the eastern oyster (*Crassostrea virginica*). He ascertained that, in general, oyster larvae are more sensitive to surfactants than clam larvae. This is in agreement with pesticide and other pollutant studies in which clams are typically more resistant. Growth of oyster and clam larvae is hindered at lower concentrations than those necessary to produce significant mortalities. Of course, diminution of larval growth rates under critical temperature conditions can be as lethal as direct toxicity, since it increases the possibilities of mortality from other causes. Fertilized eggs of clams and oysters are killed at lower concentrations than necessary to kill fully developed veliger larvae.

Since the LAS-based detergents break down rapidly when subjected to activated sludge of secondary sewage treatment, these detergents are almost completely degraded in an efficient sewage treatment plant. Therefore, most of the detergents appear in aquatic environments as degradation products, rather than as the active ingredient. Both a standard linear alkylate sulfonate (LAS) detergent and a commercial liquid LAS detergent significantly reduced the number of fertilized oyster eggs developing normally (at 0.025 mg/liter and 0.25 mg/liter, respectively). Survival and growth of larvae was decreased significantly at 1.00 and 2.50 mg/liter of the LAS and the commercial liquid detergent (Calabrese and Davis, 1967). However, there was no difference in mortality of larvae in sewage effluent containing LAS detergent degradation products and in sewage effluent without them. The toxicity of the sewage effluent, though drastic only at 200 ml/liter and higher, may have masked low-level toxic effects of the degradation products.

References

Applegate, V. C., Howell, J. H., and Smith, M. A. (1958). Use of mononitrophenols containing halogens as selective sea lamprey larvicides. *Science* **127**, 336–338.

Bearden, C. M. (1967). Field tests concerning the effects of Dibrom 14 concentrate (Naled) on estaurine animals. *Contrib. Bears Bluff Lab.* No. 45, 1–14.

Beaven, G. F., Rawls, C. K., and Beckett, G. E. (1962). Field observations upon estuarine animals exposed to 2,4-D. *Proc. Northeast. Weed Contr. Conf.* **16**, 449–458.

Bernard, F. J., and Lane, C. E. (1961). Absorption and excretion of copper ion during settlement and metamorphosis of the barnacle, *Balanus amphitrite niveus*. *Biol. Bull.* **121**, 438–448.

Bernard, F. J., and Lane, C. E. (1963). Effects of copper ion on oxygen uptake by planktonic cyprids of the barnacle, *Balanus amphitrite neveus*. *Proc. Soc. Exp. Biol. Med.* **113**, 418–420.

Bond, C. E., Lewis, R. H., and Fryer, J. L. (1960). Toxicity of various herbicidal materials to fishes. *Trans. Semin. Biol. Probl. Water Pollut. 1959* pp. 96–101.

Boyce, R., and Herdman, W. A. (1897). On a green leucocytosis in oysters associated with the presence of copper in the leucocytes. *Proc. Roy. Soc.* **62**, 30–38.

Brown, A. W. A. (1963). Chemical injuries. *Insect Pathol.* **1**, 65–131.

Butler, C. K. (1965). Field tests of herbicide toxicity to certain estuarine animals. *Chesapeake Sci.* **6**, 150–161.

Butler, P. A. (1963). Commercial fisheries investigations. Pesticide-wildlife studies: A review of fish and wildlife investigations during 1961 and 1962. *U. S., Fish Wildl. Serv., Circ.* **167**, 11–25.

Butler, P. A. (1964). Commercial fisheries investigations. *U. S., Fish Wildl. Serv., Circ.* **199**, 5–28.

Butler, P. A. (1966). Pesticides in the environment and their effects on wildlife. *J. Appl. Ecol.* **3**, Suppl., 253–259.

Butler, P. A., and Springer, P. F. (1963). Pesticides: A new factor in coastal environments. *Trans. N. Amer. Wildl. and Natur. Res. Conf.* **28**, 378–390.

Butler, P. A., Wilson, A. J., Jr., and Rick, A. J. (1962). Effect of pesticides on oysters. *Proc. Nat. Shellfish. Ass.* **51**, 23–32.

Calabrese, A., and Davis, H. C. (1967). Effects of "soft" detergents on embryos and larvae of the American oyster (*Crassostrea virginica*). *Proc. Nat. Shellfish. Ass.* **57**, 11–16.

Chin, E., and Allen, D. M. (1957). Toxicity of an insecticide to two species of shrimp, *Penaeus aztecus* and *Penaeus setiferus. Tex. J. Sci.* **9**, 270–278.

Cragg, J. B., and Vincent, M. H. (1952). The action of metaldehyde on the slug *Agriolimax reticulatus* (Muller). *Ann. Appl. Biol.* **39**, 392–406.

Davis, H. C. (1961). Effects of some pesticides on eggs and larvae of oysters (*Crassostrea virginica*) and clams (*Venus mercenaria*). *Commer. Fish. Rev.* **23**, 8–23.

Davis, H. C., Loosanoff, V. L., and MacKenzie, C. L., Jr. (1961). Field tests of a chemical method for the control of marine gastropods. Presented at the Convention of the National Shellfisheries Association, Baltimore, Maryland, August 1961. *Bur. Commer. Fish. Biol. Lab., Milford, Conn., Bull.* No. 3, Vol. 25.

deSylva, D. P. (1954). The live bait shrimp fishery of the northeast coast of Florida. *Fla. Bd. Conserv., Tech. Ser.* **11**, 1–35.

Duke, T. W., Lowe, J. I., and Wilson, A. J., Jr. (1970). A polychlorinated biphenyl (Aroclor 1254) in the water, sediment and biota of Escambia Bay, Florida. *Bull. Environ. Contam. Toxicol.* **5**, 171–180.

Fujiya, M. (1960). Studies on the effects of copper dissolved in sea water on oysters. *Bull. Jap. Soc. Sci. Fish.* **26**, 462–468.

Galtsoff, P. S. (1946). Reaction of oysters to chlorination. *U. S., Dep. Int. Res. Rep.* **11**, 1–28.

Galtsoff, P. S. (1964). The American oyster *Crassostrea virginica* Gmelin. *U. S., Fish Wildl. Serv., Fish. Bull.* **64**, 1–480.

Galtsoff, P. S., and Loosanoff, V. (1939). Natural history and method of controlling the starfish (*Asterias forbesi*, Desor). *Bull. U. S. Fish. Bur.* **49**, 75–132.

George, J. L. (1960). Some primary and secondary effects of herbicides on wildlife. *In* "Forestry Symposium," pp. 40–73. Penn. State Univ. Press, University Park, Pennsylvania.

George, J. L., Darsie, R. F., Jr., and Springer, P. F. (1957). Effects on wildlife of aerial applications of Strobane, DDT, and BHC to tidal marshes in Delaware. *J. Wild. Manage.* **21**, 42–53.

Glude, J. B. (1957). Copper, a possible barrier to oyster drills. *Proc. Nat. Shellfish. Ass.* **47**, 73–82.

Hancock, D. A. (1959). The biology and control of the American whelk tingle, *Urosalpinx cinerea* (Say), on English oyster beds. *Fish. Invest., Ser. II* **22**, 1–66.

Hanks, R. W. (1963). Chemical control of the green crab, *Carcinus maenas* (L.). *Proc. Nat. Shellfish. Ass.* **52**, 75–86.

Harry, H. W., and Aldrich, D. V. (1963). The distress syndrome in *Taphius glabratus* (Say) as a reaction to toxic concentrations of inorganic ions. *Malacologia* **1**, 283–289.

Haven, D., Castagna, M., Chanley, P., Wass, M., and Whitcomb, J. (1966). Effects of the treatment of an oyster bed with Polystream and Sevin. *Chesapeake Sci.* **7**, 179–188.

Herdman, W. A., and Boyce, R. (1899). Oysters and disease. An account of certain observations upon the normal and pathological histology and bacteriology of the oyster and other shellfish. *Lancashire Sea-Fish., Mem.* No. 1, pp. 1–60.

Hidu, H. (1965). Effects of synthetic surfactants on the larvae of clams (*M. mercenaria*) and oysters (*C. virginica*). *J. Water Pollut. Contr., Fed.* **37**, 262–270.

Hunter, A. C., and Harrison, C. W. (1928). Bacteriology and chemistry of oysters, with special reference to regulatory control of production, handling, and shipment. *U. S. Dep. Agr., Tech. Bull.* **64**, 1–75.

Jordan, H. D. (1955). Control of crabs with crude BHC. *Nature (London)* **175**, 734–735.

Lindsay, C. E. (1963). Pesticide tests in the marine environment in the state of Washington. *Proc. Nat. Shellfish. Ass.* **52**, 87–97.

Loosanoff, V. L. (1947). Effects of DDT upon setting, growth and survival of oysters. *Fishing Gaz.* **64**, 94–96.

Loosanoff, V. L., MacKenzie, C. L., Jr., and Shearer, L. W. (1960a). Use of chemicals to control shellfish predators. *Science* **131**, 1522–1523.

Loosanoff, V. L., MacKenzie, C. L., Jr., and Davis, H. C. (1960b). Progress report on chemical methods of control of molluscan enemies. Presented at the Convention of the National Shellfisheries Association, Baltimore, Maryland, August, 1960. *Bur. Commer. Fish. Biol. Lab., Milford, Conn., Bull.* No. 8.

Lowe, J. I., Wilson, P. D., and Davison, R. B. (1970). Laboratory bioassays. Progress Rept. Bur. Comm. Fish. Center for Estuarine and Menhaden Res., Pesticide Field Station, Gulf Breeze, Fla., Fiscal yr. 1969. *U. S. Fish Wildl. Serv., Circ.* **335**, 20–23.

Lowe, J. I., Wilson, P. D., Rick, A. J., and Wilson, A. J., Jr. (1971). Chronic exposure of oysters to DDT, toxaphene and parathion. *Proc. Nat. Shellfish. Ass.* **61**, 71–79.

Marks, G. H. (1938). The copper content and copper tolerance of some species of mollusks of the southern California coast. *Biol. Bull.* **75**, 224–237.

Nimmo, D. R., and Blackman, R. B. (1970). Shrimp physiology. Progress Rept. Bur. Comm. Fish. Center for Estuarine and Menhaden Res., Pesticide Field Station, Gulf Breeze, Fla., Fiscal Yr. 1969. *U. S., Fish Wildl. Serv., Circ.* **335**, 29–31.

Nimmo, D. R., Wilson, P. D., Blackman, R. B., and Wilson, A. J., Jr. (1971). Polychlorinated biphenyl absorbed from sediments by fiddler crabs and pink shrimp. *Nature (London)* **231**, 50–52.

Rawls, C. K. (1965). Field test of herbicide toxicity to certain estuarine animals. *Chesapeake Sci.* **6**, 150–161.

Rawls, C. K., and Beaven, G. F. (1963). Results of a 1962 field experiment subjecting certain estuarine animals to a 2,4-D ester. *Proc. S. Weed Conf.* **16**, 343–344.

Robbins, S. L. (1957). "Textbook of Pathology with Clinical Application." Saunders, Philadelphia, Pennsylvania.

Ryder, J. A. (1882). Notes on breeding, food and green color of oysters. *Bull. U. S. Fish. Comm.* **1**, 403–419.

Salkeld, E. H. (1950). Changes in the histology of the honey-bee ventriculus associated with the ingestion of certain insecticides. *Nature (London)* **166**, 608.

Salkeld, E. H. (1951). A toxicological and histophysiological study of certain new insecticides as "stomach poisons" to the honey bee *Apis mellifera* L. *Can. Entomol.* **83**, 39–61.

Shaw, W. N., and Griffith, G. T. (1967). Effects of polystream and Drillex on oyster setting in Chesapeake Bay and Chincoteague Bay. *Proc. Nat. Shellfish. Ass.* **57**, 17–23.

Shuster, C. N., Jr., and Garb, F. C. (1967). A note on the histopathological condition of oysters exposed to lead. Presented at the Ninth Shellfish Pathology Conference, January 27–28, 1967.

Smith, A. J. (1967). The effect of the lamprey larvicide, 3-trifluoromethyl-4-nitrophenol, on selected aquatic invertebrates. *Trans. Amer. Fish. Soc.* **96**, 410–413.

Springer, P. F. (1961). The effects on wildlife of applications of DDT and other insecticides for larval mosquito control in todal marshes of the eastern United States. Ph.D. Thesis, Cornell University, Ithaca, New York. (Abstract in *Diss. Abstr.* **22**, 1777).

Springer, P. F., and Webster, J. R. (1951). Biological effects of DDT application on tidal salt marshes. *Mosquito News* **11**, 67–74.

Ukeles, R. (1962). Effects of several toxicants on five genera of marine phytoplankton. Presented at the 1960 Convention of National Shellfisheries Association. *U. S., Fish Wildl. Serv., Circ.* **143**, p. 21 and Table 10.

Waugh, G. D. (1964). Observations on the effects of chlorine on the larvae of oysters (*Ostrea edulis* L) and barnacles (*Elminius modestus* Darwin). *Ann. Appl. Biol.* **54**, 423–440.

Waugh, G. D., and Ansell, A. (1956). The effect on oyster spatfall, of controlling barnacle settlement with DDT. *Ann. Appl. Biol.* **44**, 619–625.

Waugh, G. D., Hawes, F. B., and Williams, F. (1952). Insecticides for preventing barnacle settlement. *Ann. Appl. Biol.* **39**, 407–415.

Welsh, J. H., and Gordon, H. T. (1947). The mode of action of certain insecticides on the arthropod nerve axon. *J. Cell. Comp. Physiol.* **30**, 147–172.

Wood, L., and Roberts, B. A. (1966). Differentiation of effects of two pesticides upon *Urosalpinx cinerea* Say from the eastern shore of Virginia. *Proc. Nat. Shellfish. Ass.* **54**, 75–85.

CHAPTER V

Venom and Biotoxin Injuries

Venoms

Invertebrates may suffer damage or death from the intentional injection of poisons produced in another invertebrate. Such poisons, termed venoms, are almost always produced in special glands and often have elaborate delivery systems. Venom may be used for securing food, for defense, or for providing food and sometimes shelter for the young.

The use of venom for providing food and shelter for the young is common among the hymenopterous insects that parasitize other insects. The adult wasp paralyzes the larval host by injecting it with a toxic fluid prior to deposition of its egg on or in the host (Steinhaus, 1949). Solitary wasps use their venom both to secure food (using caterpillars or spiders) for themselves and to provide food for their developing young. Brown (1963) summarized the symptoms and effects of such wasp venoms on insects. Stung larvae quickly lose locomotor capabilities, develop a flaccid paralysis, decrease their oxygen consumption slightly, and progressively lose their responses to stimuli. Heartbeat and peristalsis continue normally, at least for several days. The toxin, which is distributed in the victim by the hemolymph, has no effect on nervous activity and apparently inhibits neither cholinesterase nor cytochrome oxidase. Since muscle contraction is not completely inactivated, it is believed that the venom acts on the neuromuscular junction rather than on the muscle.

Adult cicadas paralyzed by venom of the killer wasp, *Sphecius speciosus*, show, after approximately 1 week, histopathological changes in the nervous tissue (especially in the brain) including vacuolization and tigrolysis of the Nissl granules.

Numerous animals in addition to the insects use venom in ways similar to the insects for protection and securing food. A voluminous literature is available on the pathological effects of these venoms on man, but little research has been devoted to their effects on invertebrates. This fact is particularly interesting when one considers that most of the venoms have evolved primarily for use against invertebrates, and both the incidence of poisoning and the pathological effects are much greater among the invertebrates than the vertebrates (especially man).

Venomous Coelenterates

A distinguishing characteristic of members of the phylum Coelenterata is the possession of tentacles equipped with nematocysts to aid in capturing prey. Although there is great variation in potency (at least to vertebrates) of the nematocyst poison among various species, it is probable that all nematocysts are venomous.

Hydrozoa

Details of the morphology of *Physalia*, the Portuguese man-of-war, may be found in any textbook of invertebrate zoology. Of primary interest here are the fishing tentacles or dactylozooids, varying in number from one in *P. utriculus* to many in *P. physalia;* the tentacles contain highly venomous nematocysts located in cnidoblasts in the superficial epithelium. The toxin itself is a fluid within the nematocyst capsule that bathes the surface of the nematocyst tubule. When the tentacle touches a prey organism, stimulation of the nematocyst triggers an immediate release of the coiled nematocyst thread that is several hundred times the length of the capsule and is provided with chitinous barbs and spines and constitutes an efficient entanglement device. If the tip of the thread penetrates the prey, toxin is conveyed directly into the body through the hollow thread (Halstead, 1965). Lane (1960) observed that the thread can penetrate a surgical glove and that the number of nematocysts discharged is proportional to the size of the prey; a small copepod induces the discharge of 20–50 adjacent nematocysts, while a larger animal may elicit the discharge of several hundred thousand nematocysts.

The nematocyst toxin of *Physalia* is believed to be a highly labile protein complex (Lane and Dodge, 1958; Lane, 1960). A great deal of information is available on the effect of the isloated toxin on various vertebrates and on the gross pathology and symptoms of natural human victims, but little is known of its effect on invertebrates. It apparently paralyzes the prey and is obviously highly toxic to most invertebrates. The crude toxin, however, is nontoxic to the ciliate protozoans *Paramecium caudatum* and *Tetrahymena gelli*. Fiddler crabs, *Uca mordax*, injected with tentacular extracts exhibit clear signs of paralysis, and the venom apparently acts on the conduction or transmission of nerve impulses, since the autotomy reflex is affected, and the typical tendency to drop a leg when handled is greatly reduced. Injection of the crude toxin into fish, frogs, or mice also produces a general paralysis; it affects the nervous system, particularly the respiratory centers, and is apparently devoid of any hemolytic properties. *Physalia* toxin elicits responses of the isolated clam heart similar to those caused by acetylcholine, i.e., diastolic arrest.

Scyphozoa

Despite the fact that some of the jellyfish, especially the cubomedusae, are among the most venomous marine animals extant (Halstead, 1965), little information is available in the literature about their effects on invertebrates. No work on the toxicology or pharmacology of any schyphozoan toxin has been published, other than that of Welsh (1956), relative to the effect on the isolated horse clam heart. The most highly venomous forms, the cubomedusae, are believed to feed largely on fish, but undoubtedly attack invertebrates as well.

Anthozoa

The earliest experimentation on the effects of coelenterate venom on an invertebrate used an anemone as the source of the toxin. Cosmovici (1925a–e) injected aqueous extracts from the tentacles of the anemone *Adamsia palliata* into the legs of crabs (*Carcinus moenas*). The injections produced violent symptoms of pruritus, tetanic convulsions, paralysis, autotomy of the injected leg, cardiac fibrillation, and apparent sustained contraction of the heart muscle. The injected crabs died within one-half hour; the time of death depended on the size of the crab and the potency of the extract.

Isolated nematocyst toxin of *Metridium senile* added to seawater inhibits the ability of the snail *Littorina planaxis* to right itself and to move out of the water when placed upside down in a container. The time required for righting and withdrawal from the water depends on the concentration of the toxin; excessive doses kill the snail (Phillips and Abbott, 1957). Depression of the tendency of the fiddler crab *Uca mordax* to autotomize legs, accompanied by obvious indications of paralysis, also occurs after injection of extracts of sea anemone (*Condylactis gigantea* and *Aiptasia* sp.) tentacles (Welsh, 1956). Comparable experiments utilizing the shore crab *Hemigrapsus nudus* as the test animal and extracts prepared from the tentacles of *Metridium dianthus* produced, after injections of tentacular extracts, spontaneous autotomy of walking legs and chelae and subsequent paralysis; the extent was dependent on the dose. *Metridium* tentacular extracts produce an increase in amplitude and frequency and a tonic shortening of cardiac contractions of the isolated cardiac ventricle of the horse clam, *Schizothaerus nuttalli* (Welsh, 1956). The tentacular extracts of the jellyfish, *Cyanea*, caused an increase in frequency and more marked tonic shortening. Both extracts contain a mixture of heart excitor and inhibitor substances, with a predominance of the excitor agent, which is believed to be 5-hydroxytryptamine. Welsh suggested that relatively larger amounts of the excitor substance were indicated in *Cyanae* tentacles. Considerable information is available on the chemistry and pharmacology of coelenterate toxins and is summarized along with the effects on vertebrates in the volume by Halstead (1965).

Venomous Annelids

A number of annelids are either suspected or have been proved to possess venomous spines or fangs. Annelid biotoxins are, however, an almost unexplored area of research (Halstead, 1965). Certain species of the genera *Chloeia*, *Eurythoe*, and *Hermodice* are called bristle worms and are capable of inflicting injuries even to man by means of their pungent parapodial, bristlelike setae. For many years, Japanese fishermen have observed that flies, ants, and other insects die upon contact with dead marine polychaetes. The setae of *Eurythoe* and *Hermodice* are normally short, but when irritated, they are rapidly and remarkably extended so that the worm appears as a mass of bristles. The setae of both genera are thought to be hollow and sometimes filled with fluid, but no one has established histologically the presence of glandular elements that might produce a venom. No data are available on the effect of bristle worm stings on marine invertebrates; but the painful experience of many humans who have handled them, the above-mentioned lethal effect of dead nereid worms on insects, and tests of the only scientifically proved toxic substance isolated from an annelid [nereistoxin from *Lumbriconereis heteropodia* by Nitta (1934)] on flies, mice, and monkeys lend credence to the likelihood that some marine annelids produce a venom that serves as an effective defense mechanism.

The chemistry of nereistoxin has been studied (Hashimoto and Okaichi, 1960) and the structural formula determined (Okaichi and Hashimoto, 1962b), but little is known of its pharmacological properties except that it affects primarily the nervous system and the heart of laboratory animals. It has been shown to be more toxic to fish than to warm-blooded animals, but no information on its effect on marine inverte-brates is available.

The blood worm, *Glycera dibranchiata,* has venom glands associated with its jaws and is capable of inflicting a painful bite to man (Halstead, 1965). A number of other polychaetes have powerful jaws and also cause painful bites. Some of these, such as *Eunice aphroditois,* are sus-pected to be venomous; others, such as the giant biting polychaetes of Australia (*Onuphis*), possess powerful jaws, but whether or not a venom apparatus is present is unknown (Halstead, 1965). The role of the venom of the glycerid polychaetes has not been established, but it is probably used both offensively and defensively. The toxicity to invertebrates and the pharmacology and chemistry of the venom remain completely unknown.

Venomous Mollusks

Only two classes of Mollusca are known to contain venomous species, Gastropoda and the Cephalopoda. Venomous gastropods are placed taxonomically in the suborder Toxoglossa that contains three families: Conidae, Turridae, and Terebridae. All members of these families are believed to possess venom organs; other gastropods are known to be toxic, and some have been suspected of utilizing the poison they produce as a venom. Among the cephalopods, the octopus has been clearly shown to possess poison glands, the products of which are used as an aid in securing food.

Gastropoda

Conus spp. Members of the genus *Conus* are predaceous marine snails that inject venom into their prey prior to feeding (Kohn, 1956, 1959). The victim is paralyzed by the venom and swallowed whole; this en-ables the snail to feed on species more active than itself. The food of *Conus* consists, varying with the species, of polychaetes, other gastro-pods including other cones, pelecypods, octopods, and small fish. The morphology of the venom apparatus was described, along with an explanatory photograph and diagrams, by Kohn *et al.* (1960).

The natural food of *C. textile* is other gastropods, including other species of *Conus* (Kohn, 1959). Specimens imported from Guam and kept alive in an aquarium for periods up to 1½ years stung and fed upon a variety of gastropods collected off the coast of southern California, including *C. californicus* and several species of *Nassarius* (Kohn *et al.*, 1960).

Extracts of the venom duct of *C. textile* centrifuged and diluted with seawater and injected into the foot of *C. californicus, Nassarius tagulus*, and *N. fossatus* caused the animals to withdraw into their shells immediately after injection; often they did not reappear over a 24-hour observation period. Others extended the foot after several hours, but were unable to right themselves and died during the observation period (Kohn *et al.*, 1960). The extract of the venom duct was highly lethal to *C. californicus;* as little as 0.001 of the total venom duct extract from one *C. textile* is lethal, verifying the aquarium observations that specimens of *C. californicus* stung by *C. textile* responded similarly and suffered more than 90% mortality. Animals injected with extracts of the venom bulb and radula sheath reacted the same as controls by withdrawing into the shell immediately after injection, but reappeared within 30 minutes and fully recovered within 48 hours.

Extracts of the venom of *C. striatus* is toxic to the shore crab *Metopograpsus mesor*, killing 10 of 17 specimens injected with 0.4–10 mm^3 of venom duct contents, even though *C. striatus* feeds primarily on fish.

Cones apparently use their venom defensively in nature as well as for securing food. Kohn (1963) noted that Cummings (1936) described an incident in which an octopus attacking *C. textile* was stung. The arm (22 mm in length) was quickly withdrawn with a writhing motion. The arm was autotomized a few minutes later, and the octopus died the next day.

Symptoms in fish preceding death from cone stings consist of color change, ataxia, convulsions, or quivering and paresis. Mice injected with the venom develop ataxia, tonic spasms, dyspnea, and hyperexcitability, followed by sluggishness, paralysis, and coma prior to death. Hemorrhage or emphysema of the lungs were the only gross postmortem pathological manifestations noted (Kohn *et al.*, 1960). Death of some mice was attributed to respiratory failure followed by cardiac arrest. The toxic manifestations suggest that the principal action of the venom is interference with neuromuscular transmission, although direct action on the central nervous system is also a possibility.

Other Venomous Gastropods. The salivary glands of a number of whelks (family Buccinidae) produce a toxic secretion. These glands are not homologous to the poison gland of *Conus*, and, although their contents are discharged into the pharynx near the radular apparatus via salivary ducts, the radular teeth are not modified to purvey venom as in *Conus* and other members of the family Toxoglossa (Halstead, 1965).

Many of the whelks (*Buccinium* spp. and *Neptunea* spp.) are active, efficient predators on invertebrates, particularly other mollusks. It is possible that the salivary secretion in those forms in which it is toxic may be utilized in subduing prey. When a predaceous gastropod drills a hole through the shell of a bivalve, the victim usually gapes quickly. Although it is generally assumed that this results from the physical destruction of tissue by the predator's radula, it is equally possible that the salivary fluid paralyzes or relaxes the bivalve.

A toxic principle has been isolated from the saliva of *Neptunea arthritica;* it was identified as tetramine, one of the toxins produced by coelenterates (Asano and Ito, 1959). The concentration of the toxin in *N. arthritica* was 5–8 times that occurring in the coelenterate *Actinia equina* (Asano and Ito, 1960). The same toxin has also been reported from the salivary gland of *N. antigue* (Fange, 1960). Tetramine produces typical curarelike effects in mammals and frogs, with a drop in blood pressure, slowing of heartbeat, and a temporary paralysis of respiration. An acetylcholinelike substance has been isolated from *Buccinium undatum* and *Thais floridana* (Whittaker, 1960).

A number of species of the genus *Murex* and other members of the family Muricidae have been shown to produce a toxic secretion in the hypobranchial or purple gland. Since there is no accessory venom apparatus by which the poison may be introduced into the prey, it is difficult to accept the opinion that it is a true venom gland used for offensive and defensive purposes. However, all the muricid snails are carnivorous, typically drilling holes through the shells of bivalves, and it is not totally inconceivable that the poison is somehow introduced into the victim after the drill hole is completed.

The toxicology, pharmacology, and chemistry of the poison secreted by the hypobranchial gland have been studied in great detail by numerous investigators. Extracts prepared from the gland have been tested on leeches, crustaceans, cephalopods, and echinoderms as well as on fish, frogs, and mammals. The toxic effects are similar to those produced by curare; most organisms show immediate agitation, followed by muscular paralysis and occasionally death. Only the cephalopods, among the invertebrates tested, appeared to be unaffected by the poison (Halstead, 1965).

Cephalopoda

According to Ghiretti (1960), Salvatore Lo Bianco (1888) was the first to indicate that the octopus kills its prey by poison rather than by biting it with its chitinous jaws. Lo Bianco described the predatory activity of the octopus, noting that an octopus attacking a crab covers it with its body and tentacles in order to bring its mouth over the branchial aperature of the crab, into which it injects poison from the posterior salivary gland. Crabs taken from the octopus exhibit locomotor difficulties and violent, irregular shaking of the chelipeds and walking legs, quickly followed by immobility and death. Lo Bianco reproduced the effects experimentally by injecting octopus saliva into the gills of crabs.

MacGinitie (1942) provided additional information on the symptoms of octopus envenomation in crabs. An octopus may attack crabs larger than it is capable of surrounding with its web, in which case it hovers, tentlike, over the prey. When first seized, the crab struggles violently, often grasping the edge of the web with a cheliped. Within 20 seconds, however, the chelipeds open widely, then slowly close. Shortly thereafter, the abdomen unbends somewhat, the appendages quiver, and a brownish-colored fluid issues from the branchial canals; the crab appears to be dead within another 45 seconds. During this time, the octopus makes no attempt to use its beak on the crab.

The venom is produced in two pairs of salivary glands (anterior and posterior) and is carried to the buccal area by salivary ducts. The posterior salivary glands are much larger than the anterior and are much larger in males than females. The gross and microscopic morphology of the venom apparatus was summarized by Halstead (1965).

A number of substances have been found in the posterior salivary gland, including tyramine, histamine, acetylcholine, taurine, p-hydroxy-phenylethanolamine (octopamine), and 5-hydroxytryptamine. Each of these substances is readily diffusible, heat stable, and, at high dosages, toxic to crabs; but tyramine has been generally identified as the poison (Ghiretti, 1960). Actually, as Ghiretti pointed out, neither tyramine nor any other amine present in the salivary glands will kill a crab if injected in the quantity equal to that occurring in a lethal quantity of octopus saliva.

Octopus saliva, collected by dissection, perfusion with seawater, and electrical stimulation of the glands, is a viscous liquid containing, in addition to all the substances mentioned above, proteolytic enzymes and a hyaluronidase (Ghiretti, 1960). When a drop of saliva is injected into a crab, it produces, in succession, hyperexcitability, a quiet phase, and then paralysis. The heart and circulatory system are apparently unaffected. The saliva contains discrete factors responsible for the separate phases observed in the poisoned crab. Acetone-extracted residues (containing all the active amines of the crude saliva) injected into a crab induce the symptoms of hyperexcitability, but the quiet stage and paralysis do not occur, and the crab recovers completely within a few minutes. Injection of water-dialyzed saliva does not elicit an excitatory reaction, but the quiet and paralytic phases occur. Ghiretti (1960) isolated and purified a substance with the same paralyzing action as dialyzed saliva, which he named cephalotoxin (Ct). Its chemical composition is not yet known, but it is probably a glycoprotein.

There is considerable variation in sensitivity to cephalotoxin among different species of crustaceans; 0.1 mg of Ct per gram of tissue produces complete paralysis in 9 minutes in *Eriphia*, in 28 minutes in *Maja*, and in 65 minutes in *Squilla*. *Palinurus* and *Pagurus* are even more resistant. In all crustaceans tested, the time between injection and paralysis depends on the amount of Ct injected.

Cephalotoxin has a strong respiration inhibitory effect on crustaceans. It has also been established that Ct inhibits blood coagulation in crustaceans; the two mechanisms involved in hemostasis in Crustacea (agglutination of blood cells and coagulation of the plasma) are both inhibited in lobster blood treated with Ct.

The salivary secretions of the common European cuttlefish (*Sepia officinalis*) are purportedly toxic, as are those of the common oriental cuttlefish (*S. esculenta*), but no information is available regarding the nature of the poison (Halstead, 1965). The same is true of the squids and, perhaps, the nautiloids.

Arthropods

It is strange, considering the large number of venomous terrestrial arthropods, that no marine arthropods are known to be venomous.

Venomous Echinoderms

It has long been believed by many oyster growers and some oyster biologists that starfish apply steady suction with the tube feet while its arms surround a bivalve and also secrete a poison that enters through the opening valves, anesthetizing the mollusk and causing the adductor muscle to relax. It has been reported (Sawano and Mitsugi, 1932) that an extract of starfish stomach poured over the heart of living mollusks induced tetany and inhibited heartbeat. However, doubt has been cast on the "anesthetic" hypothesis of starfish predation by experiments showing that the effects of extracts of starfish digestive organs either poured over the heart or injected into the adductor muscle of *Mytilus edulis* were identical to the effects produced by water introduced in the same manner (Lavoie, 1956). Also, manometric measurements have shown

that starfish can exert sufficient force (up to more than 5000 g) to open
its prey by mechanical means alone. Lavoie noted that an opening of
0.1 mm between the valves was sufficient for the starfish to insert its
everted stomach and begin feeding.

Little is known of the nature of the venom apparatus or the venom of
the one known venomous starfish, *Acanthaster planci* (Halstead, 1965).
Human contact with the venomous spines, which are described by
Halstead, causes an extremely painful wound, redness, swelling, pro-
tracted vomiting, numbness, and paralysis (Fish and Cobb, 1954; Halstead, 1965). Nothing is known of the effect of the venom on invertebrates or the function in nature, though it is most likely defensive.

Many species of sea urchins possess either venomous spines or pedicellaria. Halstead (1965) reviewed historically the relatively extensive
literature on venomous pedicellaria and the clinical effects of their stings.
He also summarized what is known of the morphology of the venom
apparatus of both spines and pedicellarae.

The spines of most echinoids are solid with blunt tips and are not
venomous. Members of the families Echinothuridae and Diadematidae
have long, slender, sharp, hollow spines that penetrate deeply into most
animals contacting them and then break off in the wound. The spines
of some of the diademateds (*Diadema*) attain a length of 300 mm and
are suspected, though not experimentally demonstrated, to be venomous.
The spines of some of the echinothurida, however, are uniquely developed into venom organs. These venom organs are composed of connective and muscle tissue fibers encasing the acute tip of the spine in a
venom sac, or "poison bag." The sac contains a toxin presumably secreted
by the epithelium lining the venom sac.

Venomous pedicellariae may have the outer surface of each valve
covered by a venom gland with venom ducts emptying into the terminal
fang of the duct, as in *Toxopneuses*, or the stalk of the pedicellariae may
be encircled by the venom gland. Of the numerous types of pedicellariae,
only the globiferous ones are venomous.

Effect of the Venom on Invertebrates. The first, and one of the few,
investigations of the effects of pedicellarial venom on marine animals
(von Uexküll, 1899) noted that the small snail *Pleurobranchus meckeli*
placed on *Sphaerechinus granularis* and stung by the pedicellariae immediately rolled up into a ball and fell off the urchin. The venom, a clear
liquid that coagulates into a white mass when spread over a slide, induces a response in the octopus, *Eledone moschata*, similar to that produced by faradic (rapidly alternating currents of electricity)
stimulation.

When pedicellariae are removed from live sea urchins, macerated in seawater, and the resulting solution injected into the visceral cavity of crabs, sea cucumbers, starfish, fish, frogs, and lizards, all four types of pedicellariae are found to produce toxic substances that cause paralysis and death. The extracts from fifty pedicellariae of *Sphaerechinus* injected intravenously into an octopus caused paralysis and death. Crabs, 4–5 cm in size, die within 15–20 minutes when injected with the venom from twenty or more globiferous pedicellariae, but recover if injected with a solution containing less venom (Halstead, 1965, after Henri and Kayalof, 1906; Kayalof, 1906).

A single large globiferous pedicellaria removed with forceps from *Toxopneustes pileolus* was allowed to sting an oyster heart attached to a kymograph, and the effects were recorded (Okada, 1955; Okada *et al.*, 1955). In that experiment, the pedicellaria was observed to vigorously close its valves when the heart was contacted; this was accompanied by the ejaculation of a white, milky fluid into the heart that resulted in immediate inhibition of cardiac pulsation, followed by sustained contraction. Injections of KCl extracts of pedicellarial venom into the oyster elicits similar cardioinhibitory effects. Nothing is known of the chemistry of pedicellarial venom, and the only attempt to evaluate the pharmacological properties indicated that it contains a dialyzable acetycholine-like substance (Mendes *et al.*, 1963).

Numerous sea cucumbers (class Holothuroidea) are known to produce a toxic substance and are discussed in the next section of this chapter. Some species, however, have a protective mechanism associated with the toxin (holothurin) that is similar to a venom apparatus. Such sea cucumbers have, associated with the respiratory trees, tortuous tubular structures known as the Organs of Cuvier. Upon irritation, portions of the cuverian tubules are discharged via the anal opening. The tubules elongate and swell in seawater to form a sticky net of threads that effectively entangles predators. The Organs of Cuvier are highly toxic in some species, containing large concentrations of holothurin, a cardiac glycoside, or steroid saponin known previously only in plants (Halstead, 1965).

Very little information is available as to the toxic effects of the eviscerated cuverian tubules on invertebrates. Nigrelli and Jakowska (1960) conducted toxicity tests on crustaceans, mollusks, annelids, coelenterates, and echinoderms (including other holothurians) by placing them in a tank with one or more eviscerated sea cucumbers. They stated that a variety of reactions were elicited, ranging from simple irritability to death, but gave no details. Interestingly, the pearl fish, *Carapus bermudensis*, which lives in the cloacal and respiratory chambers of *Actin-*

opyga agassize, is highly susceptible to the holothurin produced by the eviscerated cuverian tubules of its host. The toxicology and pharmacology of holothurin is discussed in more detail in the section on biotoxins.

Biotoxins

The poisons produced by plants and animals that are not administered by a venom apparatus are also of importance in invertebrate pathology. These biotoxins may be retained in the body of the producer or released into the environment through pores of various types. Their action on the victim, therefore, may be through the gastrointestinal tract when the toxin producer is ingested or, in the case of those discharging the toxin into the environment, through the respiratory surface, general body surface, or the GI tract.

Toxigenic Algae

Widespread occurrences of toxigenic algae are found in marine, estuarine, and freshwater situations. Population explosions, the blooms or "tides," frequently cause catastrophic mortalities of invertebrates as well as fishes. Shilo (1964) reviewed selected aspects of the toxigenic algae, pointing out that the known toxin-forming algae belong to several taxonomic groups of which the dinoflagellates are the most prominent, but also includes certain blue-green algae and one species of chrysophyte.

There are a number of perplexing questions in the interrelationships of toxic dinoflagellates and invertebrates, particularly the mollusks. Some dinoflagellates, such as *Gymnodinium breve*, cause extensive mortalities of invertebrates as well as fish, apparently by the production of an exotoxin. Others, such as *Gonyaulax catenella* and *G. tamarensis*, are eaten avidly by some mollusks who store the toxin in their tissues without harm to themselves, but transvect it to their consumers; other mollusks tend to resist feeding on the toxic dinoflagellates. There is an opinion, more common in the past, the *G. catenella* is fatally toxic to mollusks in extremely heavy concentrations.

There is far more information available on the toxic effects of dinoflagellates on fish and mammals than on invertebrates. The known authenticated instances of mortalities of invertebrates caused by toxigenic algae are listed in Table V.1. The Japanese pearl oyster industry in Gokasho Bay has suffered losses as a result of plankton blooms or red tides on numerous occasions since 1893 (Nightingale, 1936). The causative organism was not identified in the earlier disasters, but Miyake

TABLE V.1

REPORTED MORTALITIES OF INVERTEBRATES FROM TOXIGENIC ALGAE

Date	Animals affected	Causative organism	Locality	Source
1891	Oysters and mussels	*Glenodinium rubrums*	Pt. Jackson, Australia	Nightingale, 1936
1893, 1904, 1910, 1926, 1933	Pearl oysters	*Gymnodinium*	Gokasho Bay, Japan	Nightingale, 1936
1901	Bottom fauna, Holothurians (*Trachostoma arenata*) Mollusks (Octopi, *Tevila crassa-teloidis*)	*Gonyaulax* sp.	California coast (San Diego to Santa Barbara)	Torrey, 1902
	Crabs (*Petrolisthes cinctipes, Cancer antennarius, Emerita analoga*)		California coast (San Diego to San Pedro)	Torrey, 1902
1907	Shellfish	*Gonyaulax polyedra*	California coast (San Pedro to San Diego)	Nightingale, 1936
1929, 1935	Oysters (*Ostrea lurida*)	*Gymnodinium splendens*	Oakland Bay, Washington	Nightingale, 1936
1933	Pearl oysters	*Gymnodinium mikimoto*	Gokasho Bay, Japan	Nightingale, 1936
1936	Oysters (*Ostrea lurida*)	*Gymnodinium splendens*	South Puget Sound	Nightingale, 1936
1946–1947	Penaeid shrimp (*Penaeus*), blue crabs (*Callinectes sapidus*), fiddler crabs (*Uca*), mud crabs, barnacles, oysters (*Crassostrea virginica*), coquinas	*Gymnodinium breve*	West Coast of Florida (Gulf of Mexico)	Galtsoff, 1949

(1935), in an unpublished summary of Japanese red water plagues, described a mass mortality in 1910 and stated that *Gymnodinium miki-moto* was probably responsible.

Although it is likely that invertebrates are frequently killed during dinoflagellate blooms, public alarm and biologists' concern over the deaths of large numbers of the more conspicuous fish usually cause the deaths of the invertebrates to go unnoticed or unrecorded. The well-known red tide in the Gulf of Mexico caused by *Gymnodinium breve*, for example, is fatal to a number of species of mollusks and crustaceans as well as fish. Reference to this fact in the literature, however, is con-fined to the general statement that barnacles, oysters, coquinas, shrimp, crabs, and horseshoe crabs are also killed by the red tide (Galtsoff, 1949; Ingle and deSylva, 1955).

The mortality pattern of invertebrates and fish from a dinoflagellate (*Gonyaulax* sp.) bloom on the California coast has been described (Torrey, 1902). The bloom was first noticed as a red streak off the entrance of San Pedro Harbor. During the next several days, as it ap-proached shore, it divided into several patches, each of which was many acres in extent. Four days after reaching shore, and 3 weeks after it was first observed, an almost unbearable sickening odor arose from the water along the beach. During that night, a large number of animals were left stranded on the beach by the outgong tide. On a 400-foot-long stretch of beach were several hundred holothurians (*Trachostoma arenata*) and several octopi, along with various fish. The fish and octopuses were dead, but many of the holothurians lived for several days. On the following day, more *Trachostoma* were left on the beach. The odor disappeared after a few days, but the water in the harbor and along the beaches was bright vermillion in color. At about this time, numerous clams (*Tevila crassatelloides*) and crabs (*Petrolisthes cinctipes* and *Cancer anten-narius*) were thrown up onto the beach. Finally, large numbers of *Emerita analoga* were left stranded. Most of the *Emerita* were alive when thrown up, but appeared debilitated and unable to burrow in their accustomed manner. Torrey believed that the *Gonyaulax* produced the lethal effects by dying in large numbers rather than through the production of a specific toxin; the resulting putrefactive changes "pol-luted" the water.

At least some of the perplexing aspects of the interrelationships be-tween toxic dinoflagellates and mollusks have been partially elucidated by recent research. Dupuy (1968) clearly showed that California mussels (*Mytilus californianus*) avidly feed on cultured *Gonyaulax catenella*, while Pacific oysters (*Crassostrea gigas*) terminate pumping and close their shells for a considerable period of time when exposed to even minimal concentrations of *G. catenella.* These data verified experimentally and explained field observations (Sparks and Sribhibhadh, 1961) that California mussels become toxic approximately 2 weeks before Pacific oysters when the two species are maintained under identical field conditions.

Recent work by Ray and Aldrich (1967) on the interactions of toxic dinoflagellates in the northern Gulf of Mexico has done even more to clarify these questions. As mentioned previously, catastrophic mortalities of invertebrates as well as fish occur relatively frequently in the northern Gulf of Mexico, but paralytic shellfish poisoning is extremely rare or nonexistent. The lack of toxic shellfish in this area is remarkable since at least two species of toxic dinoflagellates, *Gymnodinium breve* and *Gonyaulax monilata,* are endemic. Ray and Aldrich listed several possible reasons for the infrequency of shellfish poisoning in the Gulf of Mexico: "(a) failure of molluscs to feed on these dinoflagellates; (b) failure of toxin to accumulate in molluscs; (c) lack of susceptibility of higher vertebrates to the particular toxin; (d) infrequent occurrence of toxic dinoflagellate 'blooms' in areas of shellfish (mollusc) production."

The reactions and toxicity of mollusks (primarily *Crassostrea virginica,* but also including *Donax variabilis, Mercenaria campechiensis,* and *Barnea costata*) exposed to unialgal cultures of *Gymnodinum breve* and *Gonyaulax monilata* were studied in the laboratory (Ray and Aldrich, 1967). Concentrations of *Gymnodinium breve* sufficiently high to kill fish did not harm the oysters, but caused them to become lethally toxic to chicks force-fed 5 gm of oyster homogenate. In response to *Gonyaulax monilata* cultures, on the other hand, the oysters closed the valves for the duration of the 24-hour exposure to *G. monilata.* Previous experiments had shown that longer exposure to *G. monilata* caused the oysters to lose their ability to close, a condition generally indicative of impending death. As would be expected from their failure to open or pump, homogenates of the oysters challenged with *G. monilata* were nontoxic when force-fed to chicks. Of course, there is the possibility that *G. monilata* toxin is not as toxic to chicks as that of *G. breve,* a possibility that has some research validation. However, this is really immaterial since the behavioral response of the oyster precludes their transvecting the toxin. A similar response was observed in preliminary studies with the other mollusks (*Barnea costata, Donax variabilis,* and *Mercenaria campechiensis*). In each species, reduced filtering activity was indicated by direct observation or by failure of the *G. monilata* cell counts to decline. Salinity appears to be an ecological barrier to uptake of *G. breve* by oysters, since salinity levels in the estuaries where oysters are found are almost always too low to support the growth of *G. breve.*

Interestingly, *Polydora*, a polychaete inhabiting the shell of oysters, is killed by *G. monilata* cultures, but is apparently unaffected by *G. breve*.

A great deal of work has been done on the chemistry and pharmacology of the toxin responsible for paralytic shellfish poisoning. The purified toxin is a dialyzable, acid-stable, alkalilabile nitrogenous base with the molecular formula $C_{10}H_{17}O_4N_7 \cdot 2\,HCl$ and with a molecular weight of 372. Several guanidine derivatives are yielded by hydrolysis (Schantz *et al.*, 1961). The toxicity to man approaches that of *Clostridium botulinum* toxin; the lethal oral dose for man is in the range of 1–4 μg. Its effect on mammals is paralysis by neuromuscular block at myoneural junctions. The toxicity of *Gonyaulax* toxin against most invertebrates remains largely unknown, and, of course, the precise mechanism of damage in those species that are susceptible is even more enigmatic.

The toxin produced by *Gymnodinium breve* has not been studied to the extent of *Gonyaulax catenella* toxin. It has not been characterized, nor is there a procedure for its quantification (Spikes *et al.*, 1968), but it has been reported that it induces depolarization of the membranes of nerve and skeletal muscle (Sasner, 1965). There is, of course, no question that toxin from *G. breve* kills fish and, perhaps to a lesser degree, invertebrates. Most of the reported mass mortalities of marine invertebrates in other areas are reputed to have been caused by various species of *Gymnodinium*.

The toxic components of the phytoflagellate chrysomonad *Prymnesium parvum*, which have a broad spectrum of activity against many gill-breathing invertebrates (including mollusks and arthropods) and fish, have been studied and purified from algal blooms in fish ponds (Shilo, 1964). An ichthytoxin and at least one nonichthytoxic hemolysin have been shown to be separable, though they are similar in their chemical and physical properties (Shilo and Rosenberger, 1960).

TOXIC PORIFERANS

Although sponges are typically invaded and inhabited by a large variety of organisms in a wide spectrum of associations, they are only rarely utilized as food except by nudibranchs (Abeloos and Abeloos, 1932). The reasons for the sponge's relative freedom from predation have not been studied in detail, but may be related, at least in part, to the fact that many sponges produce noxious substances (Jakowska and Nigrelli, 1960).

Field observations have shown that fish, mollusks, crabs, and worms die in less than 1 hour after being placed in a bucket of seawater containing *Tedania toxicalis*, while other species of sponge observed were nontoxic under the same conditions (DeLaubenfels, 1932; Halstead, 1965, citing personal communication from DeLaubenfels). DeLaubenfels believed that *T. toxicalis* exudes a chemical substance toxic to most marine animals. A number of other species have been shown to be toxic to white mice injected intraperitoneally, including *T. ignis, Ircinia fasciculata, Spheciospongia vesparia, Dysidea etheria, Callyspongia vaginalis* (Halstead, 1965, personal communication from P. A. Zahl), and *Pseudosuberetes pseudos* (Halstead, 1965).

Toxic Coelenterates

Although coelenterates are typically venomous, there is increasing evidence that toxic substances are produced in the tentacles and other areas in addition to the venom found in the nematocyst capsule (Halstead, 1965). Thus, an invertebrate feeding on a coelenterate would not only be subject to the venomous effect of defensive nematocyst discharge, but, by absorption in the gastrointestinal tract, also to the poison present in the undischarged nematocyst capsules and other toxic materials present in the body.

As mentioned previously, some invertebrates feed on coelenterates and even utilize ingested nematocysts for their own defense. Therefore, the extent to which invertebrates are poisoned by ingestion of toxic co-elenterates is not known.

Toxic Platyhelminthes

Much of the information on the toxicity of flatworms is largely speculative, with little research having been done. Turbellarians are generally considered to have a toxic, chemical, protective mechanism to ward off predators (Halstead, 1965). It has been suggested that rhabdoidal secretions are toxic (Hyman, 1951), but laboratory confirmation is lacking. Turbellarians are also provided with primarily unicellular glands located in the mesenchyme. One in particular, the frontal gland, consists of a cluster of cyanophilous gland cells and is located near the brain ganglion; the long neck of the gland opens through a pore on the anterior tip of the body. The frontal gland is thought to be associated with the capture of food, and, therefore, it may secrete a toxin. Hyman believed the so-called "poison gland" to be a part of the sexual apparatus of the turbellarian. Although all the turbellarians apparently have toxic properties, the role of the group in invertebrate pathology remains completely unknown and awaits investigation.

TOXIC NEMERTINES

Extracts prepared from the tissues of several species of nemertine worms have been shown to be highly toxic (Halstead, 1965). As mentioned previously, there is some question as to whether a true venom apparatus exists among the nemertines. Bacq (1937) stated that the toxic material, which he called "amphiporine," is not a venom, nor is it localized in the proboscis, but is scattered throughout the tissues. Crude alcoholic extracts prepared from *Amphiporus lactifloreus, Drepanophorus cressus,* and other species and injected into crabs result in excitation, followed by prostration, and, in some cases, death. Recovery, when it occurs, requires approximately 2 hours. Bacq also isolated another toxic substance, which he called "nemertine," from *Lineus lacteus* and *L. longissimus.* Nemertine also appears to be a nerve stimulant, but is apparently less toxic than amphiporine.

There are, apparently, no records of invertebrates being poisoned by the consumption of those species of nemertine known to contain the toxic substances. Bacq also noted that amphiporine does not appear to diffuse from the worm into the surrounding water. Therefore, it is not known whether the nemertine toxins are of any significance in nature.

TOXIC ANNELIDS

Very little is known of the toxicity of annelids. In addition to the venomous jaws or pungent, bristlelike setae of some annelids described previously, a toxic substance has also been isolated from the tissues of certain polychaetes.

Japanese fishermen have known for many years that flies, ants, and other insects are killed by contact with the bodies of dead marine polychaetes (Halstead, 1965). Nitta (1934) verified the laymen's observations experimentally in one species, *Lumbriconereis heteropoda,* by testing the effects of aqueous extracts from the worm on a variety of laboratory animals. The poison, named "nereistoxin" by Nitta, has a rapid-acting anesthetic effect on a variety of insects including the housefly (*Musca domestica*), American cockroach (*Periplaneta americana*), Azuki bean weevil (*Callosobruchus chinnensis*), larvae of *Hyphantria cunea,* and the almond moth (*Ephestia catella*) (Okaichi and Hashimoto, 1962a). Apparently, the effects of nereistoxin have not been studied on any invertebrates other than insects, but it is lethal to mice and rabbits at relatively low concentrations (0.38 mg/10 gm and 1.8 mg/kg) when injected subcutaneously. Fish are even more susceptible than mammals to nereistoxin (Okaichi and Hashimoto, 1962a).

The empirical formula was postulated to be $C_5H_9NS_2$ by Nitta (1934), but the work of Hashimoto and Okaichi (1960) indicates it is actually $C_5H_{11}NS_2$. The structural formula of nereistoxin has also been worked out by these authors (Hashimoto and Okaichi, 1960; Okaichi and Hashimoto, 1962b). In view of the wide spectrum of animals to which nereistoxin is toxic, it may be assumed that nereistoxin is poisonous to some other invertebrates as well, even though supporting evidence is unavailable.

Toxic Sipunculoids

Three species, *Bonellia fulginosa*, *B. viridis*, and *Golfingia gouldi*, of sipunculoids have been shown to be toxic to a number of invertebrates. Baltzer (1925a,b) demonstrated that the skin of the introvert of the female *Bonellia* is highly toxic, but other tissues are not. Aqueous extracts of ground-up *Bonellia* introvert tissue added to aquaria in concentrations of 1:1400–1:4000 caused rapid paralysis and death in various protozoans, freshwater nematodes, and tubifex annelids.

The male *Bonellia* becomes attached to the introvert of the adult female while in the trochophore larval stage and remains as a tiny degenerate form drawing all its nourishment from the female and functioning only to fertilize the female. Baltzer believed that the extreme sexual dimorphism in *Bonellia* can be explained by the toxic effects of the female introvert tissue on the symbiont male. In his toxicity experiments, Baltzer found that an aqueous extract at a concentration of 1:3000 was lethal to adult male *Bonellia* removed from the female and placed in the aquaria. When tadpoles were kept in sublethal concentrations of *Bonellia* toxin for lengthly time periods, growth and development were inhibited, thus adding some support to Baltzer's theory. Unfortunately, Baltzer was unable to obtain enough *Bonellia* larvae to complete his investigation.

Chaet (1955) reported that the coelomocytes of *Golfingia gouldi* release a toxin when heated that is lethal to the animal. Death of the worms occurred in 90 seconds when held at 76°C. The toxin may also be released by rupturing the coelomocytes by homogenization in the absence of heat (Chaet, 1956).

Sipunculoids, presumably including the species listed as toxic, are commonly eaten by anemones, cephalopods, crabs, fish, and, in some areas, man. Whether consumption of the forms containing toxin is lethal or even harmful to the consumer is not known.

TOXIC MOLLUSKS

Mollusks are probably the most important vector of marine biotoxins to man, primarily through transvection of toxins produced by dino-flagellates (Halstead, 1965). However, they are of minor importance to invertebrate toxicology since invertebrates are apparently unaffected by consumption of toxic mollusks.

The sea hares, *Aplysia* spp., have been considered highly poisonous since Roman times; indeed, the name "sea hare" is a Roman appelation (Halstead, 1965). Modern investigations have verified the ancient beliefs, though the level of toxicity is not as high as one would expect in view of the animal's evil reputation. Contrary to laymen's beliefs, both ancient and modern, sea hares may be handled with impunity; when disturbed, however, they discharge a violet, often nauseous, fluid from their mantle. This fluid, which undoubtedly contributed to their reputation, serves a protective function, since sea anemones are the only animals known to feed on live, postmetamorphic *Aplysia* (Winkler and Tilton, 1962).

The fluid from the opaline gland of sea hares was injected into a wide variety of invertebrates: coelenterates (*Aiptasia, Anemonia, Rhopalonema, Hermiphora*); annelids (*Heteronereis* and *Asterope*); mollusks (*Nassa* and *Pleurobranchae*); arthropods (copepods and *Carcinus*); echinoderms (*Strongylocentrotus* and *Antedon*) (Flury, 1915). All animals injected developed muscular paralysis, and some were killed.

Crude aqueous and acetone extracts from the digestive glands of the California sea hare (*A. californica*) and a related species (*A. vaccaria*) have been shown to be toxic to a variety of laboratory animals, e.g., frogs, mice, chicks, guinea pigs, rats, and kittens, when injected intraperitoneally or fed via a stomach tube (Winkler, 1961). Subsequently (Winkler and Tilton, 1962), the effects of .the toxic principle, termed aplysin, on tissue preparations and living animals were studied further, and the possible chemical identity of the active molecule was suggested to be a choline ester. Aplysin initiates (displays) effects similar to acetylcholine and certain other choline esters such as succinylcholine; but aplysin is not hydrolyzed by the pseudocholinesterases, a fact that casts doubt upon the choline ester identification.

POISONOUS ARTHROPODS

Although the phylum Arthropoda contains the greatest number of species of any animal phylum, it carries, according to Halstead (1965), the dubious distinction of being the group about which the least is known of its role in biotoxicology.

POISONOUS ECHINODERMS

Asteroidea

A number of Japanese workers have studied starfish toxins, finding that several species of Japanese asteroids produce a substance that displays cardioinhibitory action on the oyster heart (Sawano and Mitsugi, 1932) and that inhibits ecdysis of fly maggots (Akiya, 1937; Ando, 1954; Ando and Hasegawa, 1955). *Astropecten, Marthasterias,* and *Echinaster,* according to Pawlowsky (1927), produce a poisonous secretion lethal to many marine animals. More recently, asterotoxin prepared from the Japanese starfish (*Asterina pectinifera*) has been shown to be a saponin and is highly toxic to the killifish, *Oryzias latipes* (Hashimoto and Yasumoto, 1960). This saponin is strongly hemolytic to vertebrates; its toxcity or mode of action on invertebrates, if any, remains uninvestigated.

An autotomizing toxin, produced by the epithelial lining cells of hepatic and aboral tissue, occurs in the coelomic fluid of *Asterias forbesi* scalded for 2 minutes in 76°C seawater (Chaet, 1962). If an arm is scalded, autotomy of that arm occurs in 8 or 9 hours; coelomic fluid taken immediately from scalded starfish and injected into normal starfish causes autotomy of the injected arm in 80% of the cases and sometimes causes the death of the recipient. The heat-stable, dialyzable toxin is also poisonous to several other starfish, a sea urchin, a marine annelid, and a horseshoe crab.

Echnoidea

Despite the large number of venomous sea urchins, few of the echinoids are reported to be poisonous. During their reproductive season, the ovaries of certain species have been reported to be toxic to man, but the group appears to be of little or no consequence in invertebrate toxicology.

Holothuroidea

Research on the toxicity of poisonous sea cucumbers was originated by Yamanouchi (Halstead, 1965) who noted, in 1929, that aqueous extracts from *Holothuria vagabunda* killed all the fish in aquaria to which they were added. Subsequently, Yamanouchi (1942, 1943a,b, 1955) investigated the chemical nature, action, and distribution of a toxic substance "holothurin" in sea cucumbers.

Aqueous extracts prepared from *H. vagabunda* are known to be toxic to a wide spectrum of invertebrates, including sea anemones (*Actinia equina, Cavenularia obesa*), crustaceans (*Pachygrapsus crassipes, Metapenaeus* sp., *Eupagarus samuelis*), mollusks (*Aplysia kurodai*), and annelids (*Eisenia foetida*), as well as marine and freshwater fish (Yamanouchi, 1955). Most animals reportedly died as a result of complete muscular paralysis. Yamanouchi found that 24 of the 27 species of Indo-Pacific and Japanese species of sea cucumbers he tested were toxic, although the level of toxicity varied with the species. The poison was found to be most concentrated in the body wall of *H. vagabunda*, but it occurred in lower concentrations in the viscera and body fluids. This is in contrast to the West Indian sea cucumber *Actinopyga agassizi*, in which most of the toxin is concentrated in the Organs of Cuvier (Nigrelli, 1952).

Holothurin has been shown (Nigrelli and Zahl, 1952) to completely inhibit growth of various species of protozoans (*Tetrahymena pyriformis, Euglena gracilis*, and *Ochromonas malhamensis*) when 2.2 mg of holothurin is added per 100 ml of nutrient liquid culture media.

Several additional reports further suggest that holothurin may have antimetabolic properties. It is lethal to planarians (*Dugesia tigrina*) in minute concentrations: between 0.00003 and 0.000001% of the crude aqueous extract (Quaglio *et al.*, 1957). Transected planarians placed in 0.01% of the crude extract for up to 1 minute exhibit normal regeneration of the anterior portions, while the posterior portions either disintegrate or fail to regenerate. Crude aqueous extracts of holothurin cause retardation of pupation in the fruit fly *Drosophila melanogaster* (Goldsmith *et al.*, 1958) and various developmental abnormalities of sea urchin (*Arbacia punctulata*) eggs, depending on the time of immersion and the dilution of the extract (Ruggieri *et al.*, 1960).

Data on the pharmacology of holothurin are scanty and apparently limited to investigations of the pharmacodynamics of the toxin on vertebrate tissue. Friess (1963) noted that purified holothurin appears to severely affect selected chemoreceptor targets and simultaneously destroys their capacity for evoking and controlling their linked physiological functions.

References

Abeloos, M., and Abeloos, R. (1932). The hepatic pigments of the nudibranch molluscs, *Doris tuberculata* Cuv., and their relations to the sponge, *Halichondria panicea*. Pall. *C. R. Soc. Biol.* **109**, 1238–1240.

Akiya, R. (1937). Starfish lees is effective for extermination of flies. (In Japanese.) *Rep. Hokkaido Fish. Lab.* No. 361, p. 10.

Ando, Y. (1954). *J. Hokkaido Fish. Sci. Inst., Ser.* 7 **11**, 29.

Ando, Y., and Hasegawa, O. (1955). Study of preventative effect of starfish lees to flies. (In Japanese.) *Rep. Hokkaido Hyg. Lab.* No. 7 p. 67.

Asano, M., and Ito, M. (1959). Occurrence of tetramine and choline compounds in the salivary glands of a marine gastropod, *Neptunea arthritica* Bernardi. *Tohoku J. Agr. Res.* **10**, 209–227.

Asano, M., and Ito, M. (1960). Salivary poison of a marine gastropod, *Neptunea arthritica* Bernardi, and the seasonal variation of its toxicity. *Ann. N. Y. Acad. Sci.* **90**, 674–688.

Bacq, Z. M. (1937). L' "amphiporine" et la "nemertine" poisons des vers nemertiens. *Arch. Int. Physiol.* **44**, 190–204.

Baltzer, F. (1925a). Ueber die giftwirkung weiblicher *Bonellia*—Gewebe auf das *Bonellia*—Männchen und andere organismen und ihre bezrehung zur bestimmung des geschlechts der Bonellienlarve. *Mitt. Naturforsch. Ges. Bern* **8**, 98–117.

Baltzer, F. (1925b). Über die giftwirkung der weiblichen *Bonellia* und ohre beziehung zur geschlechtsbestimmung der Larve. *Rev. Suisse Zool.* **32**, 87–93.

Brown, A. W. A. (1963). Chemical injuries. *Insect Pathol.* **1**, 65–131.

Chaet, A. B. (1955). Further studies on the toxic factor in *Phascolosoma*. *Biol. Bull.* **109**, 356.

Chaet, A. B. (1956). Mechanism of toxic factor release. *Biol. Bull.* **111**, 298–299.

Chaet, A. B. (1962). A toxin in the coelomic fluid of scalded starfish (*Asterias forbesi*). *Proc. Soc. Exp. Biol. Med.* **109**, 791–794.

Cosmovici, N. L. (1925a). L'action des poisons d'*Adamsia palliata* sur les muscles de *Carcinus moenas*. *C. R. Soc. Biol.* **92**, 1230–1232.

Cosmovici, N. L. (1925b). L'action des poisons d'*Adamsia palliata* sur le coeur de *Carcinus moenas*. *C. R. Soc. Biol.* **92**, 1300–1302.

Cosmovici, N. L. (1925c). Action convulsivante des poisons d'*Adamsia palliata* sur le *Carcinus moenas*. *C. R. Soc. Biol.* **92**, 1466–1469.

Cosmovici, N. L. (1925d). Autotomie chez *Carcinus moenas* provoquee par les poisons d'*Adamsia palliata*. *C. R. Soc. Biol.* **92**, 1469–1470.

Cosmovici, N. L. (1925e). Les poisons de l'extrait aqueux des tentacles et des nématocystes d'*Adamsia palliata* sont-ils detruits par l'ébullition. *C. R. Soc. Biol.* **92**, 1373–1374.

Cummings, B. (1936). Encounter between cone shell and octopus. *North Queensl. Natur.* **4**, 42.

DeLaubenfels, M. W. (1932). The marine and freshwater sponges of California. *Proc. U. S. Nat. Mus.* **81**, 1–140.

Dupuy, J. L. (1968). Isolation, culture, and ecology of a source of paralytic shellfish toxin in Sequim Bay, Washington. Ph.D. Thesis, University of Washington, Seattle.

Fange, R. (1960). The salivary gland of *Neptunea artiqua*. *Ann. N. Y. Acad. Sci.* **90**, 789–794.

Fish, C. J., and Cobb, M. C. (1954). Noxious marine animals of the central and western Pacific Ocean. *Fish Wildl. Serv.* (*U. S.*), *Res. Rep.* **36**, 21–23.

Flury, F. (1915). Über das aplysiengift. *Arch. Exp. Pathol. Pharmakol.* **79**, 250–263.

Friess, S. L. (1963). Some pharmacological activities of the sea cucumber neurotoxin. *AIBS Bull.* **13**, 41.

Galtsoff, P. S. (1949). The mystery of the red tide. *Sci. Mon.* **68**, 108–117.

Ghiretti, F. (1960). Toxicity of octopus saliva against crustacea. *Ann. N. Y. Acad. Sci.* **90**, 726–741.

Goldsmith, E. D., Osburg, H. E., and Nigrelli, R. F. (1958). The effect of holothurin, a steroid saponin of animal origin on the development of the fruit fly. *Anat. Rec.* **130**, 411–412 (abstr.).

Halstead, B. W. (1965). "Poisonous and Venomous Marine Animals of the World," Vol. I, Invertebrates. U. S. Govt. Printing Office, Washington, D. C.

Hashimoto, Y., and Okaichi, T. (1960). Some chemical properties of nereistoxin. *Ann. N. Y. Acad. Sci.* **90**, 667–673.

Hashimoto, Y., and Yasumoto, T. (1960). Confirmation of saponin as a toxic principal of starfish. *Bull. Jap. Soc. Sci. Fish.* **26**, 1132–1138.

Henri, V., and Kayalof, E. (1906). Etude des toxines contenues dans les pedicellaires chez les oursins. *C. R. Soc. Biol.* **60**, 884–886.

Hyman, L. H. (1951). "The Invertebrates: Platyhelminthes and Rhynchocoela, the Acoelomate Bilateria," Vol. II. McGraw-Hill, New York.

Ingle, R. M., and deSylva, D. P. (1955). "The Red Tide," Educ. Ser. No. 1. Bd. Conserv., State of Florida.

Jakowska, S., and Nigrelli, R. F. (1960). Antimicrobial substances from sponges. *Ann. N. Y. Acad. Sci.* **90**, 913–916.

Kayalof, E. (1906). Etude des toxines des pedicellaires chez les oursins. Thesis, Fac. Med., University of Geneve, Paris.

Kohn, A. J. (1956). Piscivorous gastropods of the genus *Conus. Proc. Nat. Acad. Sci. U. S.* **42**, 1686.

Kohn, A. J. (1959). The ecology of *Conus* in Hawaii. *Ecol. Monogr.* **29**, 42.

Kohn, A. J. (1963). Venomous marine snails of the genus *Conus.* "Venomous and Poisonous Animals and Noxious Plants of the Pacific Area," pp. 83–96. Pergamon, Oxford.

Kohn, A. J., Saunders, P. R., and Wiener, S. (1960). Preliminary studies on the venom of the marine snail *Conus. Ann. N. Y. Acad. Sci.* **90**, 706–724.

Lane, C. E. (1960). The toxin of *Physalia* nematocysts. *Ann. N. Y. Acad. Sci.* **90**, 742–750.

Lane, C. E., and Dodge, E. (1958). The toxicity of *Physalia* nematocysts. *Biol. Bull.* **115**, 219–226.

Lavoie, M. E. (1956). How sea stars open bivalves. *Biol. Bull.* **111**, 114–122.

Lo Bianco, S. (1888). Notizie biologiche reguardante specialmente il periods di maturita sessuale degli animali del Golfo di Napoli. *Mitt. Zool. Sta. Neapel* **8**, 385–440.

MacGinitie, G. E. (1942). Notes on the natural history of some marine animals. *Amer. Midl. Natur.* **19**, 213–214.

Mendes, E. G., Abbud, L., and Umiji, S. (1963). Cholinergic action of homogenates of sea urchin pedicellariae. *Science* **139**, 408–409.

Miyake, K. (1935). Conquest of the red tide, the foe of cultivated pearls. Japan. Unpublished report (from Nightingale 1936).

Nightingale, H. W. (1936). "Red Water Organisms. Their Occurrence and Influence upon Marine Aquatic Animals with Special Reference to Shellfish in Waters of the Pacific Coast." Argus Press, Seattle, Washington.

Nigrelli, R. F. (1952). The effects of holothurin on fish and mice with sarcoma 180. *Zoologica (New York)* **37**, 88–90.

Nigrelli, R. F., and Jakowska, S. (1960). Effects of holothurin, a steroid saponin from the Bahamian sea cucumber (*Actinopyga agassizi*) on various biological systems. *Ann. N. Y. Acad. Sci.* **90**, 884–892.

Nigrelli, R. F., and Zahl, P. (1952). Some biological characteristics of holothurin. *Proc. Soc. Exp. Biol. Med.* **81**, 379–380.

Nitta, S. (1934). Über nereistoxin, einen, giftigen bestandteil von *Lumbriconereis heteropoda* Marenz (Eunicidae). (In Japanese.) *J. Pharm. Soc. Jap.* **54**, 648–652.

Okado, K. (1955). Biological studies on the practical utilities of poisonous marine invertebrates. I. A preliminary note on the toxical substance detected in the trumpet sea-urchin, *Toxopneustes pileolus*. *Rec. Oceanogr. Works Jap.* **2**, 49–52.

Okado, K., Hashimoto, T., and Miyauchi, Y. (1955). A preliminary report on the poisonous effects of the *Toxopneustes* toxin upon the heart of the oyster. *Bull. Mar. Biol. Sta. Asamushi, Tohoku Univ.* **7** 133–140.

Okaichi, T., and Hashimoto, Y. (1962a). Physiological activities of nereistoxin. *Bull. Jap. Soc. Sci. Fish.* **28**, 930–935.

Okaichi, T., and Hashimoto, Y. (1962b). The structure of nereistoxin. *Agr. Biol. Chem.* **26**, 224–227.

Pawlowsky, E. N. (1927). Tiere Mit giftigen drüsen und einem besonderen verwundungsapparat. *In* "Gifttiere und inhre giftigkeit" (E. N. Pawlowsky, ed.), pp. 32–38. Fischer, Jena.

Phillips, J. H., and Abbott, D. P. (1957). Isolation and assay of the nematocyst toxin of *Metridium senile fimbriatum*. *Biol. Bull.* **113**, 296–301.

Quaglio, N. D., Noland, S. F., Veltri, A. M., Murray, P. M., Jakowska, S., and Nigrelli, R. F. (1957). Effects of holothurin on survival and regeneration of planarians. *Anat. Rec.* **128**, 604–605 (abstr.).

Ray, S. M., and Aldrich, D. V. (1967). Ecological interactions of toxic dinoflagellates and molluscs in the Gulf of Mexico. *In* "Animal Toxins," pp. 75–83. Pergamon, Oxford.

Ruggieri, G. D., Ruggieri, S. F., and Nigrelli, R. F. (1960). The effects of holothurin, a steroid saponin from the sea cucumber, on the development of the sea urchin. *Zoologica (New York)* **45**, 1–16.

Sasner, J. J. (1965). A study of the toxin produced by the Florida red tide dinoflagellate, *Gymnodinium breve* Davis. Dissertation, University of California, Los Angeles.

Sawano, E., and Mitsugi, K. (1932). Toxic action of the stomach extracts of the starfishes on the heart of the oyster. *Sci. Rep. Tohoku Univ., Ser. 4* **7**, 79–88.

Schantz, E. J., Mold, J. D., Howard, W. L., Bowden, J. P., Dutcher, J. D., Walters, D. R., and Riegel, B. (1961). Paralytic shellfish poison. VIII. Some chemical and physical properties of purified clam and mussel poisons. *Can. J. Chem.* **39**, 2117–2123.

Shilo, M. (1964). Review of toxigenic algae. *Vehr. Int. Verein. Limnol.* **15**, 782–795.

Shilo, M., and Rosenberger, R. F. (1960). Studies on the toxic principles formed by the chrysomonad *Prymnesium parvum* Carter. *Ann. N. Y. Acad. Sci.* **90**, 866–876.

Sparks, A. K., and Sribhibhadh, A. (1961). Status of paralytic shellfish toxicity studies in Washington. *Proc. Shellfish. Sanita. Workshop, Nov. 28-30 1961* pp. 266–274.

Spikes, J. J., Ray, S. M., Aldrich, D. V., and Nash, J. B. (1968). Toxicity variations of *Gymnodinium breve* cultures. *Toxicon* **5**, 171–174.

Steinhaus, E. A. (1949). "Principles of Insect Pathology." McGraw-Hill, New York.

Torrey, H. B. (1902). An unusual occurrence of Dinoflagellata on the California Coast. *Amer. Natur.* **36**, 187–192.

von Uexküll, J. (1899). Die physiologie des pedicellarien. *Z. Biol.* (*Munich*) **37**, 334–403.

Welsh, J. H. (1956). On the nature and action of coelenterate toxins. *Deep-Sea Res.* 3, Suppl., 287–297.

Whittaker, V. P. (1960). Pharmacologically active choline esters in marine gastropods. *Ann. N. Y. Acad. Sci.* **90**, 695–705.

Winkler, L. R. (1961). Preliminary tests of the toxin extracted from California sea hares of the genus *Aplysia*. *Pac. Sci.* **15**, 211–214.

Winkler, L. R., and Tilton, B. E. (1962). Predation on the California sea hare, *Aplysia californica* Cooper, by the solitary great sea anemone, *Anthopleura xanthogrammica* (Brandt) and the effect of sea hare toxin and acetylcholine on anemone muscle. *Pac. Sci.* **16**, 286–290.

Yamanouchi, T. (1942). A study of poisons contained in holothurians. (In Japanese.) *Teikoku Gakushiin Hokoku* **17**, 73.

Yamanouchi, T. (1943a). On the poison contained in *Holothuria vagabunda*. (In Japanese.) *Folia Pharmacol. Jap.* **38**, 115.

Yamanouchi, T. (1943b). Distribution of poison in the body of *Holothuria vagabunda*. (In Japanese.) *Zool. Mag.* (*Tokyo*) **55**, 87–88.

Yamanouchi, T. (1955). On the poisonous substance contained in holothurians. *Publ. Seto Mar. Biol. Lab.* **4**, 183–203.

CHAPTER VI

Pathological Effects of Ionizing Radiation

The effects of ionizing radiation on vertebrates, especially man, have been of considerable interest for many years, but concern has been greatly intensified since the advent of thermonuclear warfare and subsequent development of peaceful uses of atomic energy. However, investigations of the pathological effects on invertebrates have not been pursued with the same vigor, resulting in a paucity of knowledge in this important area.

It is beyond the scope of this volume to discuss radiation physics and radiation chemistry. The biological effect produced by exposure to an external source of radiation depends on several factors that have been summarized by Chase and Robinowitz (1967); these include the penetration or range of the radiation, the total dose received, the tissue involved, the area or volume involved, the energy of the radiation, and the time for accumulation of the dose.

In regard to the penetration of various types of radiation, alpha particles cannot penetrate deeper than about 0.1 mm of tissue and external doses would not normally penetrate even into the germinal layers of mammalian skin. Beta particles may penetrate up to a few millimeters of tissue, depending upon their energy, while gamma and X-rays are very penetrating and can cause deep-seated radiation damage. Neutrons also penetrate deeply, with tissue damage resulting from ionizations caused by recoil of the nuclei struck by the neutrons.

The dose delivered to a given volume of tissue is equal to the product of dose rate and time.

$$\text{Dose (roentgens)} = \text{Dose Rate (R/hour)} \times \text{Time (hours)}$$

Tissues vary in sensitivity to radiation. In mammals, the hematopoietic system, including the lymphatic system and bone marrow, is the most sensitive, followed in order of decreasing sensitivity by the GI system, the kidneys, and the central nervous system. Mammalian cells in order of decreasing sensitivity are lymphocytes, bone marrow cells (erythroblasts and megakaryocytes), sperm and ova, epithelial cells of the small intestine, sebaceous cells, sweat glands, eye, cartilage, liver cells, kidney cells, nerve cells, and muscle cells. There is an obvious need to establish the relative sensitivity of tissues and cells of various taxa of invertebrates to ionizing radiation, an area of research largely ignored in the past.

In regard to the area or volume involved, the smaller the area exposed to radiation, the greater the tolerance to a specific dose. Thus, whole body irradiation at a given level might be fatal, while the same dose restricted to a small area would not be lethal. Also, a given radiation dose delivered over an extended period of time produces less permanent somatic damage than the same dose delivered over a short time period.

Because the histopathological effects of irradiation and the relation to cell renewal systems is understood only in vertebrates (mainly mammals), understanding of the effects on invertebrates can best be gained by a general review of the effects on various types of mammalian tissue before considering the invertebrates.

Histopathological Effects of Irradiation on Mammalian Tissues

Most of the consequences of whole-body irradiation responsible for death in the heavily exposed mammal are mediated through a disturbance of cellular kinetics (Bond *et al.*, 1965). An irradiated organ does not fail because the remaining mature cells do not function adequately. Rather, there is no longer an adequate supply of mature functional cells, with consequent reduction in numbers below a minimum required for the organ system to function (Bond *et al.*, 1965).

Radiation lethality in mammals is due to the failure of three systems which are, in order of sensitivity, the hematopoietic system, GI system, and the central nervous system (Casarett, 1968). Thus, if an individual is exposed to a massive dose (20,000 rad*), symptoms prior to death (1–3 days) indicate that damage has occurred to the central nervous

* Units of radiation measurement are those reported by the author(s) of the cited study.

system (CNS syndrome). A smaller dose will result in lethal injury related to the GI system with survival limited to 3–5 days. Finally, the hematopoietic or bone marrow (Bond *et al.*, 1965) syndrome manifests itself after doses of 200–600 rad, and death usually occurs 3–6 weeks after irradiation.

Description of the CNS syndrome is beyond the scope of this review, and the interested reader is referred to Casarett (1968), Haley and Snider (1962), Kimeldorf and Hunt (1965), Bond *et al.* (1965), or Van Cleave (1963).

After whole-body doses that lead to the GI syndrome, early and severe changes occur in the epithelium of the small intestine. This syndrome will be described in detail (from Bloom, 1948; Bond *et al.*, 1954, 1965; Montagna and Wilson, 1956; Quastler, 1956; Ellinger, 1957; Wilson, 1959; Quastler and Hampton, 1962; Rubin and Casarett, 1968; M. F. Sullivan, 1968).

Intestinal epithelial cells arise from the division of relatively undifferentiated cells located in crypts at the villus base and then move up along the sides of the villi; the old cells are constantly being sloughed at the tips (Bloom and Fawcett, 1966). The degenerative phase that follows extremely high doses of radiation (6000 rad) results after 1 day in crypts that are shallower than normal and have only a few abnormally large cells with large, pale nuclei. Some debris is visible in the lumen, although most of it has been removed, and the villi are slightly reduced in size and have fewer cells. Two days after irradiation the crypts contain very few cells, and these are large and often have bizzare shapes. The cell population of the villi is markedly reduced and some large, abnormal cells have migrated from the crypts to the base of the villi. The tips of the villi contain cells that are irregular in size and shape, and, occasionally, whole sheets of cells are sloughed off. Three days postirradiation, occasional cell nests with no connection to the surface are all that remain of the crypts. The villi are markedly reduced in size, and the population of epithelial cells is greatly reduced. Most of the remaining cells have large, pale nuclei, and some are necrotic. By 4 days (if death has not occurred) desquamation of the villi is virtually complete; they are collapsed, and the normal architecture is lost.

Quastler *et al.* (1958) summarized the effects of irradiation on the cell renewal system. (1) Initially, irradiation interferes with the production of new cells but not with maturation rates or decay; this leads to a gradual decrease in the number of cells that form the intestinal lining. If the generative cells recover before death of the organism, they produce new cells at an abnormally rapid rate until the tissue is repopulated with a normal number of nearly normal cells. (2) In the absence of recovery, the cell population forming the intestinal lining is greatly reduced. However, the normal decay process is then modified; the remaining cells form a thin lining that persists for various lengths of time, depending on the species. (3) If mitotic arrest continues, the third phase involves the consequences to the whole organism of deterioration of the intestinal lining—complete loss is lethal within hours.

The hematopoietic or bone marrow syndrome is caused by failure of the blood-forming organs to replace blood cells either injured directly by irradiation or those that are not affected, but mature and decay normally. The following review is summarized from Bloom (1948), Jacobson (1954), Hulse (1959), Odartchenko *et al.* (1964), Fliedner *et al.* (1964), and Bond *et al.* (1965).

The primary causes of death in the acute hematopoietic syndrome are massive general infections and internal hemorrhaging. Death is caused by the loss of immune function (lymphocytes), inability to control microbial diseases (granulocytes) or to prevent hemorrhage (platelets). All blood cells are directly or indirectly affected by irradiation. Lymphocytes are killed directly in the circulation, and there is an immediate, sharp decrease in cell number. Granulocytes and/or polymorphonucleocytes are not killed in the circulation, but show a rapid initial decline due to their short life-span and the inability of the irradiated stem cells to replace lost cells. Erythrocytes and platelets are relatively radioresistant, have longer cell turnover times, and their number decreases more slowly.

Precursor cells, or stem cells, located in the bone marrow, spleen, or lymphoid tissue, are extremely radiosensitive. They are undifferentiated, have high mitotic rates, and, once irradiated, they stop proliferating. Thus, there is no replacement of any blood cell type. Platelets and red blood cells continue to function normally until they die of old age, and then their numbers also decline. Lymphocytes are killed directly and show a rapid decrease in cell number. Circulating granulocytes, erythrocytes, and platelets decrease only after existing mature cells die. No replacement of any cell type occurs because of mitotic inhibition of stem cell lines.

Cellular and Histopathological Effects of Ionizing Radiation on Invertebrates

PHYLUM PROTOZOA

Protozoa have long been favorite organisms for radiation studies because of the ease with which they can be grown in controlled cultures, thus making possible the use of large numbers of cells in essentially the same physiological conditions (Giese, 1953). Except for these animals and a very limited number of other invertebrates (insects), no other invertebrate group offers the opportunity of working with *in vitro* systems. Further advantages of using cultured protozoans are described by Giese (1967).

Reports on the effects of ionizing radiation date back to the late nineteenth century, but, until about 1930, results were often confusing or difficult to interpret for several reasons: inadequate descriptions of dose delivered, length of exposure, unit of measurement, source of radiation (even the type of radiation was often omitted), and misinterpretation of the results. Two obvious sources of confusion were the large doses necessary to kill or immobilize protozoans (LD_{50} of 350–450 kR for *Paramecium caudatum* versus 400–1200 R for most mammals), and the question of what constituted death in protozoans. Back and Halberstaedter (1945) defined the lethal dose for *Paramecium* as that dose which produced complete immobility within 10–15 minutes, but Wichterman and Figge (1954) found that immobilized paramecia ultimately became as active as controls, if examined several hours later, and subsequently divided and produced successful clones.

Studies conducted prior to 1939 will not be described due to the previously mentioned problems. Back (1939) described a delay in fission of *Paramecium caudatum* followed by normal division after exposure to low, sublethal doses of X-rays, or complete inhibition of fission and eventual death of the organism after higher doses. This study confirmed, for protozoans, what earlier workers had observed—radiation causes a delay in cell division. However, it contradicted other studies (Crowther, 1926, for example) that indicated that irradiation caused an increase in cell size and the production of "monsters." Back (1939), in fact, found that his ciliates diminished in size. It is now a well-known fact that protozoan "monsters" are caused by irradiation (Giese, 1967). Back and Halberstaedter (1945), using X-ray doses of 200–700 kR on *P. caudatum*, described the sequence of degeneration that began with altered ciliary activity, irregular locomotion, eventual cessation of movement, followed by cytolysis and death. They stated (erroneously) that complete cessation of movement served as an indication of impending death. Wichterman (1948), using *P. bursaria*, described similar alterations in ciliary action, but he also found initial size variation in surviving cells. Finally, he found, using 700 kR, a marked increase in protoplasmic viscosity, followed by irreversible coagulation. Wichterman and Figge (1954) confirmed this coagulation phenomena for *P. caudatum* after large doses of X-irradiation and thought it may have been due to functionally impaired contractile vacuoles.

Bridgeman and Kimball (1954) reported similar results after exposing *Tillina magna* and *Colpoda* sp. to X-rays. Depending on dose, they found coagulated cytoplasm, ruptured cell membranes, and morphological and behavioral alterations. Size was often diminished by extrusion of cytoplasm, normal shape was lost, and the organism assumed a spherical appearance; cyclosis was retarded, contractile vacuoles were functionally retarded, and the cilia, although intact, beat slower.

Williams (1966) described the responses of *Spathidium spathula*, which is more radiosensitive than other ciliates. With doses ranging upward from 55 kR, *S. spathula* showed one or more of the following types of behavior: (1) death of an individual cell several days after exposure—primary death; (2) death of a descendant of an original irradiated cell up to 8 days after exposure—secondary death; (3) permanent injury including one or more of the following: low division rate, giantism, loss of micronuclei, cytosome replication, heavy pigmentation, decreased motility, and the appearance of large fluid-filled vacuoles containing particulate matter; (4) temporary division retardation, followed by apparent complete recovery; (5) no apparent reaction. These changes are quite similar to those described for mammalian cells by Bacq and Alexander (1961). Williams (1966) further suggested that change in micronuclear number may be a useful measure of radiation injury in multimicronucleate ciliates. Alterations in micronuclear number, after X-irradiation, have also been reported for *Paramecium aurelia* (Geckler and Kimball, 1953), *Paramecium multimicronucleatum*, and *Paramecium caudatum* (Wichterman, 1959).

It is not practical to summarize the effects of radiation on ciliates because of the differences in sensitivity, mechanisms of repair, and organelles. The most striking cellular effects are associated with division delay, death after one to several divisions, the production of monsters, and other phenomena that are manifested when the cell cannot divide or repair itself.

Much less work has been conducted on the other classes of protozoans, and most of it is confined to ultraviolet rather than ionizing radiation. Daniels and his associates conducted many studies involving fusion of irradiated and nonirradiated *Pelomyxa* (Daniels, 1955, 1958). For a review of the effects of ultraviolet radiation on protozoans and phenomena such as photoreactivation, determination of action spectra, and the mechanism of action of UV radiation, the reader is referred to Giese (1967).

PHYLUM PORIFERA

There is only one report of the effect of ionizing radiation on sponges. Korotkova and Tokin (1965) exposed the colony-forming limestone sponges *Leucosolenia complicata* and *L. variabilis* to varying doses of beta radiation. Doses were 25, 50, and 100 krad, and observations were made 15–17 days after irradiation. They concluded that 25 krad had no influence on the rate of regeneration, while doses of 50 and 100 krad inhibited wound repair for an average of 2 or 3 days; also the number of phagocytic choanocytes decreased (due to dedifferentiation) as dose increased. Unfortunately, mortality seemed to be related to the time spent in water during irradiation and thus it seems likely that the observed changes (i.e., effects on regeneration and phagocytic activity) may be due to oxygen deprivation and subsequent anoxia rather than to the radiation. If this is true, then the report has only limited value in regard to direct radiation effects on sponges.

Phylum Coelenterata

There are few reports concerned with radiation effects in this phylum. Daniel and Park (1951) described a sequence of *Hydra* degeneration after X-irradiation that they attributed to the irradiated media rather than to direct radiation effects on cells. Daniel and Park (1953) conducted further experiments in which irradiated *Hydra* were promptly removed to an unirradiated solution. Following an 8000-R exposure, a knob appeared at the distal end of the tentacle. As disorganization proceeded, there was a loss of cells, shortening and thickening of the organ, accompanied by a loss of contractility, and eventual disappearance of the tentacle. Daniel and Park (1953) reported complete disintegration of *Hydra littoralis* after irradiation with 25,000 R. In these experiments, four hydras bearing buds in various stages of development were inadvertently included in the otherwise budless adults. The buds were only slightly damaged and continued to develop. The observation of continued development was in accord with Henshaw's (1938) finding that partially developed buds of the hydroid, *Obelia*, continued to differentiate after exposure to 25,000 R. Park (1958) found that differentiation of the hypostome, tentacles, stalk, and pedal disk occurred after irradiation with 4.5, 13, 25, and 30 kR, but initiation of buds was greatly inhibited even after 4.5 kR.

Korotkova and Tokin (1965) irradiated colonies of *Clava multicornis*, *Laomedea flexuosa*, and *Coryne loveni* that contained hydranths, stolens, and hydrorhiza; the doses of beta radiation ranged from 2 to 450 krad. They found that radiation (1) may cause disintegration of cellular systems; (2) may initiate processes of "somatic embryogenesis" (development of an entire organism from somatic cells); (3) inhibited the process of wound repair.

PHYLUM PLATYHELMINTHES

More papers have been published on the effects of irradiation on tissues in this invertebrate phylum than any others, excluding Protozoa and possibly Arthropoda. These reports are generally confined to the effects of irradiation on the planarian neoblast and regeneration. There are few papers concerned with the parasitic members of the phylum, and the information concerning effects on tissues and pathological syndromes is generally secondary to the primary observations concerned with some other phenomenon such as immunity induction. Dawes (1964) irradiated flukes (*Fasciola hepatica*) in the metacercarial stage with 3000 R and placed them in the liver of the host. Growth of irradiated flukes was greatly reduced, and Dawes attributed this to a failure of the fluke's intestine to differentiate beyond the stage at which a number of simple lateral diverticula arose on each side of the body.

Tan and Jones (1966) irradiated growing cestodes (*Hymenolepis microstoma*) *in vivo* within their hosts. They found that after extensive damage to developing proglottids of the young strobila, there was eventually complete repair, and normal strobila were regenerated. Such recovery, the authors suggested, showed that radiation is tolerated by the tissues or cells that gave rise to the recovered strobila. These resistant cells were thought to lie anterior to the more sensitive young segments.

Studies on the effect of irradiation on planarian regeneration have been conducted for more than half a century. Bardeen and Baetjer (1904) found that regeneration was completely inhibited by X-rays (doses not given). Death occurred in 20–30 days and was described as a degenerative process that began in the anterior and spread posteriorly, but no histological alterations were described. Schaper (1904) and Wiegand (1930) obtained similar results using radium. Wiegand reported that mitosis was inhibited, and disintegration of chromosomes followed. He also described "giant cells," which Dubois (1949) considered analogous to giant neoblasts filled with fragmented chromosomes that she observed in her work. Curtis and Hickman (1926) and Curtis (1928, 1936), using X-rays and radium, found that planaria continued to live for some time after irradiation, but did not regenerate. They also agreed that "embryonic formative cells" (neoblasts) were most sensitive to X-rays. Wolff and Dubois (1947) and Dubois (1949) identified the functions and abilities of the neoblasts through irradiation studies.

Dubois (1949) found that planaria receiving 8000 R remain normal for 2 weeks. Then the head thins, the eyes draw together, and the auricles become smaller. During the third week, heavily pigmented brown or black necrotic spots develop on the external edge of the auricles. Similar areas of necrosis later develop in posterior regions. The planaria swell, often become twisted, and, a few days before death, contract maximally. Disintegration of the animal occurs between the fourth and seventh week after exposure. With whole-body irradiation after amputation of the head, regeneration of the eyes (but not the head) occurs. Dubois determined that the threshold dose that inhibits regeneration lies between 2500 and 5000 R, and no regeneration occurs if the dose is greater than 8000 R. She further noted that mitosis underway at the moment of irradiation proceeds normally, but postirradiation mitoses are abnormal. Fragmentation of the chromosomes is followed by the death of the cell. Using localized irradiation, Wolff and Dubois (1947) determined that neoblasts are totipotent, have the ability to migrate (speed varies directly with temperature), and form the regenerative blastema.

Wolff and Lender (1962) have also shown that whole-body X-irradiation of planaria with doses of 5000 R prevents regeneration, and the animal dies in 3–7 weeks. Chandebois (1965a) found that the minimal dose resulting in destruction of all cells in mitosis is about 5000 R. In contrast, Hashlauer (1964) found that by subjecting planaria to lethal doses of ultraviolet light regeneration could not be completely inhibited, and small doses distributed over the entire course of regeneration significantly stimulated the regenerative growth. LeMoigne (1965) discovered that X-irradiation destroys the neoblasts during the embryonic development of the planaria and suppresses regenerative capacity without inhibiting the differentiation of the rudiments already determined at the time of irradiation.

Irradiation has played an important role in the elucidation of regeneration mechanisms in planaria. Many studies concerning regeneration and the role of the neoblast in forming the regenerative blastema have been undertaken (Wolff and Dubois, 1947; Dubois, 1949), and studies of regeneration have been reviewed by Brøndsted (1954). There is, even today, considerable controversy over the role of the neoblast in regeneration. Dubois (1949) describes the blastema as being filled with neoblasts that differentiate into epidermal cells, muscles, cerebral cells, and eye cells. Wolff and Dubois (1947) demonstrated the ability of the neoblast to migrate, using X-rays on the prepharyngeal region of a decapitated planarian. Four weeks after amputation, a new blastema was formed, and the animal regenerated normally. All animals that received whole-body irradiation died without regeneration. They explained that in the locally irradiated animals the colonization of the irradiated portion was assured by the neoblasts that migrated from the nonirradiated region and that the delay (4 weeks) was necessary for the cells to "succeed in their purpose."

Lender (1965) also described the formation of the blastema as the accumulation of neoblasts. These cells, he stated, multiply by mitosis and migrate toward the wound under the influence of a stimulus provided by the injured tissue.

Chandebois (1965b) disclaimed the "neoblast concept" of regeneration and offered instead a slightly more complex idea of a "cell transformation system." Briefly, he stated that planarians possess permanent totipotent cells in their parenchyma, but that these are fixed elements, apparently fused in a syncytium, and that they can be produced by dedifferentiation of differentiated cells. The free cells initially accumulate near the wound and show amitosis; they are finally destroyed by histolysis. They act in regeneration by providing the syncytium. Mix and Sparks (1969), in a paper to be discussed later, found that apparently normal regeneration was achieved by planaria receiving whole-body doses of ionizing radiation greater than those used by Dubois (1949) or Wolff and Lender (1962).

In a recent paper, Rose and Shostak (1968) fed vital dyes to the triclad flatworm, *Phagocata gracilis*. These dyes became concentrated and remained exclusively in the gastrodermal cells of the intact animal. Dye inclusions were identified by color and were found in 4-day postirradiation blastemas of decapitated animals that had been fed the dye. The number of inclusions increased in the parenchyma of these worms on days 1–3 after feeding, but decreased after the fourth day. Since the blastema contained only neoblasts, the authors concluded that the gastrodermal cells dedifferentiate to form neoblasts during regeneration, and the subsequent inclusions in the parenchyma of the regenerated organisms show that neoblasts originating from intestinal cells differentiated in new pathways. This paper will undoubtedly lead to a careful review of some of the older literature.

Mix and Sparks (1969) studied the gross and histopathologic effects of various levels of gamma radiation on *Dugesia tigrina*.

Gross Effects

Planarians subjected to 75 krad exhibit dramatic gross responses immediately, twisting uncontrollably and falling from the sides of the container. Within 2 hours, all negative phototropism is lost, and the irradiated animals secrete copious quantities of mucus.

A fairly definite pattern of gross pathological changes occurs in planarians subjected to radiation doses of 5 krad and above, with the time of onset dependent on the dose. Initially, abscesses appear on the body (Fig. VI.1), beginning as early as 12 hours with 75 krad, 50 hours with 50 krad, but not until day 20 with exposure to 5 krad. Subsequently, abscesses develop in the anterior regions and around the eye cups. These lesions are marked by darkly pigmented peripheries consisting of necrotic tissue (Fig. VI.2). The eyes then begin to fragment, and the head begins to lose its normal shape, eventually becoming a mass of

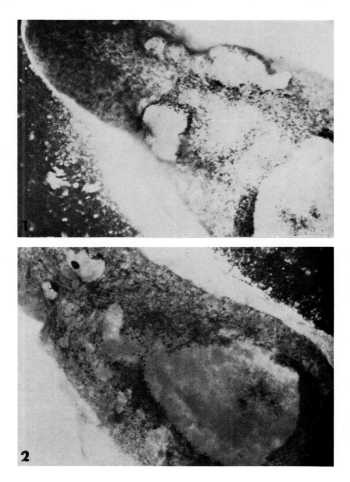

Fig. VI.1. Abscess on posterior region of planarian (*Dugesia tigrina*) subjected to gamma radiation. 2.5×. (From Mix and Sparks, 1969.)

Fig. VI.2. Abscess in middorsal region of *D. tigrina* exposed to gamma radiation. Note dark periphery consisting of necrotic tissue. 2.5×. (From Mix and Sparks, 1969.)

liquefied tissue that is subsequently shed. Occasionally, polyplike out-growths develop from the body surface.

Histopathological Effects

The first recognizable histopathological changes in planarian tissue subjected to ionizing radiation involve the neoblasts. Necrosis, character-ized by karyorrhexis and karyolysis, of the neoblasts is initiated as early as 8 hours at higher dosages and by 24 hours at all levels. Necrosis of the neoblasts causes the parenchyma to appear markedly diffuse and virtually acellular.

Radiation damage to the intestinal epithelium is dose dependent. Ex-posure to 1 and 5 krad causes no apparent effect on the gut prior to 17 days. Subsequently, the epithelium of the intestinal branches becomes intensely eosinophilic and surrounded by basophilic granular material (Fig. VI.3), but the branches retain their normal architecture. At higher levels, the epithelium is progressively reduced to a squamous cell lining or is sloughed by 30 days. The branches of the gut nearer the lateral surfaces are more severely affected at all levels than those situated more interiorly.

The parenchyma, even at low levels of irradiation, becomes markedly eosinophilic. The internal parenchyma, at 17 days, contains large amounts of an amorphous, intensely staining, acidophilic material often sur-rounded by a thin basophilic border. At higher radiation levels, struc-tures occupying the parenchyma undergo pathological changes. Dorso-ventral parenchymal muscles appear thinner than normal, and the fibrils separate after 17 days at irradiation levels of 10 krad. When epidermal–muscle borders are distinguishable at 1 month, the borders are separated by a granular, basophilic material (Fig. VI.4). The eosinophilous glands, conspicuous in the parenchyma and lateral marginal areas in the normal worm, are affected by higher doses of radiation. By 96 hours postirradia-tion with 25 krad, all eosinophilous gland cells disappear, except those at the lateral margins of the worm. A conspicuous decrease in eosino-philous glands occurs by 24 hours at 50 krad, and virtually all disappear by 150 hours after this dose. Rhabdite-producing cells, normally present in the parenchyma, are killed and frequently liberate fragmented rhab-dites into the parenchyma as they disintegrate. No rhabdite-producing cells survive beyond 17 days at doses of 10 krad and higher. Con-spicuous basophilic strands appear throughout the parenchyma; their time of appearance depends on the radiation level (24 hours at 50 krad and 150 hours at 25 krad). Eventually, the parenchyma of planarians exposed to higher levels of irradiation becomes virtually acellular (Fig. VI.5) and may appear vacuolated with large clear areas.

The response of the surface epithelium of planarians to irradiation is somewhat confusing. Some epidermal cells become hypertrophied with enlarged nuclei, while others are reduced in size.

Planarians exposed to 1 krad undergo successful regeneration. By the sixth week postirradiation, regeneration of injured tissues is nearly complete, and, by the eighth week, all tissues appear normal, and the parenchyma regains its cellular appearance. At higher levels, most animals die without apparent attempts to regenerate injured tissue.

PHYLUM ANNELIDA

Hancock (1962) exposed *Lumbricus terrestris* to X-rays and reported an $LD_{50/35}$ (50% mortality by 35 days) of 100 kR. He found no mortality in any group during the first week and no lethal effect with doses below 45 krad. The same author, Hancock (1965), stated that he induced tumors in *L. terrestris* using X-irradiation. These tumors, in Hancock's opinion, resembled myoblastomas with the neoplastic cells possibly arising from degenerating muscle fibers. "Myoblastomas" were found after 23 days in earthworms receiving 100, 400, and 1000 R. Knowing, however, that the $LD_{50/35}$ for earthworms is approximately 100 kR (Hancock, 1962) and that radiation-induced tumors are rare, it seems unlikely that 100 R (or even 1000 R) would be sufficient to induce tumor formation or to cause any deleterious effects within 23 days. Similar "myoblastomas" have since been reported by Cooper (1968) in transplant experiments and have even been found in normal worms (Cooper, personal communication).

Grass (1969) studied the effects of various doses of X-rays on the earthworm, *Eisenia foetida*. He reported that doses of 10, 50, and 100 kR produced tissue degeneration that was confined to the intestinal epithelium and body musculature. In degenerating intestinal epithelium, pycnosis, necrosis, sloughing of epithelial cells, and thickening of the basal lamina were observed. As doses increased from 10 to 100 kR, changes were qualitatively the same, but damage was more extensive. He found few, if any, changes in the stomach and attributed this fact to protection by stomach musculature. There was no repair or regeneration in any age or dose group.

PHYLUM MOLLUSCA

With the exception of LD_{50} determinations and the effects of radiation on sperm and egg cells (e.g., Sonehara, 1933), pertinent literature on histopathological effects of radiation in this phylum is restricted to oysters of the genus *Crassostrea*. Price (1962) reported the $LD_{50/30}$ for

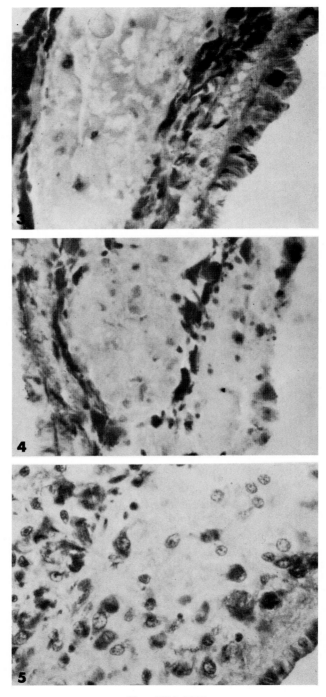

Figs. VI.3–VI.5

242

the American oyster, *Crassostrea virginica*, as 93,328 R. However, White and Angelovic (1966), using Price's data and the method of estimating 50% end points, found that the oyster was relatively resistant to radiation effects for the first 25 days, but from 25 to 50 days after irradiation they had a high mortality. They concluded that it is difficult to use $LD_{50/30}$ values as indexes of species radiation tolerance.

Investigations of the effects of ionizing radiation (gamma) on the Pacific oyster, *Crassostrea gigas*, by Mix (1970) and Mix and Sparks (1970, 1971a,b) provide information on the mechanisms of radiation injury and repair of radiation-injured tissues of oysters.

Gross Effects

Acute radiation doses above 100 krad cause mortalities, but there is a slight delay in the onset. Oysters begin dying at 5 days postirradiation, and, almost invariably, all oysters subjected to doses of 200 krad and higher succumb within 11 days postirradiation. It is clear that the threshold of acute lethal radiation lies between 100 and 200 krad, since significant mortalities do not occur below that level.

Observations of the gross pathological changes associated with acute radiation damage in oysters are obviously both subjective and, for the most part, nonspecific. Symptoms of physiological stress or impending death, such as weakening of the adductor muscle (gaping), decreased tactile response, and silt accumulation, are common to all dying oysters. More specific alterations, such as the appearance of abscesses, that appear as direct manifestations of radioinduced injury might be overlooked because of sampling problems. However, there is a relatively distinctive continuous pattern of gross response that is clearly dose dependent over a definite time interval in oysters subjected to radiation.

Oysters exposed to 1 and 5 krad exhibit no recognizable gross pathological responses. At higher dosages, the first apparent gross response is an accumulation of mucus along the periphery of the mantle edges, occurring by 48 hours after exposure to 200 and 400 krad and somewhat

---

Fig. VI.3. Necrotic intestinal branch surrounded by unidentified basophilic, granular substance in *D. tigrina* irradiated with 1 krad. 696×. Hematoxylin and eosine. (From Mix and Sparks, 1969.)

Fig. VI.4. Unidentified basophilic substance present throughout parenchyma and at epidermal–muscle borders in *D. tigrina* irradiated with 10 krad. 696×. Hematoxylin and eosine. (From Mix and Sparks, 1969.)

Fig. VI.5. Parenchyma of *D. tigrina* 150 hours after irradiation with 50 krad. Note virtual acellular appearance. Hematoxylin and eosine. 696×. (From Mix and Sparks, 1969.)

later (by 96 hours) at 10–100 krad. At the lower levels of exposure, however, mucous accumulation is transient, terminating by 144 hours postirradiation. Concomitant with mucous accumulation, tactile response decreases in oysters exposed to higher (200–400 krad) dosages, becoming apparent at 96 hours postirradiation.

The adductor muscle in oysters subjected to radiation doses of 10 krad and higher becomes progressively weakened, with the time of onset and degree of weakening directly related to the dose level. Adductor muscle weakness becomes apparent by 48 hours after exposure to 200 and 400 krad, but only after 12–14 days at radiation levels of 10–100 krad. Gross examination of the adductor muscle reveals that the muscle fibers of the white portion subjected to high levels of radiation become swollen, possibly edematous, milky colored, and easily detached from surrounding connective tissue. The dark translucent portion of the adductor muscle does not appear to be affected.

Exposure to radiation levels of 10–100 krad markedly affects the overall condition of oysters. They lose their normal color and become thin and watery with pale, rather than brown, digestive diverticula.

Abscesses, though relatively rare, may develop in irradiated oysters long after irradiation. Mix and Sparks (1970) reported four abscesses in their studies; one, adjacent to the adductor muscle, developed in an oyster subjected to 10 krad, the other three were in oysters that had received doses of 100 krad. All abscesses were large (5 mm in diameter), circular lesions of green coloration. One was situated adjacent to the heart and was present when the oyster was opened on the seventy-sixth day postirradiation, while the other two were found in the adductor muscle when the oysters were sacrificed on day 90.

Histopathology

Digestive Tubules and Collecting Ducts. The digestive tubules and collecting ducts of lethally irradiated oysters (200–400 krad) undergo early and marked histopathological changes. By 8 hours postirradiation, the distal portions of the tubule epithelial cells become swollen and then rupture, liberating acidophilic, granular debris into the lumina of the tubules. The normal architecture of the collecting duct epithelium is disrupted at 24–48 hours postinjury; nuclei are karyolytic and distal fragmentation of the cells releases cilia and cell debris into the lumina. Degeneration of the digestive tubules and collecting ducts continues, with even the basal portions sloughing after fragmenting, until (at 8 days postirradiation) most of the tubules are completely denuded. During this period, there is marked infiltration of the tubules by necrotic leukocytes.

The digestive tubules of oysters subjected to radiation levels of 20–100 krad respond similarly. Fragmentation of the distal portions of the epithelial cells begins as early as 8 hours, and, between 48 and 96 hours, most of the cells are reduced from tall columnar to low cuboidal, probably resulting from the fragmentation of the distal portions [Fig. VI.6 (normal architecture) and Fig. VI.7]. Many tubules contain conspicuous accumulations of cell debris and even entire sloughed cells. Fragmentation of tubule epithelial cells continues through 30 days postirradiation, with the severity related to dose level. Few histopathological alterations, other than decreased affinity for stains and infrequent karyolysis, occur in the collecting duct epithelium. Leukocyte infiltration of the digestive tubule and collecting duct areas (Fig. VI.8) is characteristic of the acute, nonlethal radiation syndrome. It is, however, a typical inflammatory response rather than a specific reaction to ionizing radiation. The degree of response varies from light (generally confined to the area beneath the basement membrane) to massive infiltrations into the epithelium. It is transient, beginning at about 24 hours postirradiation, increasing in intensity to 96 hours, then decreasing from 96 hours to 8–16 days, when it disappears.

Necrosis of digestive tubule epithelial cells accelerates and, by 50–60 days, presents a characteristic appearance of deeply basophilic, necrotic columnar cells (secretory absorptive cells) interspersed among pale-staining cells and apparently normal crypt cells (Fig. VI.9). Degeneration progresses, ultimately resulting in complete denudation by 60–90 days postirradiation.

The effects of low levels (1–10 krad) of radiation on the digestive tubule and collecting duct epithelium are delayed except for a nonspecific inflammatory response. By 30 days postinjury, however, the typical degeneration of the tubule epithelium begins, and denudation of the tubule occurs by 76 days.

Stomach and Gut. Localized fragmentation of the stomach of oysters irradiated with 200–400 krad begins as early as 8 hours postirradiation with rupture of the distal ends of many of the epithelial cells that releases granular, diffuse, acidophilic debris into the lumen. By 24 hours, there is some disruption in the normal architecture, with separation of cells in the basal areas and debris from fragmenting cells increasing in the lumen (Figs. VI.10 and VI.11). The normal complement of leukocytes is present in the stomach epithelium but most contain pycnotic nuclei. Leukocytic infiltration into the stomach epithelium is initiated at about 48 hours; however, virtually all the invading leukocytes are undergoing necrosis (Fig. VI.12).

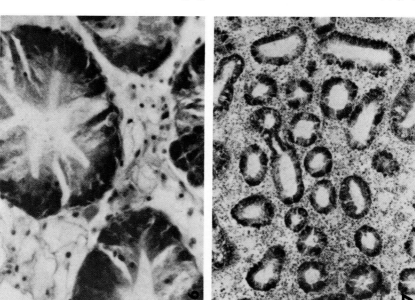

Fig. VI.6. Normal digestive tubules of the Pacific oyster, *Crassostrea gigas*. Hematoxylin and eosine. 340×. (From Mix and Sparks, 1970.)

Fig. VI.7. Digestive tubules 48 hours after irradiation with 100 krad. Note the metaplastic appearance of the tubule epithelium and the heavy leukocytic infiltration. Hematoxylin and eosine. 136×. (From Mix and Sparks, 1970.)

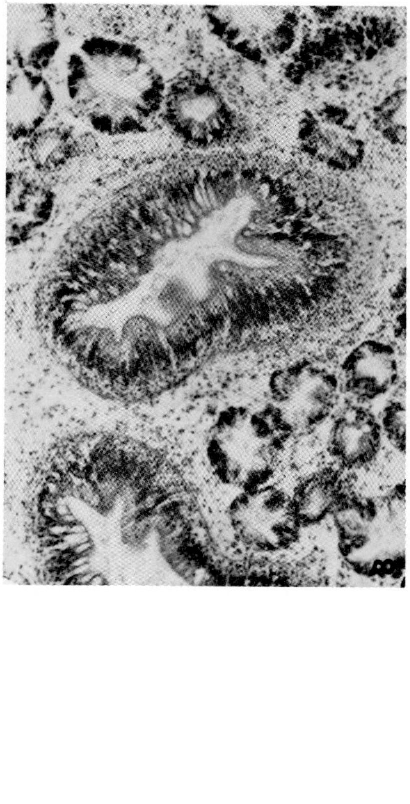

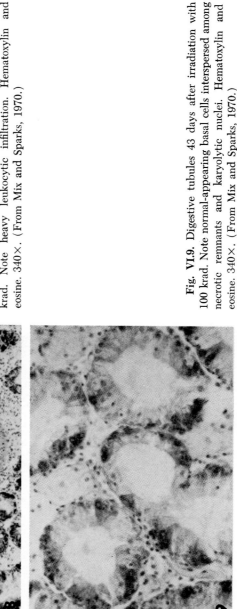

Fig. VI.8. Collecting duct 48 hours postirradiation with 100 krad. Note heavy leukocytic infiltration. Hematoxylin and eosine. 340×. (From Mix and Sparks, 1970.)

Fig. VI.9. Digestive tubules 43 days after irradiation with 100 krad. Note normal-appearing basal cells interspersed among necrotic remnants and karyolytic nuclei. Hematoxylin and eosine. 340×. (From Mix and Sparks, 1970.)

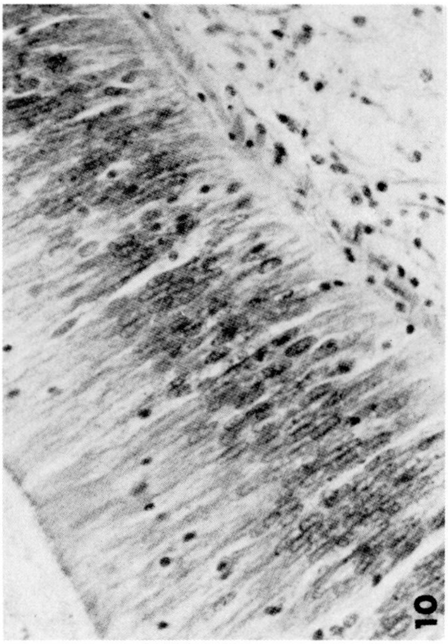

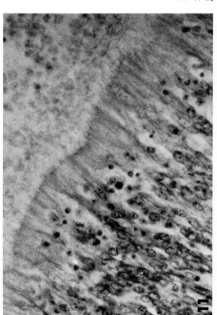

Fig. VI.10. Normal stomach of *Crassostrea gigas.* Hematoxylin and eosine. 544×. (From Mix and Sparks, 1970.)

Fig. VI.11. Stomach 24 hours after irradiation with 400 krad. Note fluid fragmentation, pycnotic leukocytes, rounded nuclei, and disrupted terminal web. Hematoxylin and eosine. 544×. (From Mix and Sparks, 1970.)

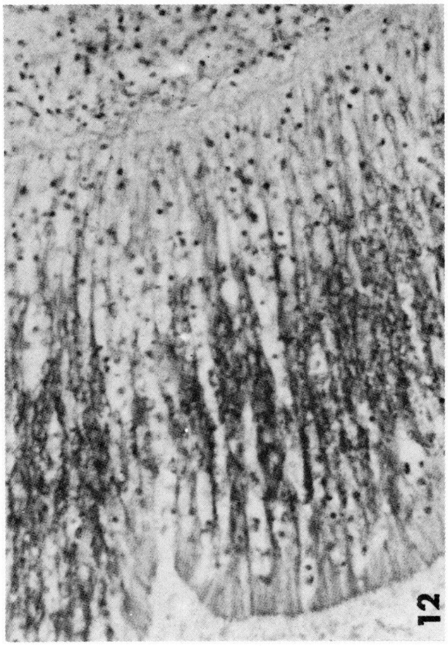

Fig. VI.12. Stomach 48 hours after irradiation with 400 krad. Note karyolytic nuclei, liquefication, and leukocytic infiltration. Hematoxylin and eosine. 340×. (From Mix and Sparks, 1970.)

Fig. VI.13. Stomach 144 hours postirradiation, 400 krad. Note complete degeneration of epithelium. Hematoxylin and eosine. 136×. (From Mix and Sparks, 1970.)

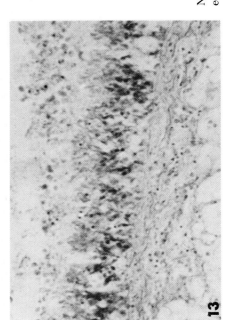

Degeneration of the stomach continues; the epithelial cells lose their affinity for stains, fragment focally, and lose virtually all semblance of architectural normality by 144 hours postirradiation (Fig. VI.13).

The gut remains normal in appearance for the first 48 hours after lethal irradiation, lacking even an inflammatory response; then it is characterized by loss of affinity for stains and karyolysis of epithelial nuclei. In marked contrast to the stomach and digestive tubules, the gut, though degenerating, retains its normal architecture through 144 hours postirradiation (Figs. VI.14 and VI.15).

Radiation damage to the stomach of oysters subjected to sublethal doses (10–100 krad) is manifested soon after exposure, with the degree of pathological response dependent on the dose received. The typical nonspecific leukocytic infiltration into the epithelium is initiated as early as 8 hours postirradiation (Figs. VI.16–VI.18). Although virtually all the leukocytes are necrotic, epithelial cell nuclei appear normal, even though some disarticulation of the basal portions does occur. By 96 hours post-irradiation, reparative mechanisms are already underway in oysters exposed to 10–50 krad, but degeneration continues at the 100-krad level, marked by a massive leukocytic infiltration of the epithelial lining of the stomach. The leukocytes, both normal and necrotic, may be interspersed uniformly among the epithelial cells or concentrated in foci resembling abscesses (Figs. VI.19 and VI.20) or ulcers (Fig. VI.21). The gut, at these levels of irradiation, exhibits similar, but less marked, pathological changes.

The stomach of oysters exposed to low levels of irradiation (1–5 krad) exhibits the typical leukocytic infiltration, but it begins somewhat later. Leukocytic foci form in the basal portion of the stomach epithelium by 48 hours (Fig. VI.20) and frequently ulcerate (Fig. VI.21), disrupting the normal architecture.

Except for a light leukocytic infiltration during the first 48 hours, no degenerative changes occur in the gut prior to 96 hours. No further damage is evident at 1 krad exposure, but the gut of oysters exposed to 5 krad develops changes similar to, but less severe than, those occurring earlier in the stomach.

Gills. Radiation doses of 200 and 400 krad cause massive damage to the gills of oysters. Leukocytic infiltration into the blood spaces and adjacent epithelium is initiated within 12 hours and accelerates through 48 hours. The gill epithelial cells rupture or slough in sheets from the filaments as the nuclei undergo necrosis. Many gill filaments become fused in the distal portions of the demibranchs after the sloughing of epithelial cells (Fig. VI.22). As many of the filaments become completely

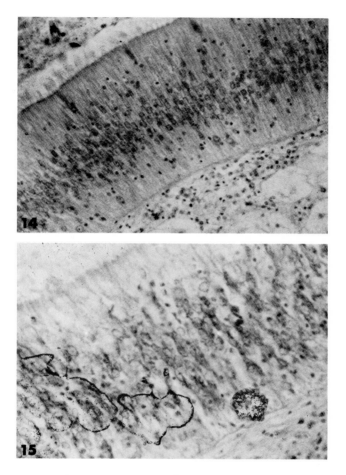

Fig. VI.14. Normal gut of Pacific oyster, *C. gigas*. Hematoxylin and eosine. 340×. (From Mix and Sparks, 1970.)

Fig. VI.15. Gut 144 hours after irradiation with 200 krad. Note normal architecture is retained, but nuclei are karyolytic. Hematoxylin and eosine. 544×. (From Mix and Sparks, 1970.)

denuded, shortening and fusion of the filaments become more pronounced (Fig. VI.23), and debris accumulates in the water tubes. By 144 hours postirradiation, pathological changes in the gills are so pronounced that total respiratory failure is probable. Complete destruction of the demibranch occurs at the 400-krad level; all epithelial cells are sloughed from the filament, leaving only the skeletal support and necrotic muscle and connective tissue (Fig. VI.24).

Sublethal (75 krad) irradiation of the gill only rarely induces acute

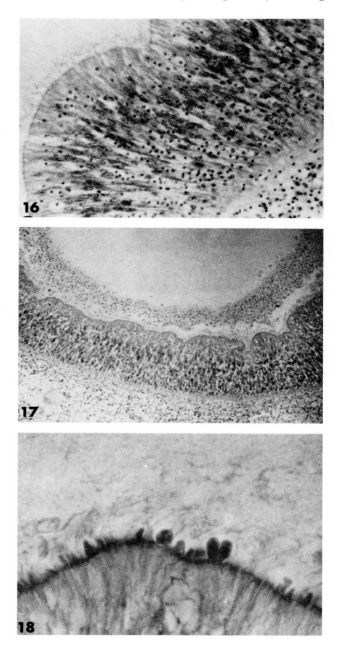

Figs.   VI.16–VI.18

cellular alterations. By 30 days postirradiation, however, gill atrophy becomes apparent and continues through 120 days. The number of lateral ciliated cells in the ordinary filament (see Galtsoff, 1964) is reduced from 3–6 cells to 2–3 cells (Fig. VI.25), and the surviving cells become hypertrophied. The ordinary filament becomes shortened, from 75–80 μm to 60–65 μm, and the epithelial cells covering the filament elongate (Fig. VI.26). The connective tissue of the gill degenerates, and the atrophied gill is infiltrated by fibroblasts that produce collagenlike fibers to replace the connective tissue.

Mantle. The primary response to irradiation in the mantle is the appearance of large numbers of fibroblasts beneath the epithelium (Figs. VI.27 and VI.28). Other nonspecific tissue changes include leukocytic infiltration into the mantle, pycnotic Leydig cells, karyolytic nuclei in Leydig cells, and sloughing of surface epithelial cells.

Gonads. Oysters of the genus *Crassostrea* are protandric hermaphrodites; the majority of *C. gigas* function first as males and then change to females during their third year (Katkansky and Sparks, 1966). The oysters irradiated by Mix and Sparks (1971b) were 3 years old and were all females undergoing gonadal development when challenged.

Twenty-four hours after irradiation at lethal levels (200–400 krad), the early developing ovocytes become pycnotic or karyolytic, and leukocytic infiltration into the gonad is initiated (Figs. VI.29 and VI.30). Leukocytic infiltration accelerates through 48 hours postirradiation, and the residual germ cells are markedly pycnotic (Fig. VI.31). More mature ova are also necrotic and are frequently sloughed into the lumen of the gonoduct. By 144 hours (Fig. VI.32), many necrotic residual cells, ovocytes, and ova slough into the lumen; the ciliated epithelium of the peripheral ducts becomes pycnotic, and the infiltrated leukocytes undergo necrosis. The germinal epithelium, gonaducts, and surrounding areas are invaded by numerous fibroblasts.

Oysters subjected to sublethal levels (1–100 krad) of ionizing radiation undergo a degenerative syndrome that follows the same sequence and differs only in time of appearance (Table VI.1).

The gonad is invaded by leukocytes within 24 hours postirradiation,

---

Fig. **VI.16.** Stomach of oyster 24 hours after irradiation with 20 krad. Note heavy infiltration of pycnotic leukocytes. Hematoxylin and eosine. 340×. (From Mix and Sparks, 1970.)

Fig. **VI.17.** Stomach 48 hours postirradiation, 50 krad. Note heavy leukocytic infiltration. Hematoxylin and eosine. 136×. (From Mix and Sparks, 1970.)

Fig. **VI.18.** Basophilic protrusions from stomach epithelial cells in an oyster 48 hours after exposure to 20 krad. Hematoxylin and eosine. 1340×. (From Mix and Sparks, 1970.)

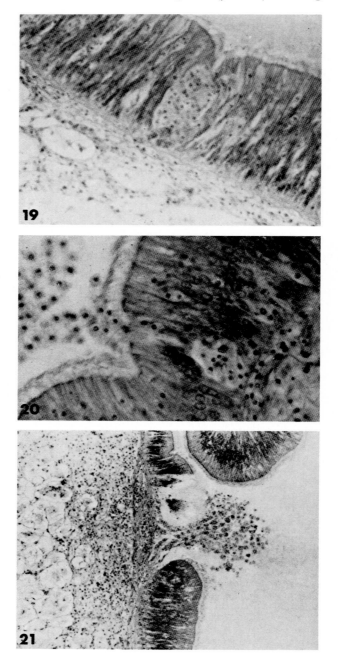

Figs.   VI.19–VI.21

and the infiltration intensifies over the next several days (Fig. VI.33). The higher dose levels inhibit maximum infiltration of leukocytes (see Table VI.1), with the response occurring later and initially much less intense than in oysters irradiated at lower doses; but the reaction increases markedly after the delay.

The developing ovocytes and residual cells undergo necrosis (Fig. VI.34) and are sloughed, leaving the germinal surfaces denuded of normal cellular covering (Fig. VI.35), but lined instead by abnormal squamous cells. Subsequently, fibroblasts invade the gonad and become the predominant cell type, eventually forming a scarlike mass consisting of fibroblasts, collagenlike material, leukocytes, and developing ova (Fig. VI.36).

The gonads contain the most actively proliferating cell population in the oyster during gonadal development; microscopic examination reveals numerous mitoses in the germinal epithelium. After all levels of irradiation, impaired cell production is demonstrated by the absence of mitotic figures by 144 hours postirradiation. Pycnotic residual cells indicate probable abortive mitoses shortly after irradiation; eventual denudation of the gonad follows irradiation with doses greater than 1 krad. Thus, two mechanisms of aplastic cytopenia occur in the germinal cells of the gonad: the death of the germinal cells due to the immediate effects of irradiation and postirradiation mitotic inhibition that eventually leads to cell death.

Tissue Repair and Regeneration

All tissues in oysters damaged by ionizing radiation exhibit varying degrees of wound repair, but only the digestive tubules have the capacity to repair and repopulate destroyed or damaged tissue, at least within 120 days postirradiation. The ultimate manifestation of digestive tubule degeneration is either complete denudation (at the 50-, 75-, and 100-krad levels) or the tubule lining being replaced with abnormal, acidophilic, squamous cells (20 krad). Mix and Sparks (1971a) described the repair

Fig. VI.19. Abscesslike collection of leukocytes in stomach epithelium 96 hours postirradiation, 100 krad. Hematoxylin and eosine. 340×. (From Mix and Sparks, 1970.)

Fig. VI.20. Large pocket of leukocytes in stomach epithelium 48 hours postirradiation with 5 krad. Note discharge of leukocytes into lumen. Hematoxylin and eosine. 540×. (From Mix and Sparks, 1970.)

Fig. VI.21. Ulceration of stomach epithelium of an oyster 8 days after irradiation with 5 krad. Note the leukocytic infiltration in underlying connective tissue, perforation of the basement membrane, and the composition of the denuded surface (homogenous, acidophilic matrix containing numerous leukocytes). Hematoxylin and eosine. 136×. (From Mix and Sparks, 1970.)

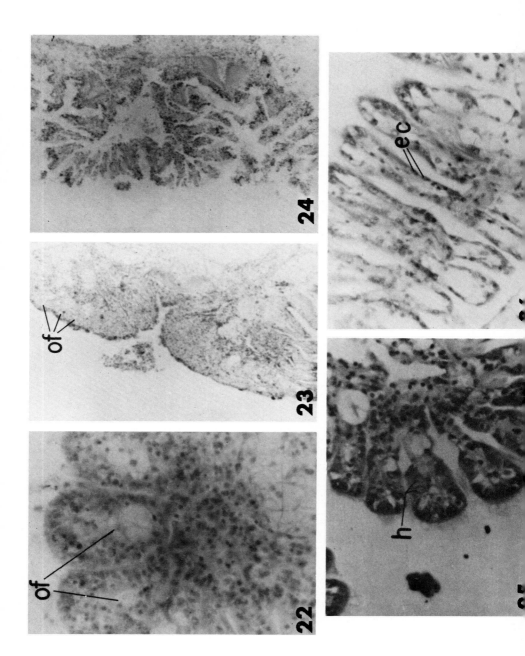

Fig. VI.22. Oyster gill filaments 24 hours postirradiation, 400 krad. Note leukocytic infiltration, early fusion of ordinary filaments (of). Hematoxylin and eosine. 480×. (From Mix, 1970.)

Fig. VI.23. Oyster gill filaments 96 hours postirradiation, 400 krad. Note complete fusion of ordinary filaments (of). Hematoxylin and eosine. 120×. (From Mix, 1970.)

Fig. VI.24. Oyster gill filaments 144 hours postirradiation, 400 krad. Note filaments are fragmented and devoid of cells. Hematoxylin and eosine. 120×. (From Mix, 1970.)

Fig. VI.25. Chronic degeneration of oyster gills after irradiation with 75 krad, 70 days postirradiation. Note decrease in number of lateral ciliated cells and hypertrophy (h) of remaining cells. Hematoxylin and eosine. 480×. (From Mix, 1970.)

Fig. VI.26. Chronic degeneration of oyster gills after irradiation with 75 krad, 80 days postirradiation. Note elongated cells (ec) covering basal portion of filament, lack of connective tissue, and decrease in number of ciliated cells. Hematoxylin and eosine. 480×. (From Mix, 1970.)

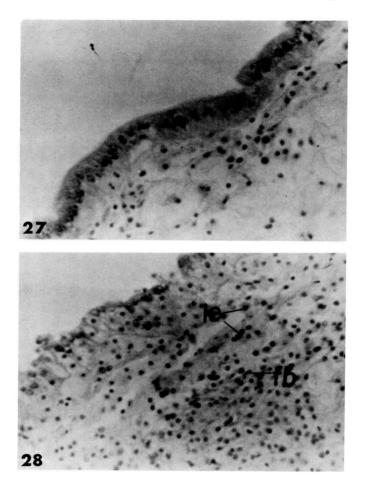

Fig. VI.27. Normal mantle of Pacific oyster. Hematoxylin and eosine. 480×.
(From Mix, 1970.)

Fig. VI.28. Oyster mantle epithelium and underlying Leydig tissue 12 days after
irradiation with 75 krad. Note infiltration of leukocytes (le) and elongated fibroblasts (fb). Mallory's triple stain. 480×. (From Mix, 1970.)

process of the digestive tubules in oysters subjected to 20 krad of ionizing
radiation.

Because of variation in the degenerative sequence at different dose
levels, in different individuals at the same dose, and even among tubules
in the same oyster, it is impossible to assign a specific time sequence to
the events constituting regeneration, except to note that it is initiated by
50–60 days postirradiation and continues through at least 90 days. Restoration of destroyed tissues requires the survival of a precursor or stem

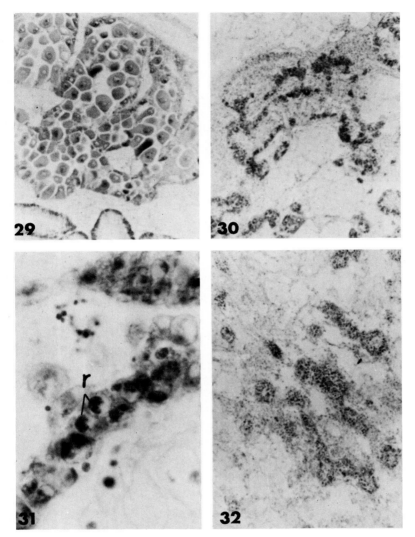

Fig. VI.29. Normal female oyster gonad with developing and maturing ova. Hematoxylin and eosine. 120×. (From Mix, 1970.)

Fig. VI.30. Female oyster gonad 24 hours after irradiation with 200 krad. Note leukocytic infiltration into the follicles. Hematoxylin and eosine. 120×. (From Mix, 1970.)

Fig. VI.31. Female oyster gonad 48 hours after irradiation with 200 krad. Note the clumped chromatin in the residual cells (r). Hematoxylin and eosine. 1200×. (From Mix, 1970.)

Fig. VI.32. Female oyster gonad 144 hours after irradiation with 200 krad. Note the necrotic leukocytes and sloughed ovocytes. Hematoxylin and eosine. 120×. (From Mix, 1970.)

TABLE VI.1

Time Sequence of Degeneration and Wound Repair (Scar Formation) in Gonads of Oysters Exposed to Various Doses of Radiation[a]

Pathological condition	1 krad	5 krad	10–20 krad	50 krad	75–100 krad
Aggregation of leukocytes and fibroblasts in blood spaces	4–12 days	4–16 days	6–16 days	6–16 days	6–16 days
Earliest period of infiltration into germinal epithelium	24 hours	48 hours	48 hours	96 hours	96 hours
Time of heaviest inflammation	144 hours	144 hours	8 days	12–30 days	12–30 days
Time germinal epithelium denuded	—	43 days	30–43 days	30–43 days	30–43 days
Degree of cicatrization					
Small gonoducts	Complete	Complete	Complete	Incomplete	Incomplete
Large gonoducts	Incomplete	Incomplete	Incomplete	Incomplete	Incomplete
Peripheral gonoducts	Complete	Complete	Complete	Complete	Incomplete
Proliferation after denudation	—	No	No	No	No

[a] From Mix (1970).

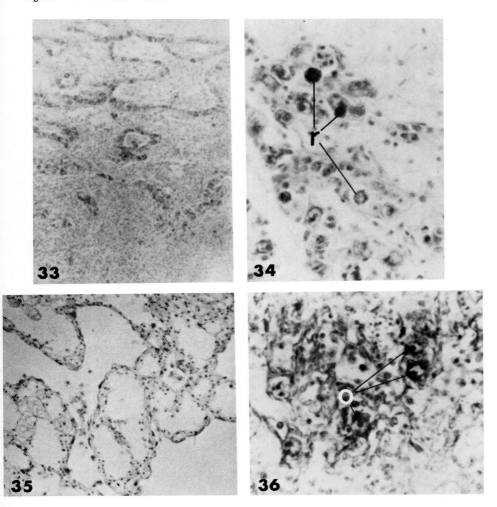

Fig. VI.33. Degeneration of the oyster gonad after irradiation with 1 krad, 96 hours post-
adiation. Note the massive leukocytic infiltration into the gonads. Hematoxylin and eosine.
0×. (From Mix, 1970.)

Fig. VI.34. Degeneration of the oyster gonad after irradiation with 1 krad, 8 days post-
adiation. Note the pycnotic residual cells (r). Hematoxylin and eosine. 480×. (From Mix,
70.)

Fig. VI.35. Completely denuded gonad of an oyster 43 days after irradiation with 50 krad.
ematoxylin and eosine. 128×. (From Mix and Sparks, 1971b.)

Fig. VI.36. Degeneration of the oyster gonad after irradiation with 1 krad, 76 hours post-
adiation. Note the fibrous scar consisting primarily of collagenlike material and the developing
a (o) amid the scar tissue. Mallory's triple stain. 480×. (From Mix, 1970.)

cell with the ability to differentiate, if necessary, and to proliferate. There are two possible sources of such cells in the radiation-injured digestive tubule. First, it is likely that some crypt cells are not injured (even though complete denudation of many tubules in the plane of section occurs in all oysters) and ultimately begin to divide. Second, it is possible that some injured crypt cells undergo subcellular repair and serve as the stem cells for reepithelization.

Digestive tubule reepithelization begins with the formation of a cell "nest" or island of crypt cells through mitosis (Fig. VI.37). Typically only one, but occasionally two, of these cell nests occurs in a digestive tubule section. Mitotic proliferation and subsequent migration continue until the entire tubule is repopulated with new epithelial cells (Figs. VI.38–VI.40). In some instances mitosis continues at the site of the original cell nests (Fig. VI.41), while in others mitoses are prominent in the cells forming the leading edge of migration (Fig. VI.42). The repopulated tubule epithelium consists entirely of large, undifferentiated crypt cells (Fig. VI.43) containing deeply staining cytoplasm and a large basal nucleus with a prominent nucleolus. Some nuclei may be distally situated if repopulation has only recently been completed. Differentiation of the epithelial cells into the normally more abundant glandular absorptive cells had not occurred by 90–120 days postirradiation when Mix and Sparks terminated their study.

Repopulation typically occurs in varying stages in an oyster, with some tubules completely repopulated, while adjacent tubules may still be denuded or in the initial stages of repair (Fig. VI.44). It is evident that all tubules are not repopulated, since oysters 90–120 days postirradiation possess only about half the normal tubule number. Therefore, acytosis of some digestive tubule populations obviously occurs.

Wound repair of radiation-destroyed gonads is basically the same at all dosage levels, but it is initiated earlier and is more complete at lower levels, especially at 1 and 5 krad. The wound repair response consists of scar formation rather than repopulation by normal cells. The cicatrization process is initiated subsequent to massive aggregation of leukocytes and fibroblasts around the periphery of the blood spaces. The aggregates form a compact capsule from which the leukocytes migrate into the injured or destroyed germinal epithelium and infiltrate the lumina of the gonoducts to form extravascular clots.

The number of leukocytes involved in this response increases as the germinal epithelium deteriorates and subsequent phagocytosis of cellular debris occurs. After the disappearance of the germinal lining, poorly defined fibers, fibroblasts, and clumped leukocytes fill the denuded gonad (Fig. VI.45) and form a scar. The smaller and peripheral gonoducts are

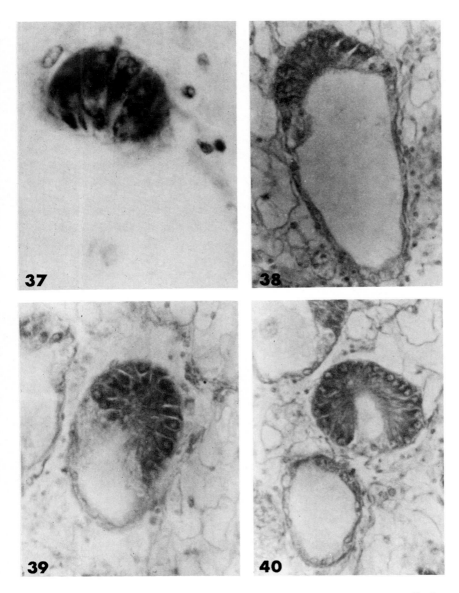

Fig. VI.37. Cell nest formed by mitotic proliferation of a surviving crypt cell after sublethal radiation. Mallory's trichrome. 1200×. (From Mix, 1970.)

Figs. VI.38, VI.39, VI.40. Subsequent migration of newly formed crypt cells around periphery of denuded tubule. Mallory's trichrome. 480×. (From Mix, 1970.)

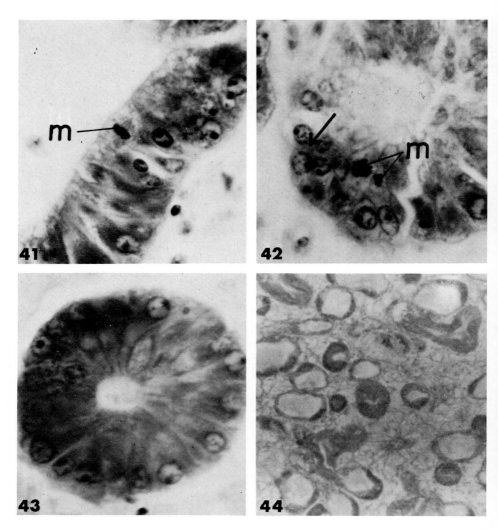

Fig. VI.41. Mitotic figure (m) at site of original cell nest in repopulating oyster digesti tubule after sublethal radiation. Mallory's. 1200×. (From Mix, 1970.)

Fig. VI.42. Oyster digestive tubule reepithelization, the leading edge of cell migration (arrow Note the large number of nuclei and the mitotic figures (m). Mallory's. 1200×. (From M 1970.)

Fig. VI.43. Completely repopulated tubule. Note that the cell population consists entirely large, undifferentiated crypt cells. Mallory's. 1200×. (From Mix, 1970.)

Fig. VI.44. Digestive tubules in various stages of repopulation 90 days after irradiation wi 75 krad. Mallory's. 120×. (From Mix, 1970.)

typically completely cicatrized at low radiation doses (Fig. VI.46), while larger ducts and those exposed to higher dosages exhibit incomplete scar formation characterized by a paucity of leukocytes and collagenlike fibers. Localized germinal proliferation is initiated at the 1-krad level after about 75 days postirradiation, with ova developing within the cicatrix.

The wound repair process of other tissues injured by ionizing radiation has not been studied in comparable detail, but in oysters irradiated with 75 krad there is progressive deposition of fibrous tissue in the gills, mantle, and beneath the GI tract. Collagen deposition is particularly conspicuous in the gills, replacing degenerated connective tissue in the filaments.

PHYLUM ARTHROPODA

Significant histopathological studies of the effects of ionizing radiation on arthropods have been confined to the insects, primarily in the development of sterile male techniques. Most adult insects are resistant to irradiation (O'Brien and Wolfe, 1964); this is attributable to the fact that most cells in adult insects are differentiated and do not undergo necrobiosis and subsequent replacement (R. L. Sullivan and Grosch, 1953). Such cells are typically much more resistant to radiation damage than dividing or undifferentiated cells (Patt and Quastler, 1963). There is, obviously, extensive proliferation in the gonad; the vulnerability of the gonia or stem cells and the resistance of most other cells to radiation damage is precisely the combination that makes the sterile male technique successful.

Additionally, there are regenerative cells in the midgut of some insects, with necrobiosis and cell replacement of the entire midgut epithelium occurring periodically throughout adult life (Snodgrass, 1935). The successful sterilization of many species of insects without reduction in longevity indicates that either cell replacement in the midguts of these insects is not greatly affected or it is not necessary for survival. However, some recent studies have demonstrated a digestive tissue syndrome after irradiation similar to that occurring in mammals. Large doses of radiation destroy the regenerative cells in the midgut of the queen honeybee (Lee, 1964), the cockroach (Mortreuil-Langlois, 1960, 1963), and the boll weevil (Mayer, 1963), leading eventually to loss of the secretory epithelium and death. Histopathological studies of the effects of ionizing radiation have not been reported from other invertebrate groups, although there have been numerous investigations of radiation effects on the gametes of echinoderms.

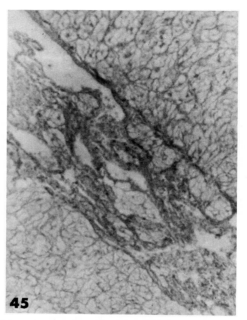

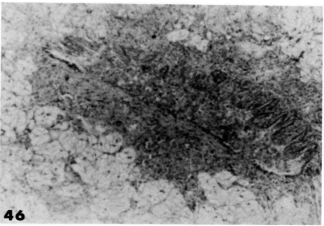

Fig. VI.45. Gonoduct 42 days postirradiation (1–5 krad). Note the gonoducts are denuded and lined with squamous cells and collagenlike fibers. Mallory's. 120×. (From Mix, 1970.)

Fig. VI.46. Gonoduct 76 days postirradiation (1–5 krad). Note the scar formed in and around large peripheral gonoduct. Mallory's. 120×. (From Mix, 1970.)

References

Back, A. (1939). Sur un type de lésions produites chez *Paramecium caudatum* par les rayons x. *C. R. Soc. Biol.* **131,** 1103–1106.

Back, A., and Halberstaedter, L. (1945). Influence of biological factors on the form of roetgen-ray survival curves. Experiments on *P. caudatum. Amer. J. Roentgenol. Radium Ther.* **54,** 290–295.

Bacq, Z. M., and Alexander, P. (1961). "Fundamentals of Radiobiology," 2nd ed. Pergamon, Oxford.

Bardeen, C. R., and Baetjer, F. H. (1904). The inhibitive action of the roentgen rays on regeneration in planarians. *J. Exp. Zool.* **1,** 191–195.

Bloom, W., ed. (1948). "Histopathology of Irradiation from External and Internal Sources," Nat. Nucl. Energy Ser. IV-221. McGraw-Hill, New York.

Bloom, W., and Fawcett, D. W. (1966). "Textbook of Histology," 8th ed. Saunders, Philadelphia, Pennsylvania.

Bond, V. P., Silverman, M. S., and Cronkite, E. P. (1954). Pathogenesis and pathology of post-irradiation infection. *Radiat. Res.* **1,** 389–400.

Bond, V. P., Fliedner, T. M., and Archambeau, J. O. (1965). "Mammalian Radiation Lethality: A Disturbance in Cellular Kinetics." Academic Press, New York.

Bridgeman, J., and Kimball, R. F. (1954). The effects of X-rays on division rate and survival of *Tillina magna* and *Colpoda* sp. with an account of delayed death. *J. Cell. Comp. Physiol.* **44,** 431–445.

Brøndsted, H. V. (1954). Planarian regeneration. *Biol. Rev.* **30,** 65–126.

Casarett, A. P. (1968). "Radiation Biology." Prentice-Hall, Englewood Cliffs, New Jersey.

Chandebois, R. (1965a). Variations du pouvoir régénérateur de planaires irradiées (*Dugesia subtenaculata*) en fonction de la dose de rayons X. *C. R. Acad. Sci.* **260,** 4834–4837.

Chandebois, R. (1965b). Cell transformation systems in planarians in regeneration. *In* "Regeneration in Animals and Related Problems," pp. 131–143. North-Holland Publ., Amsterdam.

Chase, G. P., and Robinowitz, J. L. (1967). "Principles of Radioisotope Methodology," 3rd ed. Burgess, Minneapolis, Minnesota.

Cooper, E. L. (1968). Multinucleate giant cells, granulomata, and "myoblastomas" in annelid worms. *J. Invertebr. Pathol.* **11,** 123–131.

Crowther, J. A. (1926). The action of X-rays on *Colpidium colpoda. Proc. Roy. Soc., Ser. B* **100,** 390–404.

Curtis, W. C. (1928). Old problems and a new technique. *Science* **67,** 141–149.

Curtis, W. C. (1936). Effects of X-rays and radium upon regeneration. *In* "Biological Effects of Radiation" (B. M. Duggar, ed.), Vol. 1, pp. 411–458. McGraw-Hill, New York.

Curtis, W. C., and Hickman, J. (1926). Effects of X-rays and radium upon regeneration in planarians. *Anat. Rec.* **34,** 145–146.

Daniel, G. E., and Park, H. D. (1951). The effect of x-ray treated media on hydra tentacles. *J. Cell. Comp. Physiol.* **38,** 417–426.

Daniel, G. E., and Park, H. D. (1953). Glutathione and x-ray injury in hydra and *Paramecium. J. Cell. Comp. Physiol.* **42,** 359–367.

Daniels, E. W. (1955). X-irradiation of the giant amoeba, *Pelomyxa illinoisensis.* I.

Survival and cell division following exposure. Therapeutic effects of whole protoplasm. *J. Exp. Zool.* **130**, 183–197.

Daniels, E. W. (1958). X-irradiation of the giant amoeba, *Pelomyxa illinoisensis*. II. Further studies on recovery following supra-lethal exposure. *J. Exp. Zool.* **137**, 425–442.

Dawes, B. (1964). A preliminary study of the prospect of inducing immunity in fascioliasis by means of infections with X-irradiated metacercarial cysts and subsequent challenge with normal cysts of *Fasciola hepatica* L. *Parasitology* **54**, 369–389.

Dubois, F. (1949). Contribution a l'étude de la migration des cellules de régénération chez les planairies dulcicoles. *Bull. Biol. Fr. Belg.* **83**, 215–278.

Ellinger, F. (1957). "Medical Radiation Biology." Thomas, Springfield, Illinois.

Fliedner, T. M., Andrews, G., Cronkite, E. P., and Bond, V. P. (1964). Early and late cytological effects of whole body irradiation on human marrow. *Blood* **23**, 471–487.

Galtsoff, P. S. (1964). The American oyster *Crassostrea virginica*. *U. S. Fish Wildl. Serv., Fish. Bull.* **64**, 1–480.

Geckler, R. P., and Kimball, R. F. (1953). Experiments with *Stentor coeruleus* on the nature of the radiation-induced delay in fission in the ciliates. *J. Protozool.* **5**, 151–155.

Giese, A. C. (1953). Protozoa in photobiological research. *Physiol. Zool.* **26**, 1–22.

Giese, A. C. (1967). Effects of radiation upon protozoa. *In* "Research in Protozoology" (T. Chen, ed.), Vol. 2, Pergamon, Oxford.

Grass, J. C. (1969). The pathological effects of ionizing radiation on the earthworm, *Eisenia foetida*. M.S. Thesis, University of Washington, Seattle.

Haley, T. J., and Snider, R. S. (1962). "Response of the Nervous System to Ionizing Radiation." Academic Press, New York.

Hancock, R. L. (1962). Lethal doses of irradiation for *Lumbricus*. *Life Sci.* **11**, 625–628.

Hancock, R. L. (1965). Irradiation induced neoplastic and giant cells in earthworms. *Experientia* **21**, 1–4.

Hashlauer, J. (1964). The effects of visible and UV light on the regenerative growth of planaria. *Strahlentherapie* **125**, 604–630.

Henshaw, P. S. (1938). Radiosensitivity and recurrent growth in *Obelia*. *Radiology* **32**, 466–472.

Hulse, E. V. (1959). The depletion of the myelopoietic cells of the irradiated rat. *Brit. J. Haematol.* **5**, 369–378.

Jacobson, L. O. (1954). The hematologic effects of radiation. *In* "Radiation Biology," (A. Hollaender, ed.), Vol. I, Part 2. McGraw-Hill, New York.

Katkansky, S. C., and Sparks, A. K. (1966). Seasonal sexual pattern in the Pacific oyster *Crassostrea gigas* in Washington State. *Fish. Res. Pap.* **2**, 80–89.

Kimeldorf, P. J., and Hunt, E. L. (1965). "Ionizing Radiation: Neural Function and Behavior." Academic Press, New York.

Korotkova, G. P., and Tokin, B. P. (1965). Responses of sponges and coelenterates to beta irradiation. *Radiobiology* **5**, 40–50.

Lee, W. R. (1964). Partial body radiation of queen honeybees. *J. Apicult. Res.* **3**, 113–116.

LeMoigne, A. (1965). Effets des irradiations aux rayons x sur le développement embryonnaire et le pouvoir de régénération a l'eclosion de *Polycelis nigra* (Turbellarie, triclade). *C. R. Acad. Sci.* **260**, 4627–4629.

Lender, T. (1965). La régénération des planaires en régénération. *In* "Regeneration in Animals and Related Problems," pp. 95–112. North-Holland Publ., Amsterdam.

Mayer, M. S. (1963). Biological and histopathological effects of gamma radiation on three life stages of *Anthonomous gradis Boheman*. Unpublished Ph.D. Thesis, Texas A & M University, College Station.

Mix, M. C. (1970). The histopathological effects of ionizing radiation on the Pacific oyster, *Crassostrea gigas*. An examination of generative syndromes, cellular reparative mechanisms, and their relation to normal cell renewal systems. Ph.D. Thesis, University of Washington, Seattle.

Mix, M. C., and Sparks, A. K. (1969). Histopathological effects of ionizing radiation on the planaria, *Dugesia tigrina*. *Nat. Cancer Inst., Monogr.* **31**, 693–707.

Mix, M. C., and Sparks, A. K. (1970). Studies on the histopathological effects of ionizing radiation on the oyster *Crassostrea gigas*. I. The degenerative phase involving digestive diverticulae, stomach and gut. *J. Invertebr. Pathol.* **16**, 14–37.

Mix, M. C., and Sparks, A. K. (1971a). Repair of digestive tubule tissue of the Pacific oyster, *Crassostrea gigas*, damaged by ionizing radiation. *J. Invertebr. Pathol.* **17**, 172–177.

Mix, M. C., and Sparks, A. K. (1971b). The histopathological effects of various levels of ionizing radiation on the gonad of the oyster, *Crassostrea gigas*. *Proc. Nat. Shellfish. Ass.* **61**, 64–70.

Montagna, W., and Wilson, J. W. (1956). A cytologic study of the intestinal epithelium of the mouse after total body X-irradiation. *J. Nat. Cancer Inst.* **15**, 1703–1735.

Mortreuil-Langlois, M. (1960). Effet des rayons X sur l'intestin moyen de *Blabera fusca* Br. *C. R. Soc. Biol.* **154**, 1769–1770.

Mortreuil-Langlois, M. (1963). Etude histopathologique de l'intestin moyen de *Blabera fusca* Br. (Orthoptera) au cours d'une période prolongée de post-irradiation. *Bull Soc. Zool. Fr.* **98**, 539–546.

O'Brien, R. D., and Wolfe, L. S. (1964). "Radiation, Radioactivity, and Insects." Academic Press, New York.

Odartchenko, N., Cottier, H., Feinendegen, L. E., and Bond, V. P. (1964). Mitotic delay in more mature erythroblasts of the dog, induced *in vivo* by sublethal doses of X-rays. *Radiat. Res.* **21**, 413–422.

Park, H. D. (1958). Sensitivity of hydra tissues to X-rays. *Physiol. Zool.* **31**, 188–193.

Patt, H. M., and Quastler, H. (1963). Radiation effects of cell renewal and related systems. *Physiol. Rev.* **43**, 357–396.

Price, T. J. (1962). Accumulation of radionuclides and the effects of radiation on molluscs. "Biological Problems in Water Pollution," 3rd Semin., Environ. Health Ser. Water Supply and Pollution Control, U. S. Public Health Service, Washington, D. C.

Quastler, H. (1956). The nature of intestinal radiation death. *Radiat. Res.* **4**, 303–320.

Quastler, H., and Hampton, J. C. (1962). Effects of ionizing radiation on the fine structure of the intestinal epithelium of the mouse. I. Villus epithelium. *Radiat. Res.* **17**, 914–931.

Quastler, H., Sherman, F. G., Brecher, G., and Cronkite, E. P. (1958). Cell renewal, maturation and decay in the gastrointestinal epithelial of normal and irradiated animals. *Proc. U. N. Int. Conf. Peaceful Uses At. Energy, 2nd, 1958* Vol. 22, pp. 202–205.

Rose, C., and Shostak, S. (1968). The transformation of gastrodermal cells to neoblasts in regenerating *Phagocata gracilis* (Leidy). *Exp. Cell Res.* **50**, 553–561.

Rubin, P., and Casarett, G. W. (1968). "Clinical Radiation Pathology," Vols. I and II. Saunders, Philadelphia, Pennsylvania.

Schaper, A. (1904). Experimentelle Untersuchungen uber den Einfluss der Radium-strahlen und der Radiumemanation auf embryonale und regenerative Entwick-lungsergange. *Anat. Anz.* **25**, 398–414 and 326–337.

Snodgrass, R. E. (1935). "Principles of Insect Morphology." McGraw-Hill, New York.

Sonehara, S. (1933). Studies on the effects of X-rays upon the development of a pond snail *Lymnea* (Radix) *japonica* Jay. *J. Sci. Hiroshima Univ., Ser. B, Div. 1,* pp. 151–169.

Sullivan, M. F. (1968). "Gastrointestinal Radiation Injury," Monogr., Nucl. Med. Biol., Ser. No. 1. Excerpta Medica Found, Amsterdam (Printed by Reidel, Dordrecht, Netherlands).

Sullivan, R. L., and Grosch, D. S. (1953). The radiation tolerance of an adult wasp. *Nucleonics* **11**, 21–23.

Tan, B. D., and Jones, A. W. (1966). X-ray induced abnormalities and recovery in *Hymenopolis microstoma. Exp. Parasitol.* **18**, 355–373.

Van Cleave, D. D. (1963). "Irradiation and the Nervous System." Roman Little-field, New York.

Wiegand, K. (1930). Regeneration bei Planarien und *clavelina* unter dem Einfluss von Radiumstrahlen. *Z. Wiss. Zool.* **136**, 255–318.

White, J. C., and Angelovic, J. W. (1966). Tolerances of several marine species to Co 60 irradiation. *Chesapeake Sci.* **7**, 36–39.

Wichterman, R. (1948). The biological effects of X-rays on mating types and conjugation of *Paramecium bursaria. Biol. Bull.* **94**, 113–127.

Wichterman, R. (1959). Mutation in the protozoan *Paramecium multimicronucleatum* as a result of x-irradiation. *Science* **129**, 207–208.

Wichterman, R., and Figge, F. H. J. (1954). Lethality and the biological effects of x-rays in *Paramecium*; radiation resistance and its variability. *Biol. Bull.* **106**, 253–263.

Williams, D. B. (1966). Effects of X-rays on cell lethality and micronuclear number in the ciliate *Spathidium spathula. J. Protozool.* **13**, 272–277.

Wilson, S. G., Jr. (1959). Radiation-induced gastrointestinal death in the monkey. *Amer. J. Pathol.* **35**, 1233–1251.

Wolff, E., and Dubois, F. (1947). Sur une méthode d'irradiation localisée permettant de mettre en évidence la migration des cellules de régénération chez les planaires. *C. R. Soc. Biol.* **141**, 903–906.

Wolff, E., and Lender, T. (1962). Les neoblastes et les phénomènes d'induction et d'inhibition dans la régénération des planaires. *Ann. Biol.* **38**, 499–529.

CHAPTER VII

Tumors and Tumorlike Conditions in Invertebrates

A tumor is defined as any swelling or abnormal mass of tissue. Through medical usage, however, the term is now virtually restricted to neoplasia (new growth), which is characterized by autonomy—growth of cells or tissues independent of the normal laws of growth of the organism. Growth of the tumorous tissue is not only independent and in excess of the normal tissue, but it persists at the same rate and pattern after termination of the stimulus that initiated the accelerated growth. In a definition slightly more restrictive than the original, Scharrer and Lochhead (1950) defined tumors as the result of abnormal cell proliferation. In the latter context, invertebrate tumors are of relatively common occurrence, but, since many invertebrate tumors arise in reaction to injury or in response to parasitic invasion, tumors in the vertebrate sense are rare in invertebrates.

Tumor nomenclature in vertebrates has evolved over many years of intensive study, and, even though it does not follow a consistent scheme, it is useful in describing the origin, location, and microscopic or gross characteristics of the growth. As Scharrer and Lochhead (1950) pointed out in their review of invertebrate tumors, a number of serious difficulties are encountered in any attempt to assess the relative abundance and mechanisms of tumorous changes in invertebrates. A major problem is the fact that most invertebrate zoologists who are familiar with invertebrate gross and microscopic anatomy are inexperienced in tumor diagnosis, and pathologists who specialize in tumor research are generally unfamiliar with invertebrate anatomy, physiology, and reactions to injury. Sparks (1969) noted that except for the insects tumors have neither been sought nor studied in invertebrates to the extent that they have been investigated in the vertebrates, particularly in mammals. Scharrer and Lochhead also wisely suggested that terminology developed almost exclusively for use in mammalian pathology should not be applied to invertebrates until their comparative relationships are better understood. Dawe (1968) elaborated on this theme by pointing out that

> the art-science of correlating histopathological characteristics of neoplasms with their biological characteristics has been developed to a high degree of perfection for neoplasms of man and of many domestic and laboratory animals . . . through the accumulation of vast amounts of pathological materials in centers where it became possible for pathologists to acquire broad and detailed experience. This process of accumulating the requisite materials and experience has not yet taken place in the field of invertebrate neoplasia.

He supported Scharrer and Lochhead's opinion that it is dangerous to attempt to extrapolate from knowledge of human and mammalian disease to invertebrate pathology, and he proposed the establishment of centers for the collection and study of anomalous growths and forms in invertebrates. As Dawe pointed out, one such center, the Registry of Tumors of the Lower Animals, has already been established in the Smithsonian Institution in Washington, D. C. Other centers of invertebrate pathology, pathobiology, and comparative pathology have been established or are in the process of being organized (Sparks, 1970).

In time, as Dawe remarked, a sound body of knowledge and experience will be developed that will resolve much of the confusion now existent in invertebrate oncology. Considerable progress has been made recently through awakening interest in invertebrate tumors. The Registry of Tumors of the Lower Animals has contributed greatly to the impetus; it sponsored a symposium on Neoplasms and Related Disorders of Invertebrate and Lower Vertebrate Animals in June, 1968. Numerous original papers were presented and are discussed in this chapter. Dawe (1969) discussed phylogeny and oncogeny; Pauley (1969) critically reviewed neoplasia in mollusks; and Sparks (1969) reviewed the status of knowledge of neoplasia in the major invertebrate taxa not discussed at the symposium.

Despite the danger of using vertebrate terminology and nomenclature, there are some basic definitions and terminologies of vertebrate tumors that should be understood by the invertebrate pathologist and evaluated in studies of invertebrate tumors. One of the basic classifications of tumors is based on the number of cell types that make up the tumor. Most vertebrate tumors consist of only one cell type and are called simple; some are composed of two cell types, typically derived from the same germ layer, and are called mixed tumors; and tumors composed of cells from more than one germ layer are called compound tumors or teratomas. All types contain a nonneoplastic connective tissue stroma associated with the neoplastic cells. The most common classification of mammalian tumors divides them into benign and malignant. Recognition of malignancy in a mammalian tumor is, of course, the most important criterion of the neoplasm. There are a number of recognizable criteria by which such a distinction can be made in vertebrate pathology, and most, though not all, vertebrate tumors can be differentiated into one of the two types. Table VII.1 summarizes the comparative characteristics of benign and malignant tumors of vertebrates. Malignancy in invertebrate tumors, however, is not well documented, and the criteria are not established with any certainty.

Most of the described tumors of invertebrates appear to be benign by the criteria listed in Table VII.1. It should be mentioned that many of the structures described in the literature as tumors are actually hyperplasia or unusual proliferation of typical cellular components in response to injury or parasitic invasion and would not be considered tumors by the vertebrate pathologist. A second confusing point is the fact that

TABLE VII.1

COMPARISON OF THE CHARACTERISTICS OF BENIGN AND MALIGNANT TUMORS[a]

Characteristics	Benign tumor	Malignant tumor
Structure and differentiation	Structure often typical of the tissue of origin	Structure often atypical, i.e., differentiation imperfect
Mode of growth	Growth usually purely expansive; capsule formed	Growth infiltrative as well as expansive so that strict encapsulation is absent
Rate of growth	Growth usually slow; mitotic figures scanty, and those present are normal	Growth may be rapid with many abnormal mitotic figures
Progression of growth	Usually progressive slow growth that may come to a standstill or retrogress	Growth rarely ceases; often rapid and usually progressive to a fatal termination
Metastasis	Absent	Frequently present

[a] From Robbins (1959) modified from Willis.

there is considerable lack of agreement among workers as to the neoplastic nature of many of the described tumors. The point at which
hyperplasia or hypertrophy become neoplasia is perhaps debatable and
must await further study in most instances. Since the vast majority of
invertebrate tumors and related growths described in the literature are
found in nature, it is usually impossible to determine whether the proliferation of tissue would persist at the same rate and in the same pattern if the stimulus initiating the proliferation were removed.

If we ignore for the moment those cellular proliferations obviously
initiated in response to injury and wound repair and consider the so-
called spontaneous tumors that, as Scharrer and Lochhead (1950) correctly pointed out, are actually growths of unknown cause, the typical
invertebrate tumor has certain characteristics. Although the tumor may
be a polyp or pedunculous mass, the cellular makeup usually consists of
relatively normal, recognizable cell types. The tumor is sometimes experimentally induced, but the lack of mitotic figures, especially abnormal
ones, indicates relatively slow growth. It is seldom possible to ascertain
whether growth of the tumor has stopped or retrogressed, or whether
it will or has already progressed to a fatal termination. Finally, there
are, to my knowledge, no authenticated examples of metastasis in any
invertebrate tumor, exclusive of the insects.

As Dawe (1969) so succinctly pointed out, "At present, it is far more
important that biologists characterize the neoplastic diseases or related
disorders of growth and form they observe than that they apply 'correct'
names." As a guide to encourage this, he compiled a tabulation of
parameters to be considered. This is valuable to the biologist who is not
an experienced oncologist (Table VII.2). Obviously, not all of the
parameters can be evaluated when a neoplasm is first encountered. It
should, however, serve as a guide for as complete a description as possible of the growth and as an outline for further study.

Phylum Protozoa

Opalinid Nuclear Abnormalities

Metcalf (1928) reported several abnormal specimens of opalinids, a
group of parasitic flagellates inhabiting the large intestines of frogs and
toads. These opalinids (*Opalina abtaigona*) contained, in addition to the
typical numerous normal nuclei, many abnormal nuclei that Metcalf
thought closely resembled certain mammalian cancer cell nuclei. Among
the diagnostic features of mammalian cancers are multipolar and other
atypical mitoses, amitotic division of chromatin, degenerating nuclei,

TABLE VII.2

<small>Parameters of Tumors—An Aid in Description and Study[a]</small>

I. Macroanatomic Characteristics

Anatomic location	Boundary features
Size	Necrosis
Shape, surface properties	Ulceration
Color	Invasion, gross
Consistency	Metastasis, gross

II. Microanatomic Characteristics

1. Tissue and cell type of origin
2. Stromal characteristics and vascular supply
3. Degree and kind of histologic atypicality
4. Unicentricity versus multicentricity
5. Invasiveness
   a. *In situ*
   b. Local invasion—fat, fascia, bone, etc.
   c. Lymphatic invasion
   d. Blood vascular invasion (or hemocoel)
   e. Perineural invasion
   f. Other
6. Metastases—cf. primary
7. Cytochemical characteristics
   a. Pigments
   b. Enzymes
   c. Intracellular storage (glycogen, mucin, keratin, hormones, fats, etc.)
8. Extracellular products (collagen, reticulin, cartilage, bone, amyloid, shell, chitin, etc.)
9. Karyotype
   a. Modal number and distribution pattern of chromosomes
   b. Qualitative abnormalities (breaks, deletions, additions, translocation, fusions, marker chromosomes, etc.)
10. Ultrastructure
    a. Changes of organelles, endoplasmic reticulum, mitochondria, plasma and nuclear membranes, lysosomes, Golgi, tonofibrils, myofibrils, microtubules, desmosomes, etc.
    b. Infectious agents (viruses, bacteria, fungi, parasites)
11. In tissue or cell culture
    a. Cell morphology (fibroblastoid, epithelioid, amoeboid)
    b. Structure-forming capability (papillae, glands, fascicles, etc.)
    c. Cell adhesiveness
    d. Cell cohesiveness
    e. Contact inhibition of movement
    f. Contact inhibition of division
    g. Pinocytotic activity
    h. Membrane properties (zeiosis, microfilopodia, etc.)
    i. Nuclear properties (pulsation, rotation, amitosis, etc.)

TABLE VII.2 (*Continued*)

III. Macrobiological Characteristics

Growth continuous or intermittent?	Transplantability
Growth reversible?	Progression
Lethality	Radiosensitivity
Responsiveness (hormones)	Drug sensitivity
Dependency (hormones)	Host age, sex developmental stage
Growth rate	Functional activities (pigment
Growth in ascites form	changes, endocrine target effects)

IV. Microbiological Characteristics

1. *In vivo*
   a. Relation to host's genomic constitution (hereditary control)
   b. Transmissibility by infectious agent
   c. Antigenic profile (histocompatibility antigens, fetal antigens, virus-specific antigens, paraproteins, etc.)
   d. Metabolic properties (functional products such as protein and steroid hormones, serotonin, histamine, etc.)
   e. Response to antimetabolites, hormones
   f. Cell kinetics (life-span, mitotic time, intermitotic time, size of nonproliferating pool, etc.)
2. *In vitro*
   a. Nutritional requirements
   b. Growth rates
   c. Cloning efficiency
   d. Gain or loss of antigens
   e. Gain or loss of transplantability
   f. Inducible enzymes, e.g., transaminases
   g. Other properties as studied *in vivo*

---

a From Dawe (1969).

and enlarged cells. Metcalf observed nuclei whose chromatin was arranged in a coarse-meshed cap on a clear, glassy-appearing body; he also noted that the granular achromatic substance was organized into well-developed and fairly regular lines of granules resembling a mitotic spindle. The chromatin material, however, never formed distinct chromosomes. The chromatin-capped spheres were usually single, but two or three such masses of chromatin were sometimes found in a single abnormal nucleus.

All these phenomena, in Metcalf's opinion, indicated amitotic division of the chromatin. The division of the chromatin was often unequal, and subsequent stages of nuclear degeneration resulted in gradual disappearance of the nuclear contents and wall. The chromatin-capped spheres and chromatin masses were the most persistent structures, but they ultimately disappeared, leaving empty spaces in the cytoplasm, often with remnants of chromatin masses. Metcalf further noted that the degenerating nuclei were larger than normal, again resembling nuclei of some neoplasms.

Additional study of other opalinids indicated, in Metcalf's opinion, that the abnormal nuclei were derived from cells in which amitosis followed abnormal mitosis. In *Protoopalina caudata,* which normally contains two nuclei, he reported occasional cells whose nuclei divided without cytoplasmic division; others contained nuclei whose chromosomes had divided to produce double the normal complement without either nucleus or cytoplasm dividing. Metcalf noted that in all the abnormalities, cells and nuclei were greatly enlarged, and nuclear degeneration and disappearance also occurred occasionally along with degeneration of cytoplasm in the enlarged cells. He summarized those abnormal characteristics in opalinids which he believed corresponded cytologically to mammalian cancer.

> 1) enlarged cells with enlarged nuclei arising by division of the nucleus or by fusion of nuclei. Such nuclei have too many chromosomes; 2) unequal division of such enlarged nuclei causing still further distortions in kinds as well as numbers of chromosomes present; 3) amitotic division of the chromatin without division of the nucleus; 4) degeneration of nucleus and of cell.

Most modern workers tend to discount Metcalf's findings simply as examples of abnormal mitoses. The characteristics he described do resemble those frequently associated with neoplasia in higher animals, and some of the mechanisms concerned may quite possibly be similar. Metcalf's report is better understood when viewed in the perspective of his own time, for he was stimulated to fit his observations to the theory of Boveri (1914). This theory in turn was based on the observation of unequal nuclear division after polyspermic fertilization of sea urchin eggs, coupled with the demonstration still earlier by von Hansemann (1890) that unequal division of chromatin occurred in human neoplasms.

Mottram (1940) demonstrated that benzo[a]pyrene, which increases cell growth when applied to vertebrate tissues, increased growth rates in *Paramecium* cultured in various concentrations of the substance. Paramecia of abnormal shape developed in concentrations of 1 and 0.5 ppm after prolonged culture (122 and 62 days).

The experimental cultures never underwent conjugation, and abnormal specimens removed and placed in normal media as single-cell cultures developed into populations of abnormal paramecia. Some were many times normal size, some were midgets, others were "Siamese twins" or triplets, and some were apparently normal. Offspring of the abnormal forms were often normal and outgrew the abnormal, but some abnormal forms bred true through many generations. Long exposure to the carcinogen, therefore, results in the production of polymorphic cells that are strikingly similar to those found in mammalian tumors initiated by benzo[a]pyrene. They occur among cells that are stimulated for extended periods to accelerated growth; they occur only after long exposure to the hydrocarbon; the abnormal cells continue to be produced after the stimulus is removed, and the cell population is highly polymorphic.

Wolman (1939), using 3,4-benzopyrene and methylcholanthrene, reported increased proliferation in cultures of paramecia, as did Spencer and Melroy (1940). However, Tittler (1948), working with the predaceous ciliate, *Tetrahymena geleii,* found no observable effect on the growth curve of *Tetrahymena* exposed to carcinogenic hydrocarbons. Tittler believed that the positive results of earlier workers were due to faulty culture techniques with too many variables (including the influence of food organisms in the form of bacteria) for valid results. Moewus (1959) later demonstrated that benzedine, another carcinogen, stimulates the mitotic activity of *Polytoma uvella,* a phytomonadinid flagellate, at a dosage of 0.1 μg/ml.

Sonneborn (1954) compared populations of ciliates containing certain cytoplasmic particles (κ, λ, μ, π) to neoplastic cells. The current knowledge of the genetics and biochemistry of these cytoplasmic particles was summarized by van Wagtendonk (1969) along with his concept of their neoplastic nature.

Van Wagtendonk believes that these peculiar cytoplasmic components of *P. aurelia* can be considered as neoplastic equivalents. As he pointed out, the essential difference between "killers" and "sensitives" is that the killers possess large numbers of rod-shaped particles in their cytoplasm that are absent in the cytoplasm of the sensitives. The particles multiply in the cytoplasm of those strains that carry a certain dominant nuclear gene, and they have never been seen to arise *de novo.* Under special conditions, the particles are capable of "infecting" paramecia of the proper genetic background even when cultured *in vitro.* In these respects, according to van Wagtendonk, the particles are comparable to the viruses associated with certain tumors, especially with the agent found in mammary tumors of mice. He further postulated that the particles may have evolved from prototypic infectious agents into cytoplasmic components that conferred ecological advantages, and, again, they can be considered to be neoplastic equivalents because they evolved into a new growth of tissue resembling more or less the tissue from which they arose or which they invaded.

Phylum Porifera

I know of no reports of tumors or related neoplastic growths in the sponges. It is not known whether this is because the group lacks such processes, or because investigators have not sought, recognized, or reported neoplastic abnormalities in the sponges.

Phylum Coelenterata

Reports of neoplasms in coelenterates are infrequent, and their authenticity has often been questioned. Korschelt (1924) reported strange "neoplasms" rather than normal buds in the polyps of the hydroid *Syncorybe decipiens* during senescence.

Squires (1965a) described three abnormal corallites from the only known specimen of *Madrepora kauaiensis*. Since the specimen was dried and the polyp tissue was destroyed prior to Squires' examination, soft tissues could not be studied, and only abnormalities of the skeleton could be described. Coral growth is accretionary, and developmental sequences are preserved in the structure of the skeleton; therefore, more light is shed on the probable structure of the soft tissues in the corals by study of the skeleton than in most invertebrate groups.

Although malformed corals are relatively common, the majority of abnormalities can be directly attributed to the effects of predation or reaction to injury. Those not readily explainable by these phenomena have been called abnormal and were not studied further. In the description of *Madrepora kauaiensis,* at least one unusual corallite was noted and interpreted as an individual of another species growing attached to *M. kauaiensis*. Squires disagreed with that interpretation because the skeletal elements of the abnormal corallites were organically connected to the colony and, thus, were formed asexually by polyps lower on the branch. The fact that the abnormal polyps were conspicuously larger than their "siblings" suggested a greater growth rate, an assumption supported by structural features indicative of rapid growth.

Madrepora characteristically possess solid, thick septa with minutely dentate margins and sides. The pathological corallites had septa that were highly fenestrate and lacerate, a condition characteristic of mussid corals and leading to the previously mentioned suggestion that a second species was present.

The skeletal arrangement of the abnormal polyps suggested that the normal division and insertion of the mesenteries was followed until the normal polyp formation was exceeded. Then mesentery formation and septal insertion became disordered and chaotic, destroying the normal symmetry of the polyp. The pathological polyps arose from normal individuals and gave rise to normal polyps.

Squires noted that any unusual environmental condition initiating the abnormal growth *should* have affected more than 3 of the 239 polyps in the colony, but that other polyps of similar age were unaffected. Although corals are preyed upon by fish and predaceous invertebrates, the regenerative capabilities are high, and repair is relatively rapid and complete. Occasional duplication of septa occurs along the margins, but it is the result of duplication of only a single mesenterial pair. Since the disturbance of growth in an injured coral is minimal, Squires did not think that the abnormal corallite resulted from predation or accidental injury. This conclusion was based largely on the unusually large size of the polyps. It was his feeling that the natural growth-regulating mechanism was removed, allowing the abnormal increase in size. Although the actual nature of the growth-regulating mechanism in corals is not known, it is thought to be related to the attainment of sexual maturity, because growth rates decrease sharply after sexual maturity is reached.

White (1965) and Soule (1965) both took issue with Squires' interpretation. White agreed that the growths were neoplasms in the general definition of aberrant new growths, but warned against the possible inference that they represented malignant neoplasms or cancers. He noted that malignant neoplasms usually arise originally as an individual altered cell that gives rise to additional abnormal cells resulting in a malignant growth, and while reversion to normal growth occurs, it is of rare occurrence. Thus, the reversion to production of normal polyps by all three of the abnormal corallites would be unlikely. White also offered an alternative theory of their origin by pointing out the extraordinary resemblance in structure and distribution to the "galls" common on plants resulting from various sedentary predators. These occur in aquatic environments, for example on the green alga *Vaucheria,* as a result of the presence of a rotifer. The abnormal growth resulting in galls is initiated by "growth hormones" secreted by the resident predator.

Soule (1965) believed that the abnormal corallites were actually colonies of a cyclostomata ectoproct belonging to the genus *Tichenopora,* which is so similar to the appearance of the mineralized portions of *Madrepora* that many species of the genus were originally described in the genus *Madrepora.*

Squires (1965b) agreed with White that gall formation was probable in the corals because they are hosts to large numbers of organisms that cause galls in other organisms, but was of the opinion that the abnormal corallites described originated from other stimuli since examination of the specimens gave no evidence of coral overgrowth of other organisms. In reply to Soule's assumption that the growths were bryozoan colonies, Squires noted that although he had no *Tichenopora* to study X-ray diffraction patterns differ in bryozoans and corals, and the diffraction patterns in the abnormal corallites were similar to those of normal corallites from other portions of the colony.

More recently, Lenhoff *et al.* (1969) encountered some phenomena in a mutant hydra that allowed an experimental approach to the study of a possible neoplasia analog and provided some insight into the mechanisms by which neoplastic growths might develop in the hydroids. It has been possible only in recent times, with the development of mass culture methods, to obtain sufficient numbers of zygotes for fruitful search for mutants.

The above authors have isolated a number of mutants with morphological aberrancies. One *Chlorohydra viridissima* was particularly unusual in that it did not produce asexually by budding; its size, dry weight, and protein content were about 10 times greater than normal. It could not be induced to form gonads under the conditions causing gonad development in normal animals; it frequently developed multipolar body forms, and these multipolar forms arose from regenerating animals, demonstrating aberrancies in the organization of the animal.

When a normal hydra is bisected, the apical portion regenerates another basal disk, and the basal portion regenerates another head. The mutant, however, regenerates a second head, giving rise to a bipolar animal (Fig. VII.1). The basal portion regenerates even more abnormally; as in a normal hydra, another head regenerates at the site of the section, but, simultaneously, numerous small protrusions develop randomly along the basal portion of the animal (Fig. VII.2). Each protrusion subsequently develops into a tentacle, with as many as 50 or more occurring on a single individual. The tentacles cluster in groups of 6–10 (Fig. VII.3); a mouth and hypostome develop in the center of each cluster, which emerges as a small hydranth along the body tube (Fig. VII.4). The individual hydranth is capable of feeding and the appearance after 2–3 weeks is that of a colonial marine hydroid. Within the next 2 or 3 weeks, each hydranth separates as a monopolar animal (Fig. VII.5).

Thus, the mutants not only regenerate, but, through atypical regeneration, increase in numbers. Transection is not, however, required for the mutant to initiate this unusual type of asexual reproduction, as can be seen in Fig. VII.6. The offspring from a single monopolar mutant over a 3-month period are shown. As Lenhoff and his associates pointed out, the asexual reproduction in the mutant differs greatly from the budding process of the normal hydra. Also, the random formation of tentacles in the mutant never occurs in the region of the distal head (upper one-fourth to one-third of the body column).

Lenhoff *et al.* (1969) postulate that the head end of hydra in some

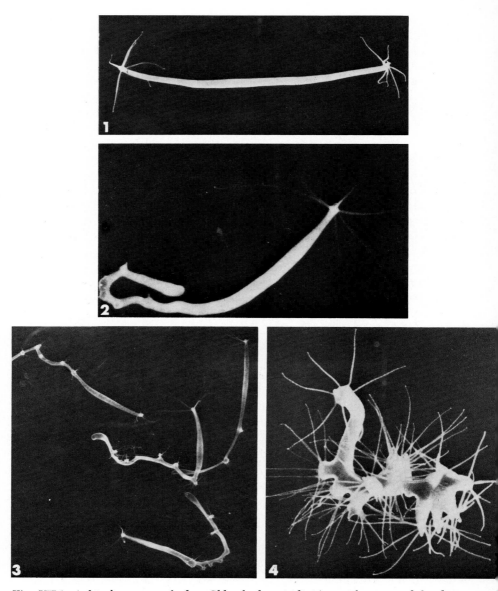

Fig. VII.1. A bipolar mutant hydra, *Chlorohydra viridissima*, with a second head regenerat[ed] rather than a basal disc. (From Lenhoff *et al.*, 1969.)

Fig. VII.2. Basal portion of a mutant hydra with a regenerated head. Note small protrusi[on] developing along basal portion. (From Lenhoff *et al.*, 1969.)

Fig. VII.3. Mutant hydra in which the protrusions form tentacles in clusters. (From Lenh[off] *et al.*, 1969.)

Fig. VII.4. Mutant hydra with clusters of tentacles forming small hydranths along the body tu[be] (From Lenhoff *et al.*, 1969.)

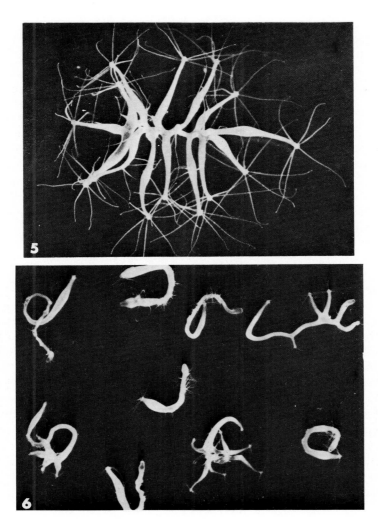

Fig. VII.5. Multipolar mutant hydra derived from atypical regeneration of the basal portion of a bisected mutant hydra. Note this animal has gone through the successive stages illustrated in Figs. VII.2–VII.4. (From Lenhoff *et al.*, 1969.)

Fig. VII.6. A clone of mutant hydra derived from a single monopolar mutant over a period of 3 months. Note each hydranth, such as those shown in Fig. VII.5, separates as a monopolar animal. (From Lenhoff *et al.*, 1969.)

unknown manner controls the polarity of the animal. In normal animals, this control apparently extends over the entire animal, while in the mutant it extends over only part of the apical portion. An alternative suggestion by Lenhoff *et al.* is the possibility that the mutant may have a defect in the budding process, and, being unable to channel its growth into the formation of buds, grows to a size it cannot regulate. Thus, the atypical regenerative properties and organization of the basal portion may merely reflect pleomorphic effects stemming from its large size.

When the above authors grafted the apical half of a mutant hydra to the basal half of a normal individual, budding was initiated in the mutant portion within 1 day and continued at a relatively constant rate for 13 days. The budding process was apparently normal, but all the progeny were mutant, i.e., failed to bud. Budding from the graft was sporadic after the thirteenth day and ceased by the twenty-third day postgrafting. Subsequently, the basal (normal) half of the animal was removed and replaced with another normal basal portion. The mutant portion again began budding; some of the new progeny were mutant while others appeared normal.

The normal-looking buds developed into animals that had all the characteristics of normal hydra, and they budded rapidly, but their buds developed into both normal and mutant individuals. "Thus, mutant somatic cells appear to proliferate in a host 'normal' individual to segregate, come off as a bud, and express their phenotype in those buds." This, as Lenhoff and his associates pointed out, raises the question of whether the phenomenon observed in the normal-looking heterocytes is analogous to neoplasia in higher organisms. Obviously, populations of somatic cells of a different genotype survive and multiply in a host organism. Eventually a mass of these cells accumulates, but they form a bud and leave the host as normal cells do rather than remaining and producing an abnormal growth as in higher animals. In view of the lack of complex vascular and nervous systems in hydra, there is little chance of masses of these cells obstructing critical areas, but deleterious effects do occur in that the heterocytes invariably transform into individuals of the mutant phenotype.

Phylum Platyhelminthes

Tumors among the flatworms have been reported only in the class Turbellaria. This is most probably related to the high regenerative capacity of the turbellarians. There is either an intrinsic potential for neoplastic growths when the regenerative processes are misdirected, or, because of the great number of investigations of planarian regeneration, there is an increased probability of observation of both spontaneous and experimentally produced tumors. Dubois (1949) demonstrated that regeneration in planarians is accomplished through the activities of undifferentiated totipotent cells, the neoblasts. When a wound occurs, surrounding neoblasts quickly differentiate to form a new epidermis, while other neoblasts migrate to the wound area and differentiate into the other cell types necessary to replace the damaged or lost tissue.

Several workers (Goldsmith, 1939; Stéphan, 1960; Lange, 1966) have reported spontaneous tumors in planarians. In apparently the first record of such an occurrence, Goldsmith (1939) observed a specimen of *Dugesia tigrina* in a stock culture maintained in his laboratory for 3 years, with two dorsal pigment stripes rather than one. Detailed examination revealed that the abnormal animal possessed two pharynges. After it was isolated, the planarian developed a small outgrowth from the dorsal surface of the postpharyngeal region after 4 weeks, and two additional outgrowths appeared within 2 months. Attempts to produce a clone from this animal, by making a transverse cut anterior to the pharynges in the abnormal and subsequent planarians derived from the operation, resulted in the production of eight animals at the time of reporting. Interestingly, the transverse cuts were not followed by the typical regeneration of a single head at the anterior end of the posterior piece and a tail at the posterior end of the anterior piece. Instead, multiple heads, as many as eight in one instance, with vesicles and pigment cups, and multiple tails developed similar to those produced by the splitting operation in previous regeneration studies. Also, an additional dorsal outgrowth developed in one of the animals produced by the transverse cuts. Although no figures were presented and the histology of the growths was not discussed, Goldsmith noted that the outgrowths resembled those produced experimentally by a variety of stimuli that will be discussed.

Stéphan (1960) described what appeared to be similar spontaneous tumors in the same species. Again, no figures were included, but the author presented a description of the tumors' histology, noting that they appeared to be massive infiltrations of the parenchyma by glandular cells. Secretions accumulated in open pockets developing in the base of the dorsal outgrowths. The rupture of the pockets or pouches, followed by disintegration of the crown, produced on the dorsal surface an appearance identical to that of the ventral surface of the worm.

Lange (1966), working with cultures of *Dugesia etrusca* and *Dugesia ilvana,* found high incidences of spontaneous tumor development in both species. Fifty-four tumor-bearing animals were studied, and, since no discernible difference between the tumors of the two species was observed, they were described together.

Most tumors originated as reddened swellings or lumps, or group of lumps, on the posterior tip of the animal, stretching the epidermis and distorting the tails (Fig. VII.7). As the growth slowly developed, the tail thickness was greatly increased, impeding movement. The tumors became darkly pigmented and eventually developed clear white tubes from which mucus was occasionally extruded (Fig. VII.8).

The tumor-bearing animals continued to divide even after the tumor became the largest part of the body. The anterior portions gave rise to normal individuals in which no tumors were observed over a 1-year period, while the posterior portions retained the tumors. The abnormal posteriors regenerated new heads at normal rates, then divided again, leaving a smaller posterior portion still bearing the tumor. Eventually the posterior portion and the tumor lysed.

Only one specimen, a *D. etrusca,* developed tumors on both the anterior and posterior regions (Fig. VII.9). Fission did not take place in this animal, and the midbody shrank rapidly until lysis of both the unaffected portion and the tumorous anterior and posterior regions occurred.

Wound repair and regeneration were abnormal in the tumors. A cut in normal tissue results in muscular contraction, which pinches off the wounded area and prevents cellular loss to the exterior. This is followed by differentiation of local neoblasts to form a new epithelium over the

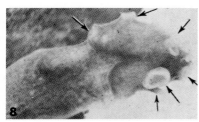

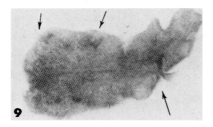

Fig. VII.7. Posterior portion of tumor-bearing planarian (*Dugesia* sp.) showing first stage in development. Arrow shows the dorsal epithelium stretched over the lump. (From Lange, 1966.)

Fig. VII.8. Differentiated stage of the growth. Note (arrows) clear white tubes. (From Lange, 1966.)

Fig. VII.9. Final stage of the tumor, with growths on both anterior and posterior portions of the animal (see arrows). (From Lange, 1966.)

wounded surface, and finally by emigration of neoblasts from distant nests into the area to regenerate the missing or damaged tissues. In contrast, incisions into tumorous tissue induced no muscular contraction. The wound remained open with cells streaming out of the body until all the tumorous tissue and large quantities of normal tissue were lost. Eventually contraction did occur, but well beyond the original boundary of the tumor; normal regeneration followed.

When Lange reduced the food to one feeding per week in his tumorous populations, no new tumors developed except in two worms in which tumors first appeared 1 week after feeding had been reduced.

Although most tumors originated at the posterior tip of the worms, in two instances a tumor developed in the dorsal midregion just posterior to the mouth. The tumor in each animal assumed a wide craterlike appearance with a clear ring of tissue forming a rim around the apex of the mound. Neither of these animals divided, and additional tumors developed on the posterior tips of the animals. The posterior tumors began to lyse, and the animals were fixed for histological study.

Histologically, tumorous animals exhibited increased tail thickness, consisting mainly of papillary and cryptlike structures covered and lined by mucus-producing epithelium (Figs. VII.10 and VII.11) and a broken or perforated basement membrane. The craterlike tumors arising anteriorly contained large masses of highly differentiated tissue, including nervous, parenchymal, and epithelial tissues, and crypts lined with epithelium (Figs. VII.12 and VII.13). The top of the tumor was lined with an epithelium that resembled that of the ventral body surface, as did the epithelium lining the crypts, and large pockets of neoblasts were scattered around the tumor. Lange postulated that the ventral-type epithelium lining the tumor crater may have resulted from the bursting of a crypt. He further stated that the cells and tissues comprising the tumors were not in themselves abnormal, but that their relationship to surrounding tissues caused the abnormality.

Foster (1963) reported nodular growths in *Dugesia dorotocephala* resulting from treatment with the carcinogens 3,4-benzopyrene, 3-methyl-4-dimethylaminoazobenzene, and 1,2-benzanthracene. The three carcinogens induced growths in 5–9% of treated worms. Growths did not occur in untreated control animals or in controls treated only with alcohol. Nine percent of the *Dugesia* challenged with 1,2-benzanthracene developed nodular growths, and 21% of regenerating animals, sectioned after 2 months of treatment and observed for another month, were affected. The nodules began to appear in the whole animals after 3 months.

Benzopyrene is highly fluorescent, therefore its retention in the planarians could be observed by fluorescent microscopy. Preliminary experi-

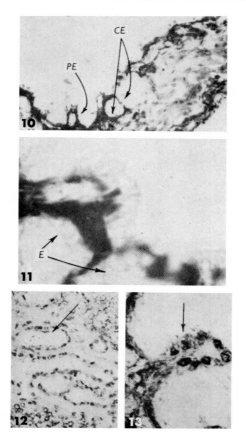

Fig. VII.10. Posterior portion of sagittal section of tumorous planarian. CE, epithelium-lined crypt; PE, epithelium-lined papillary structure. Methyl-green pyronine. Approximately 200×. (From Lange, 1966.)

Fig. VII.11. Enlargement of portion of Fig. VII.10. E, epithelial lining of crypt. Approximately 900×. (From Lange, 1966.)

Fig. VII.12. Crypts lining anterior craterlike tumors. Compare epithelial lining with Fig. VII.11. Hematoxylin. Approximately 250×. (From Lange, 1966.)

Fig. VII.13. Higher magnification of epithelial lining of crypts. Hematoxylin. Approximately 600×. (From Lange, 1966.)

ments demonstrated that benzopyrene treatment caused an intense silver-blue fluorescence in the treated animals which was retained for an average of 28 days and a maximum of 60 days. Since the fluorescence did not diminish appreciably during the first week after treatment, weekly treatments were adopted. High retention of 3-methyl-4-dimethylaminoazobenzene caused a yellowish coloration of planarians treated with that compound. Regeneration blastemas exhibited benzopyrene fluorescence, strengthening the theory that the blastema is formed by totipotent neoblasts migrating into a wound area. Examination of numerous nodular growths with a fluorescence microscope demonstrated higher concentrations of benzopyrene in the growths than in the adjacent normal tissue. The carcinogen was also retained by the mucous coat secreted by the cyanophilous glands, and numerous crystals, which the author assumed were crystals of the carcinogens, were seen in histological examination of the digestive tracts of treated planarians.

The lesions typically arose as one, two, or occasionally multiple nodules on the pigmented dorsal surface of treated worms; they markedly altered the pigmentation and never developed on the unpigmented ventral surface. When treated animals were removed from the carcinogens and bisected, nodules developed rapidly between 10 and 20 days after sectioning. The growths progressively increased in size until they encompassed the entire animal, with no recognizable head or tail remaining. Then the organism underwent necrosis and died. Death occurred in the interval between 10 days and 1 month after the nodule appeared.

Foster felt that these growths histologically exhibited similarities to mammalian tumors. Heavily basophilic cells, with an increased nucleocytoplasmic ratio, invaded the intestine in treated animals, but did not occur in controls. Mitotic figures, indicating rapid cellular proliferation, were common in the intestine, but were only rarely seen in other regions, including the regeneration blastema. Some aberrant mitotic figures were observed, a common characteristic of mammalian neoplasms. The bulk of the nodules consisted of histological elements typical of the intestine. The muscle fibers and connective tissue that normally separate the channels of the intestine as they pass along the sides of the body were invaded and disrupted by elements of the digestive system. Multinucleated giant cells occurred in the subepidermal tissue overlying the intestine and were surrounded by accumulations of pigment, suggesting a type of melanotic transformation. Many large cells in the area contained large globules and smaller fragments of pigment, and the normally dendritic pigment cells were rounded, enlarged, and irregularly distributed around the nodule of intestinal cellular elements. In the region of hyperplastic nodules, the epidermis was also hyperplastic, with four or five layers of cells present in contrast to the normal single cell layer. However, the epidermis was usually lost directly over the nodule.

As noted previously, Foster considered these nodules to be neoplastic rather than simply hyperplastic because of the gross morphology, histo-

logical structure, and progressive nature of the growths. Although the
fatal termination at 10 days to 1 month is unusually rapid for a tumor,
the small size of the animal makes it possible for the entire organism to
be involved in a short time. She noted that the growth of these nodules
resembled the rapid proliferation and infiltration of some *Drosophila*
tumors. The terminal phase of the growths described by Foster is similar
to that described by Henderson and Eakin (1961) in planarians treated
with metabolic antagonists in the form of purine analogs, in that the
growth in both cases is largely proliferative. Because the animal becomes
a rounded mass of loosely adherent cells that break up at any attempt
to transfer the organism into fixative, histological study of the terminal
phase is difficult. Therefore, it was not possible to determine the cell
type most severely affected; however, epidermal, subepidermal, and intestinal tissues all exhibited growth abnormalities. Because of the involvement of several cell types, Foster suspected the growths to be proliferating teratomas, but the occurrence of nodules only on the pigmented
backs of treated animals and the gross and microscopic evidence of pigment cell involvement make a melanotic type of transformation con-
ceivable. Because of the rarity of mitotic divisions in planarians, the
occurrence of numerous mitotic figures, including aberrant ones, in the
intestine assumes special importance and strengthens Foster's contention
that the growths are truly neoplastic. It is not yet known whether the
hyperplastic epidermis covering the nodules is lost through necrosis or
whether the nodule perforates the epidermis as a result of mechanical
stress; the extensive hyperplasia of intestinal elements certainly make the
latter possible. The fact that the progressive growth of the lesions results
in the death of the affected organism lends credence to the assumption of
the neoplastic nature of the tumors.

Foster (1969) presented additional results of her investigations on
malformations and lethal growths in planarians treated with carcinogens
at the Symposium on Neoplasms and Related Disorders of Invertebrate
and Lower Vertebrate Animals held at the Smithsonian Institution,
Washington, D. C., June 19–21, 1968. The study was originally designed
to investigate the effects of carcinogens on adult planaria, but numerous
offspring were produced allowing attention to be focused on the abnormalities of the progeny of carcinogen-treated adults.

Adult *Dugesia dorotocephala*, a species that has not been reported
to exhibit spontaneous abnormalities other than black-pigmented spikes,
were treated for 24-hour periods at weekly intervals with either 3-methyl-
chloranthrene (MCA) or benzo[a]pyrene (BP) in a concentration of
4 ml/liter of BVT media of a saturated acetone solution of the carcinogen.
After 2 months, the worms were sectioned transversely into 3 segments,

and 10 days after sectioning, when regeneration was virtually complete, they were subjected to another 24-hour treatment in the carcinogen.

Nine percent of the MCA- and 7% of the BP-treated animals developed tumors. A typical tumor (Fig. VII.14) was approximately one-third the length of the worm and arose from two stalks on the dorsal surface of the tail. Histologically, the tissue appeared disorganized, and there was considerable vacuolization of the cells. There was extensive epithelial hyperplasia and hypertrophied muscle overlying a large area composed entirely of heavily pigmented cells. The center of the tumor contained gland cells and unidentified cells with pycnotic nuclei and large globules of pigment. Nests of basophilic neoblasts were present, leading Foster to believe that at least part of the growth had arisen from neoblasts. Peculiarly, the normal single-cell layer of gastrodermal cells was entirely replaced by cells containing many pigment granules.

Because of the small numbers of tumors produced by the treatment, Foster devised a method for implanting pellets of the carcinogen mixed with paraffin in the body through a small slit on the dorsal surface (Fig. VII.15). Tumor incidence was markedly increased by the implantations; 5 of the 10 MCA-implanted animals developed heavily pigmented growths, and 4 of 10 of the BA-implanted worms developed nonpigmented nodular growths. All were lethal to the worms, but none were studied histologically. "Some MCA-implanted animals developed heavily pigmented growths with small, secondary, bulbous, nonpigmented

TABLE VII.3

LETHAL GROWTHS AND DEVELOPMENTAL ANOMALIES IN THE OFFSPRING OF
PLANARIA TREATED WITH CARCINOGENS[a]

Treatment of parents	Number of offspring	Growths in offspring	Malformations in offspring
MCA, regenerated	40	12 Papilliform tumors	6 Eyespots poorly developed; 4 enlarged heads and eyes; 3 fused eyespots
MCA, whole	None		
BP, regenerated	75	2 Nodular tumors	3 Enlarged heads; 2 small heads and eyes; 2 fused eyespots
Acetone, regenerated	65	1 Nodular tumor	Variation in pigmentation; 2 enlarged heads and eyes
Untreated controls, whole	77	None	1 Small head and eyes; 1 fused eyespot
Untreated controls, regenerated	None		

[a] From Foster (1969).

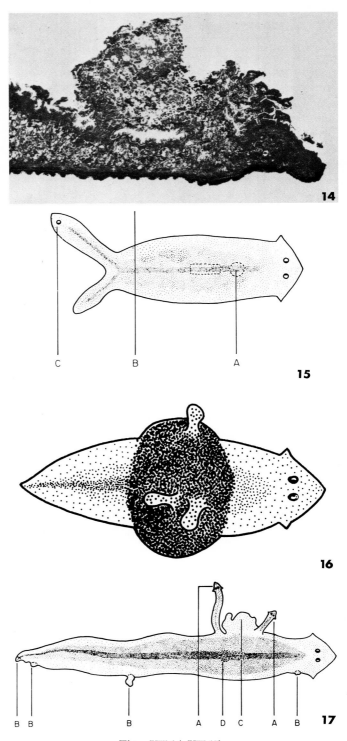

Figs. VII.14–VII.17

growths on their surfaces" (Fig. VII.16), and "one BA-implanted animal showed multiple nodular growths and a large, nodular growth bordered by two normal secondary heads" (Fig. VII.17).

Interestingly, there were far more lethal tumors produced in the off-spring of MCA-treated worms than in the treated worms themselves (30% versus 7%). Twelve papilliform tumors (Figs. VII.18 and VII.19) and thirteen malformations were recorded (Table VII.3). As shown in the table, fewer tumors and malformations developed in the offspring of BP-treated worms, and the controls had none.

The twelve offspring of MCA-treated adults that developed heavily pigmented knobs on their dorsal surfaces quickly surpassed the normal progeny in size. Pigment cells were lost from the head and tail regions, suggesting migration of pigment cells into the tumor (Fig. VII.18). Histo-logical examination 14 days after capsule insertion showed the knob to consist of a covering epithelium over pigment cells and a central core of neoblasts. The animals were centrally thickened, and the gastrodermal cells in that part of the worm associated with the tumor were hyperplastic (Fig. VII.20). In the gastrodermis, there were large, intensely basophilic nuclei surrounded by a thin area of deep blue (hematoxylin and eosine) cytoplasm. Many of these basophilic cells occurred in "tabs" of tissue extending abnormally into the digestive tract (Fig. VII.21). After 2 weeks, the worms with the papilliform growths and gastrodermal ab-normalities began to regurgitate food, developed large bulbous tumors on the dorsal stalks by the twentieth and twenty-first days, and the ten remaining unsacrificed animals died.

While it is not clear what portion of the cellular abnormalities in the carcinogen-treated planarians represent response to injurious effects (necrosis, inflammatory response, and repair of the injured areas) and what portion is neoplasia, the production of typical lethal growths in the offspring of carcinogen-treated worms further strengthens the likelihood that they are truly neoplastic.

Fig. VII.14. Sagittal section of a planarian, *Dugesia dorotocephala*, treated with 3-methylcholanthrene. Note the extensive epithelial hyperplasia and increased pig-mentation. The center of the tumor consists of nests of neoblasts. (From Foster, 1969.)

Fig. VII.15. Sketch illustrating benz[*a*]anthracene implantation technique. A, Implantation site; B, region of fission after 1½ months; C, head development in normal tail development region. (From Foster, 1969.)

Fig. VII.16. Sketch of planarian with heavily pigmented growth and several nonpigmented secondary growths developed after 3-methylcholanthrene implantation. (From Foster, 1969.)

Fig. VII.17. Sketch illustrating the effect of benz[*a*]anthracene implantation. A, Secondary heads; B and C, multiple nodular, nonpigmented growths; D, approximate implantation site. (From Foster, 1969.)

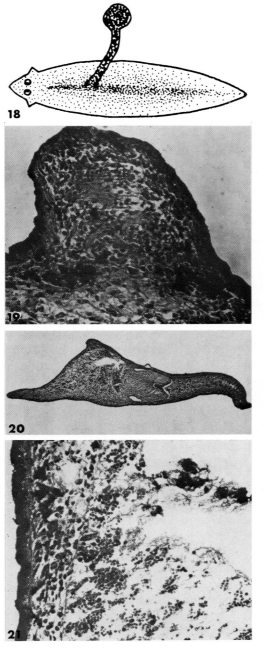

Fig. VII.18. Sketch illustrating papilliform tumor on offspring of 3-methylcholanthrene-treated adult planarian. Note the heavy pigmentation of the tumor. At 21 days, extensive depigmentation of remainder of animal had occurred. (From Foster, 1969.)

Fig. VII.19. Photomicrograph of papilliform tumor on offspring of 3-methylcholanthrene-treated adult. Histologically, the growth consists of a covering epithelium, numerous pigment cells, and a central core of neoblasts. (From Foster, 1969.)

Fig. VII.20. Sagittal section of offspring of 3-methylcholanthrene-treated adult. Note the extensive gastrodermal hyperplasia below lumen of digestive tract. (From Foster, 1969.)

Fig. VII.21. "Tab" of tissue extending abnormally into gastrodermis of offspring of 3-methylcholanthrene-treated adult. Note the large, basophilic nuclei. (From Foster, 1969.)

Phylum Annelida

Class Polychaeta

Reports of tumors in annelids are relatively rare and confusing. Thomas (1930) studied large numbers of polychaetes, *Nereis diversicolor,* that possessed globular, whitish, tumorlike structures at various sites, predominantly on the dorsal or ventral surface of the middle third of the body. Microscopically, the growths were seen to develop either around degenerating oocytes or degenerating bristles. Thomas made the point that the two types of growths were of the same nature, but the initial aspect varied because they developed at the expense of different organs.

Thomas was aware of the possible error in designating them tumors, since they developed as a result of irritation provoked by degeneration of oocytes and bristles. He contended, however, that voluminous amounts of new tissues were formed, and, therefore, they were neoplastic structures for which the term "tumor" was completely appropriate.

The growths arise as slightly hypertrophied parapodia with intensified coloration and stretched integument. As growth continues, the affected parapodia become more conspicuous, though still small and globular and with the free extremities still pointed. Eventually the entire base of the parapod becomes involved, causing the free extremity to become atrophied to a buttonlike structure resting on the swollen base. Thomas noted that he had observed one *N. diversicolor* with fifteen of these tumorous buttons. A tumorous parapod may remain isolated, or the contiguous bases may become involved.

Tumors of *Nereis diversicolor* may also arise on the segments independent of the parapodia, varying from small, whitish, barely perceptible spots up to voluminous tumors completely engulfing one or several segments. The tumors appear whitish because of loss or lack of the normal heavy vascularization, particularly of the parapodia.

Microscopic study revealed that many of the tumors arise around degenerating oocytes. At the time of origin, a group of oocytes becomes tightly encapsulated by a connective tissue sheath. Thomas felt there was a distinct possibility that each degenerating oocyte was individually encapsulated, because he frequently observed a single egg surrounded by a dense sheath of four or five rows of connective tissue cells. Subsequently, another sheath is produced to encapsulate a group of the individually encapsulated oocytes. Amoebocytes are numerous throughout the tumorous tissues, but are particularly abundant around those oocytes in advanced stages of degeneration, where the cytoplasmic contents of the amoebocytes clearly show that they are phagocytizing the degenerating oocytes.

Thomas also noted some structures, whose origins were enigmatic, among the degenerating oocytes. He was finally able, after studying a great deal of fresh material, to demonstrate that the bristles of the parapodia degenerated and were encapsulated and phagocytized in the same manner as the oocytes.

However, encapsulated oocytes and degenerated bristles occur both in normal worms and in healthy areas of tumorous worms. In the normal tissue, the encapsulated structures are isolated or grouped into very small foci, whereas in the tumors the degenerated oocytes occupy large areas. In an ecological study of the occurrence of tumors, Thomas was able to show an apparent relationship between frequency of tumors, numbers of degenerating oocytes, and salinity; oocyte degeneration and tumor formation increased with decreasing salinity.

Thomas presented a detailed account of the degeneration of the oocytes, noting the pycnosis and subsequent karyorrhexis of the nucleus and the intense reaction, which he called inflammation, to this degeneration. Amoebocytes infiltrate the area, form a plasmodium surrounding the inclusion, and develop the encapsulating sheath mentioned previously. Active multiplication in the lining of the blood vessels and the coelomic epithelium produces the amoebocytes. These "neoformed" amoebocytes organize into connective tissue around reactional foci and probably reproduce themselves amitotically. The tumorous mass may infiltrate between the integument and the body musculature, especially into the interior parts of the dorsal musculature. Thomas contended that the mass of degenerated material, the free elements, and the neoformed tissue constitute a true benign tumor.

Class Oligochaeta

On several occasions, tumors have been reported in earthworms. Gersch (1954) described earthworm epithelial tumors induced by application of benzopyrene and remarked that his experiments demonstrated the value of earthworms as experimental models in the study of tumors. Stolk (1961a,b) reported on pharyngeal tumors in the earthworm and described giant nuclei exhibiting marked polymorphy in the cells making up most of the tumorous tissue. Various enzymatic activities (three dehydrogenase systems) of the tumor were studied that were said to be "in general in good agreement with those obtained in some human tumors . . . ."

However, Hancock (1961a) reported abnormal giant nuclei in the esophageal epithelium of nontumorous earthworms (*Lumbricus terrestris*) in which the nucleolus was frequently of bizarre shape, and the nuclei appeared polyploid. Subsequently, Hancock (1961b) described several neoplasticlike lesions in earthworms subjected to X-rays (1000 rad). The most common lesion resembled a granular cell myoblastoma of vertebrates, a muscle tumor that may arise from degenerated, adult, striated muscle fibers. In a later paper, Hancock (1965) described this tumor more fully.

Granular cell myoblastomas in vertebrates contain acidophilic granules and either large vesicular or small hyperchromatic nuclei. Hancock noted that these features can be demonstrated in the earthworm lesions. A few worms receiving 1000 rad 23 days prior to sacrifice contained masses of cells exhibiting the characteristics of myoblastoma cells; the granular cells contained acidophilic material, some of the cells contained large vesicular nuclei, others had small hyperchromatic nuclei, and the striated muscle was no longer present. The seminal vesicles no longer contained spermatogenic cells; these were replaced by a solid mass of cells resembling myoblastoma cells except for a lack of acidophilic granules and the presence of only hyperchromatic nuclei. A worm receiving an intermediate level of irradiation, 400 rad, had, when sacrificed at 23 days, the coelom completely filled with myoblastoma cells, and most of the muscle tissue was missing. At a low level of irradiation, 100 rad, one earthworm sacrificed after 23 days contained myoblastoma cells that had invaded the coelom and were also present in the lateral muscle tissues. Hancock (1965) reported (Fig. VII.22) giant nuclei in the myoblastoma-like areas in the altered seminal vesicles.

Hancock (1965) also reported, for the first time, multinucleated giant cells in irradiated earthworms (Fig. VII.23). These cells were usually present in the coelom after varying doses and time after irradiation. Seven days after irradiation with 1000 rad, earthworms contained giant cells with as many as 40 nuclei. In a worm sectioned 12 days after receiving 600 rad, the muscle exhibited degeneration and hyalinization (Fig. VII.24), and giant cells appeared to be embedded in a region of fibroblasts that had replaced the muscle fibers. Hancock suggested that such regions in the muscle tissues may be the sites of origin of the myoblastomas.

Cooper (1969) questioned the relationship of myoblastoma and multinucleate giant cell formation in *Lumbricus* to irradiation since the same amount of irradiation did not produce these structures in his experiments. Furthermore, he noted that he routinely observed multinucleate giant cells in close association with obviously normal nephridial tubules, and that myoblastomas are common in worms collected in the wild. Some

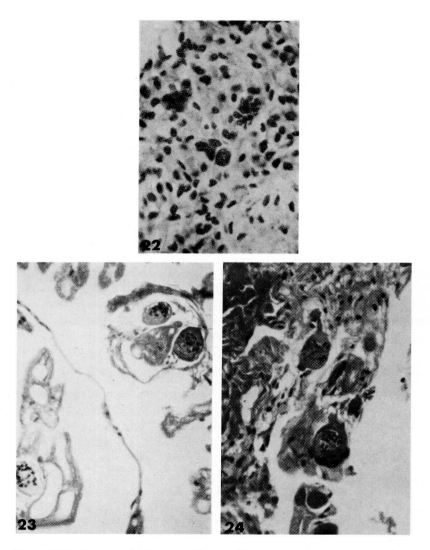

Fig. VII.22. Giant nuclei in a myoblastomalike mass in the seminal vesicle of an irradiated earthworm, *Lumbricus terrestris*. (From Hancock, 1965.)

Fig. VII.23. Multinucleated giant cells in the coelom of an irradiated earthworm. (From Hancock, 1965.)

Fig. VII.24. Multinucleated giant cells in the muscle tissue of an irradiated earthworm. Note the degeneration and hyalinization of the muscle and the fibroblasts surrounding the giant cells. (From Hancock, 1965.)

of the wild worms "completely riddled with tumors" are generally debilitated and die, while others with only one or a few tumors appear healthy, and the tumors may even regress.

Concurrently with an investigation of tissue-transplantation immunity in earthworms, Cooper (1968) studied granulomata, giant cells and myoblastomas in normal and grafted *Lumbricus terrestris* and *Eisenia foetida*. He proposed that giant cell and tumor formation may be related components of the defense system or products of the defense system that have escaped its influence.

In the latter paper, Cooper (1968) described each of these pathological conditions in some detail. To fully appreciate the relationship of the myoblastoma to the body wall, a brief review of the normal architecture of that portion of the worm is necessary (Fig. VII.25). As Cooper pointed out, the epithelium consists of two predominant cell types that produce albumin and mucus, respectively, and less numerous basal epithelial cells. The epithelium is separated from the underlying muscle layers by a basement membrane. Just beneath the basement membrane is a band of circular muscle that contains isolated pigment granules and pigment enclosed in discrete cells between the bundles of muscle fibers. A longitudinal layer of muscle lies immediately below the circular muscle layer, and it, in turn, is underlain by a double layer of peritoneal epithelium. Connective tissue cells are interspersed between the muscle bundles in both layers. Blood vessels are enclosed in the peritoneum and send branches, via the connective tissue, to the muscle layers and epithelium.

At some locations in a typical myoblastoma, only the muscle layers are involved (Fig. VII.26), but at others both the muscle layers and the epithelium are affected. The longitudinal muscle bundles remain essentially intact except that numerous connective tissue cells invade the spaces between the bundles. In contrast, muscle bundles in the area of circular muscle are rare, being largely replaced by tumor tissue. Acidophilic coelomic cells accumulate in large numbers at the base of the tumor and invade the spaces between longitudinal muscle bundles making it difficult to distinguish between the connective tissue cells and other cells from the coelom. The nuclei of the connective tissue cells and the infiltrating acidophilic coelomic cells are similar, but the connective tissue cells are more elongated than the coelomic cells. Individual chloragogen cells in small numbers also emigrate into the area.

The nuclei of the acidophilic cells are small, eccentric or central in position, with relatively prominent nucleoli and a paucity of chromatin. The cytoplasm is uniformly granular and acidophilic; the size ranges from 4 to 5 μm. A larger coelomic cell (4–9 μm) that frequently occurs in the tumor is less acidophilic and contains large, uniformly distributed vacuoles. These two coelomic cell types often coalesce to form a large syncytium. Where these cells penetrate between muscle bundles, apparent muscle degeneration occurs, evidenced by the absence of fibers, especially in the longitudinal muscle layers (Fig. VII.27).

Involvement of the epithelial layer, when it occurs, results in destruc-

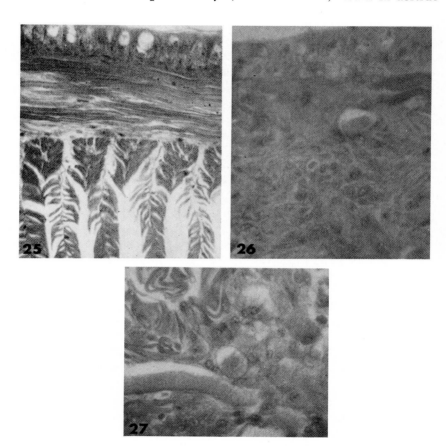

Fig. VII.25. Section of normal earthworm integument showing the epithelium and circular and longitudinal muscle layers. Approximately 400×. (From Cooper, 1968.)

Fig. VII.26. A typical region in a "myoblastoma" in an unirradiated earthworm. Note that the epithelium is essentially normal, but little muscle tissue remains. Approximately 400×. (From Cooper, 1968.)

Fig. VII.27. A myoblastoma containing prominent blood vessels. Note the numerous acidophilic cells present at the base of the longitudinal muscle bundles. Approximately 640×. (From Cooper, 1968.)

tion of the basement membrane and modification of epithelial cells to cuboidal rather than tall columnar that lack mucous and albuminous granules. Mucous granules can be found occasionally interspersed among the degenerated components of the muscle regions. Although most epithelial cells appear viable, occasional pycnotic nuclei appear, especially where the basement membrane is destroyed.

As mentioned previously, Cooper (1968) reported giant cells apparently identical to those described by Hancock (1961a) in all worms, both with and without grafts (Figs. VII.28 and VII.29). Thus, these giant cells are ubiquitous in the coelomic cavity of earthworms, occurring most frequently in association with the nephridial tubules. They are quite large (14–60 μm) and are similar in appearance to acidophilic coelomic cells except that in the multinucleate condition the nuclei are more intensely basophilic and the cytoplasm is more acidophilic and

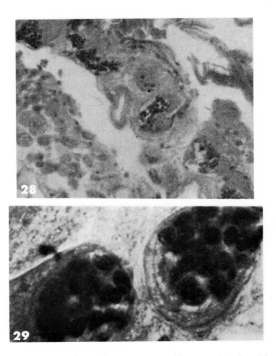

Fig. VII.28. Numerous multinucleate giant cells associated with the nephridial tubules of an unirradiated earthworm. (From Cooper, 1968.)

Fig. VII.29. Higher magnification of multinucleate giant cells of the earthworm. Note the thick cell membranes and relatively small amount of cytoplasm; the numerous small, round or oval, dark-staining nuclei are grouped eccentrically in the cells. Approximately 1000×. (From Cooper, 1969.)

less vacuolated. The nuclei are eccentrically located, and the number varies from a minimum of 5–10 to a maximum of 40–50.

Since giant cells of various origins are common in vertebrates, it is not surprising that they occur in annelids, and they can be expected in other invertebrates. The question at this point, however, is whether the earthworm giant cells are consistent with foreign-body giant cells of vertebrates or whether they are tumor giant cells. Cooper believes, and I concur, that the presence of large numbers of nuclei in the earthworm giant cells more closely resembles foreign body giant cells of vertebrates than true anaplastic cells. He (Cooper, 1968) does point out, however, that the occasional giant cells found in association with the earthworm "myoblastomas" may be tumor giant cells.

The factors that determine whether the coelomic cells encountering foreign material will develop into tumors, granulomata, or giant cells are still unresolved (Cooper, 1968). Cooper suggests that two populations of coelomic cells may be involved in coelomic defense, with one genetically determined to become giant cells upon encountering foreign material and the other proliferating to become granulomatous or tumorous.

Cooper (1969) also investigated the possibility of inducing tumor formation in earthworms by applying a vertebrate carcinogen to the body wall. Painting with MCA (3-methylcholanthrene) saturated acetone solution every other day for 12 and 24 days resulted in necrosis and denudation of the epithelium and underlying muscle layers, formation of scar tissue in the less severely affected regions, marked epidermal thickening in the least severely affected areas, and moderate to intense coelomic cell response. Although there was some increased interstitial cellularity of circular muscle of myoblastoma cell type in the 12-day experiment, Cooper stated that myoblastomas could not be induced. MCA crystals placed under autografts produced only extensive granulomas. Acetone, used as a control, produced essentially the same response as MCA-saturated acetone except for the above-mentioned interstitial cellularity. Finally, worms subjected to 1000 rad total-body irradiation were normal 40 days postirradiation.

Phylum Sipunculida

On two occasions, tumors have been reported in *Sipunculus nudus*. Hérubel (1906) described a small tumor on the side of a specimen of this species. He was particularly interested in the abnormality because it was the first observed in the more than 1000 sipunculids he had dissected over a number of years at the Roscoff Laboratory. The tumor was on the right posterior third of the animal and consisted of an oval base approximately 1 cm in diameter and an elevated hump projecting outward approximately 5 mm. The tumor consisted of numerous cells that were compressed into a hollow cavity within the skin, limited on the top by the epidermis and cuticle and on the bottom by the circular muscles of the integument. Careful microscopic study of sectioned material convinced Hérubel that the muscle fibers were degenerating, leading him to diagnose the structure as a muscle tumor. He did not attempt to ascertain the cause of the condition, but assumed it was of parasitic origin. The cells in the mass were, according to the author, unquestionably amoebocytes or leukocytes carried there by the cutaneous canals. He further noted that some of the amoebocytes in the section were in the process of phagocytizing degenerating muscle fibers.

Unfortunately, Hérubel provided no illustrations, so it is impossible to determine whether the abnormality actually was a tumor or, as seems much more likely from his own description, a heavy leukocytic response to an injury. Hérubel's description of a hollow cavity filled with cells is more a description of an abscess than a tumor.

Ladreyt (1922) described a tumor that he considered malignant in the dorsal canal of Poli in S. *nudus*. A chance dissection revealed the neoplasm originating from the endothelium of the anterior third of the esophageal tube and appearing, at the opening of the coelom, like a large pinkish-gray pea projecting into the coelomic cavity. Normally, the cells covering the internal and external surfaces of the canals of Poli are rather flattened and closely resemble the endothelial cells of the lymphatic system of vertebrates. Like the peritoneum that it resembles, the remainder of this vascular epithelium contains ciliated and nonciliated cells that differentiate into the urn cells, the red blood corpuscles, and the chloragogen cells. In the tumor-bearing sipunculid, all the cells of the vascular endothelium were modified at the beginning of their neoplastic evolution. Laydret stated that this observation demonstrates the generality of the phenomena of dedifferentiation, which supposedly characterizes the beginning of the cancerous process in the higher vertebrates. This "precancerous" stage in the sipunculid was, according to Laydret, indicated by marked hypertrophy, amitosis, frequent multipolar mitoses, and nuclear fragmentation in cells in the walls of the canal of Poli.

The cells were thought to have developed progressively into fusiform or epitheloid variants, so that the fully transformed atypical cancer cells were greatly elongated, with free extremities more or less swollen. They contained large nuclei, irregular or globular in shape, that were rich in chromatin. The tumor consisted of a massive aggregate of one stratum or several strata of these cells, greatly elongated in the anterioposterior direction and surrounding the vascular axis.

Laydret noted that the functional capacity of the endothelium seemed to be closely related to the degree of morphological dedifferentiation. He believed the endothelium was functional in the early stages of this change, but progressively lost function so that no traces of respiratory pigment or excretory activity remained in the atypical cells.

Laydret posed the question of whether this growth was a cancerous tumor and, if so, to what histological type was it related. He noted that hyperplasias associated with inflammation are common in invertebrates and are most frequently granulomas of parasitic origin, quite obviously different from the tumor in question. In concluding that the tumor was cancerous and malignant, he listed the following supportive points. There was a precancerous stage characterized by morphophysiological dedifferentiation and intense multiplication of the parietal cells resembling precancerous stages in the digestive tube of higher vertebrates. A cancerous stage followed, in which the endothelial elements were clearly atypical, with the malignant cells crowding into the general body cavity and its appendages (cutaneous canals). Its malignant nature was indicated by the tumor's effect on the respiratory and excretory functions and by regression and degeneration of the muscular and nervous systems. The morphology, development, and effects on the organism seemed to Ladreyt so similar to the comparable properties of endotheliomas and epitheliomas in higher animals that he considered it a complex malignant tumor, which he called an endoepithelioma.

Other authors have not considered the evidence sufficient for a positive diagnosis of malignancy, and, as Scharrer and Lochhead (1950) pointed out, the lack of photomicrographs or any illustrations of the tumor makes conclusive interpretations difficult, if not impossible. I cannot help wondering how Laydret was able to ascertain so much about the origin and development of the tumor when only one example, and that apparently well developed, was available for study.

Phylum Mollusca

Though certainly rare by vertebrate standards, tumors or presumed tumors appear to be more common in mollusks than in most invertebrate phyla. Whether this is because molluscan tissue tends more toward neoplasia than that of other invertebrate groups or because deviations from normal growth is more obvious is unanswerable at present. A third, and quite possibly major, reason for the apparent greater incidence of tumors in mollusks is that more mollusks are examined both grossly and histologically than any other noninsect invertebrate group.

Several authors have reviewed, with varying degrees of completeness and wide differences of opinion, the literature on tumors of mollusks as individual papers or as a part of general review articles on tumors in invertebrates. Pauley (1969), however, has completely and critically reviewed the past work in a presentation restricted to mollusks.

CLASS AMPHINEURA

There have been no reports of tumors in the chitons.

CLASS GASTROPODA

The earliest record of a possible tumor in a gastropod is found in a description by Simroth (1905) of what he called a "hump-backed snail" because of a humplike curvature of the posterior part of the mantle shield of a slug of the genus *Arion*. Strangely, Simroth did not consider it a malformation, though he did call it abnormal, but a new type. Even though the hump-backed slug was the only abnormal member of a brood of otherwise normal animals, Künkel, the breeder, subsequently described it as a new species, *Arion simrothi*, which, though never collected again, was perpetuated in the literature as a valid species until at least 1957.

Boettger (1956) reported a second example of apparently the same abnormality in one of two slugs, *Arion ater*, that were collected and placed in a terrarium. The abnormality appeared suddenly as a humplike curvature on the back after 1 month in captivity; the growth reached maturity after 3 days and maximum size after 16 days. The tumorous slug was sluggish, had little appetite, and died about 3 weeks after the tumor became apparent. Boettger noted some minor differences in the gross appearance of his specimen from that described by Simroth, but gross dissection revealed that the relationships of organs within the tumorous area were quite similar. Boettger noted that the anterior portion of the hump was filled with a brownish watery exudate, as was the curved portion of the mantle, while the posterior portion of the hump was entirely filled with organs normally found in the foot.

Histological examination, which was unfortunately neglected in Simroth's specimen, of a series of sections of Boettger's slug revealed a remarkable reduction and atrophy of the mucous gland cells of the external epithelium. The subepidermal connective tissue, however, had undergone considerable proliferation, resulting in an overall increase in skin thickness from the normal of approximately 1.6–1.8 mm to 2 mm. Szabó and Szabó (1934) had previously noted that the same phenomenon occurs regularly in aged specimens of *Limax* and *Deroceras* and also found it in a tumor of *Limax,* demonstrating that the mucous gland cells of these gastropods degenerate with advancing age and certain pathological conditions. Boettger found no changes in the circular musculature, but the connective tissues associated with the musculature cords contained cysts of a gregarine. He assumed that the parasitic infection had initiated connective tissue proliferation and that mechanical and, particularly, chemical stimuli acted on the snail's tissue to cause premature alteration of the skin through atrophy of the mucous gland cells of the epidermis.

The third case this abnormality was reported by Frömming *et al.* (1961). It occurred in a different species, *Deroceras reticulatum,* collected along with a number of apparently normal individuals of the same and other species. Although the hump-backed snail appeared lively in the terrarium, the authors sacrificed it for histological study after first photographing it in the live state.

Histologically, Frömming *et al.* were unable to detect tumorous tissue. The musculature of the swollen portion of the slug was poorly developed; the individual bands were thin, and the typical three muscle systems making up the membrane muscle sac that covers the hump were difficult to distinguish. The epithelium was normal, but the mucous cells of the underlying connective tissue were reduced in size and number. The inside of the sac contained viscera of normal appearance. The authors interpreted the hump-backed condition as a deformity caused by hatching, whereby the young snail, in forcing its way through an abnormally small opening in the shell, had its visceral organs displaced upward and backward. A hernial sac resulted that retained the visceral organs in their unnatural position. Temporary deformations in hatching slugs caused by the pressure of emerging through a hole of insufficient size are not rare; the abnormality in this case, and presumably in the two discussed previously, is herniation of the tissues allowing the visceral organs into the spongy network of muscle bundles making up the membrane muscle sac and formation of a hernia sac to contain the viscera in the abnormal location. This appears to be a correct interpretation since the only characteristic known to be common to all three cases is the displaced viscera. It is unfortunate that Boettger's hump-backed snail was infected with a gregarine that apparently initiated some confusing pathological changes.

There can be no question, however, of the neoplastic nature of a tumor in *Limax flavus* described by Szabó and Szabó (1934). The slug possessing the neoplasm was raised in the laboratory and was 3¾ years old when the tumor was first noticed. The tumor grew rapidly, collapsed, then grew quickly again—a pattern often repeated in the 3 months the tumor was observed. The authors surgically removed part of the tumor for microscopic study and fixed the still-living animal at approximately 4 years of age, since the maximum age for the species was approaching.

Grossly, the tumor consisted of a colorless structure measuring 15 mm in width, 8 mm in length, and 5 mm in height, resting on the mantle edge and against the tentacles (Fig. VII.30) and originating from the body by a thin stalk. Histologically, the tumor was covered by an epithelium varying from cuboidal to squamous rather than the columnar epithelium found on the mantle of a normal animal. Immediately beneath the surface epithelium, the stroma of the tumor consisted primarily of epithelial cells grouped to produce a glandular appearance, hollow cavities surrounded by a layer of cells, and supported by muscle and connective tissue cells. Primarily alveolar, but also tubular, glands were imitated. The epithelium lining these glandlike structures was one layer in thickness, usually columnar or cuboidal, but occasionally squamous in type. It is not clear whether these "glands" were functional, but the authors noted that small amounts of mucoid secretions and necrotic debris could be seen in the hollow cavities. Internally, the stroma of the tumor consisted primarily of dense connective tissue, rich in nuclei and containing round glandular cells of a type found in much greater numbers in the normal body wall. There also appeared to be considerable proliferation of connective tissue in the normal tissue near the tumor and of the epithelial tissue at the base of the tumor.

As the authors pointed out, this tumor must be considered a definite case of neoplasia for several reasons.

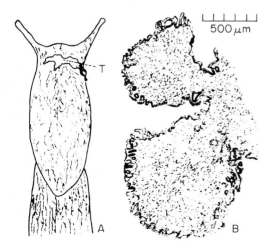

Fig. VII.30. A, Sketch of tumor (T) on slug (*Limax flavus*). B, Microscopic appearance of tumor. (From Szabó and Szabó, 1934.)

1. There was a tendency toward uncontrolled growth and proliferation.
2. The histological structure of the tumor differed from all normal tissue, since the epithelium never forms glandular structures in the normal animal.
3. It occurred where no tissue normally exists.
4. It was sharply delineated and differed from the normal tissue in structure.
5. There was no apparent injury, eliminating an inflammatory response as the initiator of the proliferation and growth.
6. Hypertrophy may be eliminated since the new tissue failed to agree histologically with normal tissue, and the tendency toward uncontrolled growth and proliferation eliminates hyperplasia.

Two apparently similar tumors of slugs (*Arion rufus*) were described by Frömming *et al.* (1961) in the same paper in which the hump-backed snail was discussed. The authors correctly pointed out that their tumors were unrelated to the epithelial tumor of *Limax* described by Szabó and Szabó, even though both sets of authors referred to them as flaplike tumors of the mantle shield.

One of the two tumors occurred on a young slug of 3½ months and consisted of a faint vesicular swelling with a smooth glossy external appearance, while the other appeared on an older slug 18 months old. The latter was much larger in size and contained on open lesion on its dorsal surface, which the authors termed an open tumor. It seems clear that the lesion had undergone trauma after proliferation, rather than the trauma being the cause of the proliferation. The two growths were considered to be identical by the authors, differing only in size. They were defined as tumorlike hyperplasias of the mucous cells of the membrane muscle sac. The mucous cells were greatly enlarged, so the process of hypertrophy was as conspicuous as the hyperplasia. The hypertrophy of mucous cells caused compression of muscle layers and connective tissue, in some instances displacing these tissues and even the intestine. The hyperplasia and accompanying hypertrophy of mucous cells in the larger tumor displaced the connective tissue and muscle tissue lying in between their origin and the membrane, and the neoplastic mucous cells then arched over the upper part of the membrane. There were no indications of true invasive growth and no mitotic figures, so the authors excluded the possibility of malignancy. The initial cause of the abnormalities could not be established; no parasites or bacteria were found, and, as mentioned previously, the character of the inflammatory response of the ulcerated portions makes it clear that the trauma followed the proliferation rather than initiating it.

Gersch (1950) described a large growth in the "lung" or mantle cavity of the pulmonate land snail *Helix pomatia* that he said "must be labeled a tumor." Since the neoplasm replaced a large amount of normal tissue through uncontrolled proliferation, this diagnosis is apparently correct. The increased proliferation of connective tissue cells, resembling the Leydig cells of bivalves, filled approximately one-third of the respiratory organ. Histologically, there appeared to be an altered nucleocytoplasmic ratio of the proliferative connective tissue cells; the center of the growth consisted of a mass of these cells with reduced cytoplasm. The nuclei of these cells could be distinguished from those of normal connective tissue cells only by the absence of typical nucleoli. However, Gersch did not consider the possibility that the growth under consideration resulted from hyperplasia. The lack of mitotic figures, especially bizarre patterns of division, argues against the anaplastic nature of the tumor.

Subsequently Nolte (1962) described two tumors on a snail of the same species. The larger consisted of a whitish, nodular mass approximately 1 cm^3 in size located on the thick-walled blind sac of the sheath, inside the loose connective tissue. The second growth, though smaller, was similar in gross and microscopic appearance and occurred on the lower part of the salivary gland. Sections showed that both tumors consisted of vesicular, proliferated epithelial tissue surrounded by normal connective tissue. In the salivary gland tumor, the vesicles lay close to one another and crowded out or replaced part of the normal glandular tissue. The vesicles consisted of an irregular, multilayered epithelium with necrotic centers. Three recognizable layers occurred in the vesicle walls: an outer layer consisted of small, light cells of various sizes and shapes with deeply staining nuclei. A middle zone of smaller spindle-shaped cells with basophilic cytoplasm was arranged parallel to the vesicle wall. Numerous indications of nuclear constriction and fragmentation were seen in the middle layer, but no mitotic figures could be found. The inner vesicular wall consisted of large round cells with apparently coagulated cytoplasm and remarkably large, round nuclei. All cells in this layer appeared necrotic, with pycnotic nuclei, but a gradual transition could be seen from the more nearly normal cells on the outer zone of the layer to the lumen of the vesicle that contained only remnants of decomposed cells. Nolte apparently was able to find additional vesicles just forming, each consisting of an accumulation of large cells similar to those of the inner layer of the mature vesicle surrounded by smaller cells forming a still indistinct vesicular wall. Small necrotic centers had already developed in some of the developing vesicles.

P. H. Fischer (1954) described a pedunculated fibrous tumor originating from the mantle of a gastropod, *Pleurobranchus plumula*. The tumor consisted of two unequal spheres joined together and connected to the mantle by a nonretractable stalk. Histologically, the tumorous mass and the stalk consisted of fibrous connective tissue forming a compact, irregularly arranged core covered by an unbroken epithelial covering. The nodular portion of the growth contained abnormal, deeply basophilic fibers radiating out from the center. Numerous connective tissue cells were also present in the nodular portion and were, in Fischer's opinion, the probable source of the basophilic fibers. The lack of mention of mitotic figures or invasiveness indicates that this, too, was a benign lesion.

CLASS SCAPHOPODA

There are, to my knowledge, no reports of tumors in the class Scaphopoda.

CLASS PELECYPODA

Reports of tumors in bivalves are relatively common in comparison to other invertebrate groups. The possibility that true neoplasia, in the vertebrate sense, does not occur in this group has been considered (Sparks et al., 1964a), because a number of so-called tumors have been shown subsequently to be hyperplastic reactions to injury rather than true neoplasias. The difficulty of distinguishing between neoplasia and hyperplasia was pointed out by Scharrer and Lochhead (1950), leaving one to sometimes feel that neoplasia in the bivalves is hyperplasia of unknown cause. However, as Pauley (1960) noted, recent studies of reaction to injury and wound repair in oysters have done much to clarify the injury response in mollusks and to facilitate distinction of neoplasia from it, even though borderline cases exist that are difficult to distinguish because of their superficial resemblance to neoplasia.

Apparently, the earliest report of a tumor in pelecypods was by Ryder (1887); he described a benign mesenchymal tumor originating in the pericardium of the American oyster, Crassostrea virginica. Although no figures were presented, an excellent description of the gross appearance of the tumor was given. It consisted of a soft, pliable, nodular mass 1 inch long and ½ inch wide located in the pericardial cavity just anterior to the adductor muscle. The pericardial cavity was enlarged and the heart displaced forward by the tumor. Subsequent authors (Smith, 1934; Butros, 1948; Sparks et al., 1964a) interpreted Ryder's description of the histology of the growth to designate it as a mesenchymal tumor. Pauley (1969), after careful study of Ryder's paper, concluded that the tumor was of epithelial origin, originating from the tall columnar epithelium lining the pericardial cavity. Pauley believed that Ryder failed to recognize the epithelial origin because of neoplastic changes. Since no figures of the histology of the growth were provided, it remains impossible to be sure of either its cellular makeup or its origin.

Smith (1934) described and illustrated a benign, nodular tumor in the same species that he stated "was not unlike the one found by Ryder." However, this statement is not correct in a number of ways, as Pauley (1969) pointed out. Smith's tumor was $1\frac{1}{4} \times 1 \times \frac{1}{3}$ inch in size and lay outside the pericardium to which it was connected by a narrow stalk. The tumor and its stalk were composed of mesenchymal tissue indistinguishable from normal Leydig cells, and it was covered by a single layer of ciliated columnar epithelium. An intense inflammatory response was evidenced by a heavy leukocytic infiltrate just beneath the surface epithelium.

A remarkably similar tumor was described by Sparks *et al.* (1964a) from a Pacific oyster, *Crassostrea gigas* (Fig. VII.31). Grossly, this growth consisted of an ovoid mass approximately 20 mm in length and 15 mm in width, lying in a slight depression just dorsal to the rectum and slightly anterior and dorsal to the adductor muscle outside the mantle. The tumor was connected to the body in the area of the rectum, from which it appeared to have originated, by a stalk of apparently identical composition. On cut section, the tumor appeared to consist of typical connective tissue covered by a thin capsule.

Microscopically (Fig. VII.32), there was a stroma of relatively typical Leydig cells, but special staining demonstrated increased collagenlike material, particularly in the center and underlying the basement membrane of the epithelial covering. The capsule was covered by an epithelium, primarily tall columnar, much like that found on the free edges of the mantle, but conspicuously convoluted with deep folds or crypts. Other areas of the epithelium consisted of low cuboidal cells characteristic of normal mantle epithelium. Beneath the epithelial covering there was, in many cases, a heavy leukocytic infiltrate surrounding and internal to the previously mentioned collagenlike fibers. The stroma was heavily vascularized, and there was a conspicuous leukocytic infiltration throughout much of the Leydig cell area. The stalk of the tumor appeared to have arisen from the dorsal portion of the mantle slightly anterior to the adductor muscle. Study of sections of the apparent origin strongly indicated that the connective tissue of the stalk was continuous with that of the mantle. Through the courtesy of Dr. Victor Loosanoff, the authors were able to directly compare the histology of the tumor in *C. gigas* with that of Smith's tumor in *C. virginica* by study of Smith's original slides and concluded that the two were identical in structure except for more conspicuous epithelial convolutions in *C. gigas*. Sparks *et al.* were reluctant to give a name to the two tumors, but noted that they were probably benign neoplasms of Leydig cells or hamartomas since they consisted of cells of normal appearance in an abnormal location. The marked inflammatory reaction undoubtedly resulted from repeated trauma to the tumor by movements of the shell.

Several months after the tumorous Pacific oyster was given to the above authors, the same commercial packer found and gave them an-

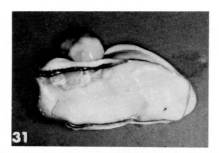

Fig. VII.31. A Pacific oyster, *C. gigas,* with a mesenchymal tumor situated just dorsal to the rectum and dorsoanterior to adductor muscle. (From Sparks *et al.,* 1964a.)

Fig. VII.32. Microscopic appearance of the tumor. Note the deep crypts in the columnar epithelial covering and the heavily vascularized area containing numerous collagenous fibers beneath the epithelium. Gomori's trichrome. 43×. (From Sparks *et al.,* 1964a.)

other oyster with an apparently identical growth (Sparks *et al.,* 1964b) (Fig. VII.33). During gross examination, however, the rectum, which normally passes over the dorsal surface of the adductor muscle, could not be located, leading the authors to strongly suspect that the apparent stalked tumor was actually the expanded rectum. This suspicion was confirmed by histological examination.

Gross examination revealed the abnormal tissue to consist of an ovoid mass measuring $24 \times 18 \times 16$ mm. Its surface was smooth, rather shiny, yellow-brown in the fixed state, and somewhat fluctuant. It was located in a depression on the dorsal side of the oyster, lying dorsoanteriorly to and depressing the anterior portion of the adductor muscle. It lay outside the mantle, and, as mentioned above, the rectum could not be located in the usual position along the dorsal surface of the adductor muscle. On cut section, the growth appeared to consist of a central, brown, laminated mass averaging 13 mm in diameter surrounded by a capsule ranging from $\frac{1}{3}$ to 3 mm in thickness.

Histologically, the central part of the abnormal growth appeared as a granular, eosinophilic, laminated mass, with colonies of bacteria and fungi scattered through it (Fig. VII.34). Necrotic leukocytes with pycnotic nuclei were observable at the periphery of the mass. The glandular, ciliated, tall columnar epithelium surrounding the laminated mass and forming the inner border of the capsule was normal in appearance, even though unusually large numbers of leukocytes appeared to be moving across this epithelial border. Surprisingly, the epithelium did not appear to be compressed by the large fecolith, but, rather, the tissue directly beneath the epithelium appear to be forming a fibrous wall, composed of collagenlike fibers, flattened Leydig cells, and leukocytes. The cells in this wall contained round, pycnotic nuclei rather than the elongated nuclei typical of developing fibrosis in oysters. There was a pronounced inflammatory reaction in the capsule, characterized by heavy leukocytic infiltration, edema, vascular dilation, and congestion of the smaller blood vessels. However, much of the capsule was composed of apparently normal Leydig cells. The periphery of the capsule consisted of a collagenlike wall, overlain by necrotic and fragmented cuboidal epithelial cells. A similar abnormality in the intestine of an oyster was seen by the above authors in a slide loaned to them, along with slides of Smith's (1934) mesenchymal tumor, by Dr. Victor Loosanoff. In the latter case, however, it was not as conspicuous, since it was located in an area of the intestine surrounded by the visceral mass.

Although the fecal impaction described above fits the broadest of all definitions of tumor, i.e. swelling, it probably cannot be diagnosed as a neoplasm. The cellular architecture of the rectum was fairly normal ex-

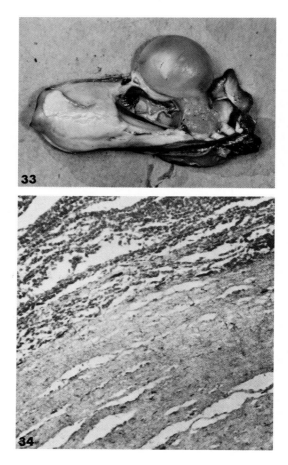

Fig. VII.33. A Pacific oyster, *C. gigas,* with a tumorlike fecal impaction. Note the gross resemblance to Fig. VII.31. (From Sparks *et al.,* 1964b.)

Fig. VII.34. Microscopic appearance of the fecal impaction. Note the laminated character of the central mass and the band of necrotic leukocytes surrounding it. Harris hematoxylin and eosine. 228×. (From Sparks *et al.,* 1964b.)

cept for the tremendous increase in the size of the organ. It is quite likely that the epithelium of the rectum had undergone hyperplasia to form the greater circumference; the same is probably the case with the other cellular elements.

This abnormality serves an important function in demonstrating the ease with which an abnormality can be mistakenly diagnosed as a tumor. If the authors had not carefully studied the gross appearance and, especially, the relationships of the abnormal structure to other organs, it is possible that the rectum would not have been recognized in the microscopic study. The final result could have conceivably been the description of an unusual tumor of intestinal origin. (Note the similarities in gross appearance in Figs. VII.31 and VII.33.)

A second mesenchymal tumor in *C. gigas,* described by Sparks *et al.* (1969), resembled in many ways the tumors previously discussed, but differed considerably from them in other characteristics. It was an elongate ovoid mass, measuring 30 × 13 × 15–20 mm, lying, in its normal position, to the left of and anterior to the adductor muscle and covering the pericardial sac (Fig. VII.35). It was soft and pliable in some areas and semisolid but still flexible in others. The surface was irregular and nodular in appearance, and the growth superficially seemed to originate in the area of the pericardial sac. Closer observation revealed it had originated ventrally from the main body mass, had grown into the suprabranchial chamber, then penetrated the mantle, and had grown dorsally to occupy the position dorsal to the body.

The tumor was studied microscopically in five different areas as shown in Fig. VII.36, and, since the microscopic structure varied somewhat, a brief description of each is presented. Section 1 possessed cryptlike depressions on that portion of the surface covered by tall columnar epithelium (Fig. VII.37) containing numerous mucous cells. The major portion of the surface in this area was smooth and covered by a single layer of cuboidal epithelium. One portion of the surface was traumatized, and the surface epithelium was missing. Beneath the surface epithelium was an area of inflammation that was highly edematous, infiltrated by leukocytes, and contained both muscle and collagenlike fibers. Beneath this inflammatory area were heavy deposits of collagen that formed the central core of the tumor (Fig. VII.38).

Section 2 contained some epithelial crypts. Within the body of the tumor there were several conspicuous concentric layers of muscle and collagen fibers (Fig. VII.39) circularly arranged with much finer fibers running perpendicular to them. This portion of the neoplasm contained some extremely interesting structures that appeared under low magnification to be digestive tubules (Fig. VII.40). However, higher magnification revealed that they were not epithelial structures, but were composed of collagenlike fibers and what appeared to be leukocytes (Fig. VII.41). These structures were called "pseudotubules" by the authors. Section 2 was, in general, more cellular than section 1 and had many Leydig cells throughout. Edematous areas and heavy collagenous deposits beneath the epithelium indicated some trauma had also occurred at this location in the tumor.

Section 3 was similar to section 2, but possessed a more convoluted

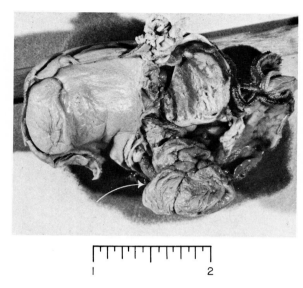

Fig. VII.35. *Crassostrea gigas* with pedunculated mesenchymal tumor. Tumor pulled down to show origin from body beneath adductor muscle. (From Sparks *et al.*, 1969.)

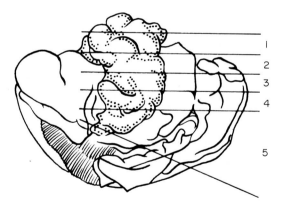

Fig. VII.36. Sketch of above tumorous oyster showing normal position of the tumor and the 5 areas of the tumor sectioned for microscopic examination. (From Sparks *et al.*, 1969.)

surface (Fig. VII.42) and a more pronounced inflammatory reaction beneath the surface epithelium (Fig. VII.43). Section 4 was similar to sections 2 and 3 with concentric layers of muscle and collagen still present. However, both the central and peripheral areas were more edematous than in the previous sections.

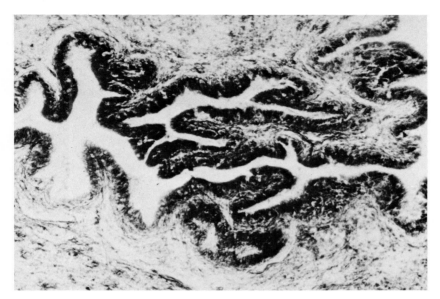

Fig. VII.37. Cryptlike depressions on portion of surface in section 1. Note tall columnar epithelium and numerous mucous cells. Mallory's trichrome. (From Sparks *et al.*, 1969.)

Section 5 had an almost normal Leydig cell area around the periphery. The peripheral area still contained pseudotubules and the concentric bands of collagenlike fibers and muscles, although the bands were fewer and thinner than in the other areas. Collagenlike deposits were interspersed among the Leydig cells near the surface epithelium, and the central area was highly edematous with only a few scattered Leydig cells.

Grossly, this tumor was similar to the large benign tumors observed on *Anodonta* (Butros, 1948; Collinge, 1891; Williams, 1890) and *Crassostrea* (Smith, 1934; Sparks *et al.*, 1964a). However, this growth differed microscopically from the other lesions in the extensive collagenlike areas and core, extensive edematous areas, concentric bands of muscle and collagenlike fibers, and pseudotubules. The inflammatory reaction noted appears to be a common feature among benign lesions in bivalves.

This growth was diagnosed as a benign tumor, since all cell types present were normal, even though the peculiar pseudotubules represent previously undescribed structures, and there were neither indications of invasiveness nor mitotic figures noted.

Pauley and Sayce (1967) provided gross descriptions of two tumorous growths from *C. gigas,* but were unable to study them microscopically

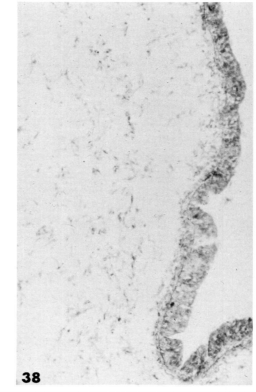

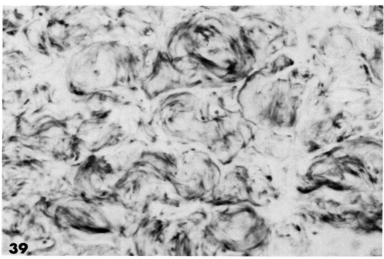

Figs. VII.38 and VII.39

because of improper fixation. One of these was a smooth textured, dark-green growth arising from the mantle, while the other consisted of an apparent polypoid or papillary neoplasm emerging from a bifurcation in the adductor muscle.

Among the most interesting tumors of mollusks, and for that matter of all the invertebrates, are the papillary epitheliomas occurring in the Sydney rock oysters, *Crassostrea commercialis,* in restricted localities in Australia. Two tumorous oysters (Figs. VII.44 and VII.45) were found by Mr. Peter Wolf during an inspection of an oyster shucking plant on the Hawkesbury River, New South Wales, Australia (Wolf, 1969). The operator of the plant reported that similar growths were noted occasionally, but not in great numbers. Additional tumorous oysters were subsequently obtained by the packer, but were unsuitable for histological study because of poor fixation. Grossly, however, the tumors appeared identical to the first two collected. Still later, Wolf located another source of tumorous oysters, with similar if not identical lesions, near Nowra, 100 miles south of Sydney, Australia.

With one exception, which will be discussed later, all the tumors originated on the ventral side of the oysters, arising on the inner mantle adjacent to the inner lobe and anterior to the inhalant chamber. In the earliest stage found to date (Fig. VII.44), the tumor is rather small, approximately 2 mm in diameter, spherical, and has a slightly grooved surface. Additional specimens appear to demonstrate progressive growth to 5–7.5 mm in diameter (Fig. VII.45), becoming oval with a deeply ridged surface (Fig. VII.46). In the most advanced tumor yet found (Fig. VII.47), the growth attained a size of 39 mm in length and 22 mm in width (Wolf, 1971), approximately half the length of the oyster. In the largest tumor, the deep ridges have progressed to the point that fingerlike projections are presented at its outer periphery. This lesion, in contrast to all others found, is situated on the dorsal side of the oyster, but still originates from the mantle lobe (P. Wolf, personal communication). At least one oyster (Fig. VII.48) has been found with a well-developed tumor on one mantle lobe and an early developing tumor on the mantle lobe on the opposite side, with no tumorous tissue connecting the two.

Histologically, all of Wolf's tumors are remakably similar; therefore, the description of the larger of the two tumors described by Wolf (1969) is typical of the neoplasm. The growth is epithelial in origin, apparently

Fig. VII.38. Smooth surface of *C. gigas* tumor covered by single layer of cuboidal epithelium in section 1. Note edematous area beneath epithelium. Hematoxylin and eosine. (From Sparks *et al.,* 1969.)

Fig. VII.39. Heavy deposits of collagenlike material in section 1 of tumor. Mallory's trichrome. (From Sparks *et al.,* 1969.)

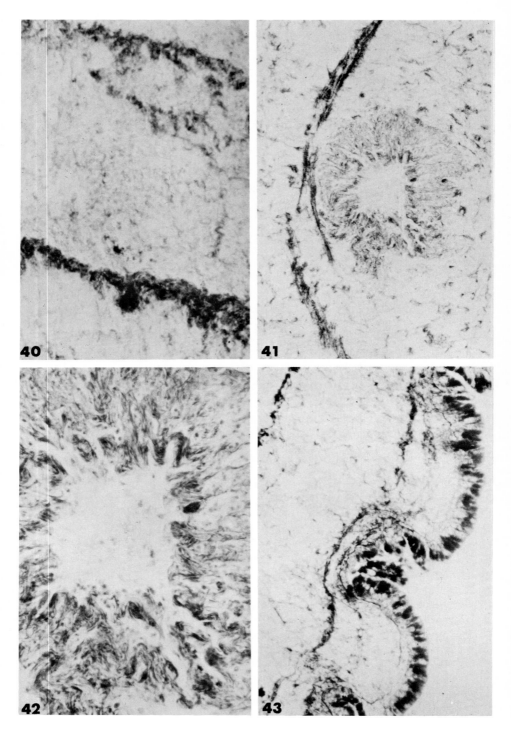

Figs. VII.40–VII.43

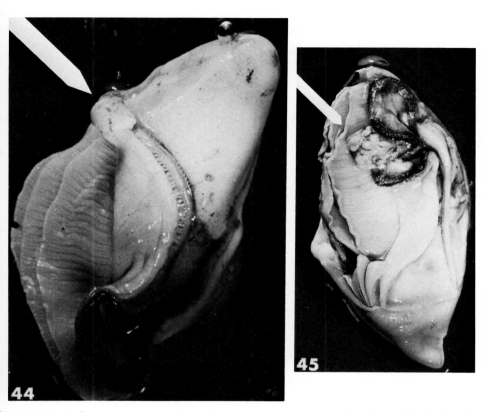

Fig. VII.44. Sydney rock oyster with small tumor on inner mantle lobes. From Hawkesbury River, Australia. (From Wolf, 1969. Courtesy of P. H. Wolf.)

Fig. VII.45. Sydney rock oyster with larger tumor on inner mantle lobes. From Hawkesbury River, Australia. (From Wolf, 1969. Courtesy of P. H. Wolf.)

Fig. VII.40. Mesenchymal tumor in *C. gigas.* Concentric layers of muscle and collagenlike fibers within the body of the tumor in section 2. Note the finer fibers perpendicular to the layers. Mallory's trichrome. (From Sparks *et al.,* 1969.)

Fig. VII.41. "Pseudotubule" in section 2 of tumor. Note similarity under low power magnification to appearance of digestive tubules. Mallory's trichrome. (From Sparks *et al.,* 1969.)

Fig. VII.42. "Pseudotubule" at higher magnification. Note that the structure is composed of collagenlike fibers and apparent leukocytes. Mallory's trichrome. (From Sparks *et al.,* 1969.)

Fig. VII.43. Section 3 of tumor. Note convoluted epithelial surface. Mallory's trichrome. (From Sparks *et al.,* 1969.)

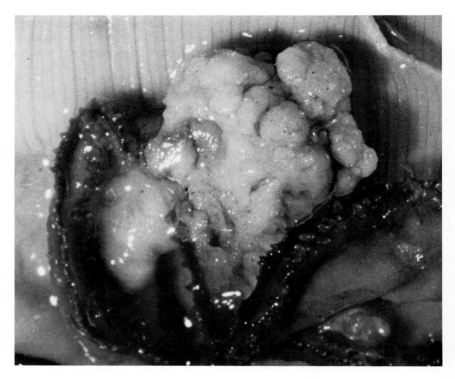

Fig. VII.46. Higher magnification of the tumor in Fig. VII.45. (From Wolf, 1969. Courtesy of P. H. Wolf.)

arising from the middle fold, which it closely resembles histologically at its outer periphery. It is composed of columnar and cuboidal epithelium interspersed with numerous mucous cells (Figs. VII.49 and VII.50). The number of mucous cells is greatly reduced in the interior of the tumor, and most of the cells, which vary from cuboidal to columnar, are ciliated (Fig. VII.51). Numerous cavities are formed by ridges and folds of the proliferating epithelium and are filled with leukocytes, indicating an intense inflammatory response (Fig. VII.52). Although growth is expansive from the middle mantle lobe outward, producing the grossly recognizable tumorous mass, it is also invasive, penetrating the mantle by way of the middle lobe. In the tumor's invasion of underlying tissue, Leydig cells are replaced by infiltration rather than by displacement and distortion. Muscle tissue is also surrounded and invaded (Fig. VII.53 and VII.54) by the proliferating neoplastic tissue. Frequent mitotic figures (Fig. VII.55) are indicative of rapid growth of the lesion.

Subsequently a more advanced stage of the tumor was described (Wolf, 1971). The expansive growth basically consisted of closely com-

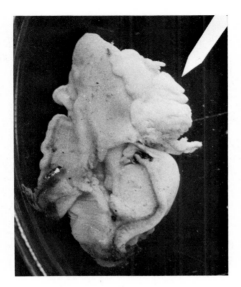

Fig. VII.47. Sydney rock oyster with advanced tumor. From Nowra, Australia. Note the tumor is situated on the dorsal side of the oyster and is approximately half the length of the oyster. (From Wolf, 1971. Courtesy of P. H. Wolf.)

pressed epithelial villi or papillae supported by branching cores of connective tissue. The resulting crypts and clefts clearly communicated with the mantle cavity and were filled with masses of necrotic cells resulting from exfoliation of epithelium from the villous processes.

The invasive portion of the neoplasm extended through the muscle and connective tissue of the mantle into the gonads. The ovarian tissue was not destroyed, and invasion of normal tissue stopped just short of the digestive diverticula.

At the base, the neoplastic epithelium displaced the normal mantle epithelium for a short distance, and the junction of normal and tumor epithelium was abrupt. Cytologically, the tumorous epithelium exhibited cellular atypia with large nuclei and reduced cytoplasm, causing increased cellularity in the tumorous tissue. The increased nucleocytoplasmic ratio and increased basophilia of the cytoplasm produced a pronounced hyperchromatism of the lesion. Mitotic figures were moderately numerous, particularly toward the periphery of the growth.

Most of the criteria of vertebrate tumor malignancy occur in Wolf's tumor. Although grossly recognizable by its expansive growth, the tumor is also invasive, apparently very much so, infiltrating the mantle in all instances and through the gonads almost to the digestive diverticula in the most advanced case found to date. Growth is apparently rapid, as shown by the increased mitotic activity, and evidence of anaplasia is

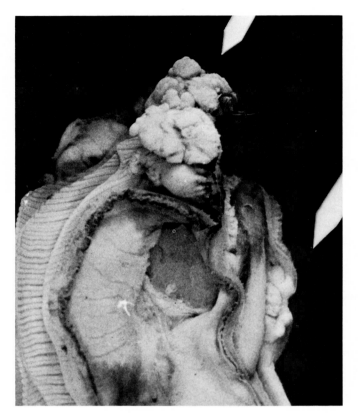

Fig. VII.48. Sydney rock oyster with 2 unconnected tumors. From Nowra, Australia. (Courtesy of P. H. Wolf.)

present in the form of altered nucleocytoplasmic ratio, increased cellularity, and increased cytoplasmic basophilia. Metastasis has not been proved or even hinted at by Wolf, but the occurrence of two tumors on at least one oyster with no tumorous tissue connection between them is at least suggestive of the possibility of the smaller tumor having arisen metastatically. The final criterion of malignancy, death of the afflicted organism, has not been established, but it is highly probable that invasion of the digestive tubules and other vital organs would have followed and led to a fatal termination in the oyster bearing the largest tumor if the oyster had not been sacrificed.

It is enigmatic indeed that two populations of oysters developing apparently identical tumors have arisen in confined areas separated by more than 100 miles of coastline. Even more perplexing is the agent responsible for the tumorgenesis. Pollution, industrial (chemical and radioactive) and domestic, can be eliminated. While there is a paper mill about 6

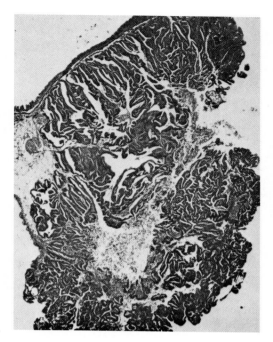

Fig. VII.49. General histological structure of Sydney rock oyster tumor. (Section from tumor in oyster shown in Fig. VII.45.) Approximately 15×. (From Wolf, 1969.)

miles upstream from the Nowra site, oysters between the mill and the affected bed do not develop tumors. The tumorous oysters from Nowra have all been collected from a bed naturally set on the bottom. All other commercial production in the area and virtually everywhere else in Australia is off the bottom (on sticks or in elevated trays). This would lead one to suspect that a carcinogenic substance is present in the mud of the oyster bed; however, the tumorous oysters from the Hawkesbury River area were grown off the bottom. Wolf (1971) lists three possible explanations for the fact that the twelve tumor-bearing oysters from the Nowra area were from the same shellbed.

(1) A genetic mutation occurred which produced a population of tumor-susceptible oysters and the tumors appeared spontaneously, (2) the infectious agent, which may be a virus, is localized in this shellbed, and (3) the infectious agent is ubiquitous but only causes tumors in oysters rendered tumor-susceptible, as by a genetic mutation.

Wolf has arranged for electron microscopic study of the tumors and is setting up transplantation and infectivity studies in the hope of discovering the cause of the tumors.

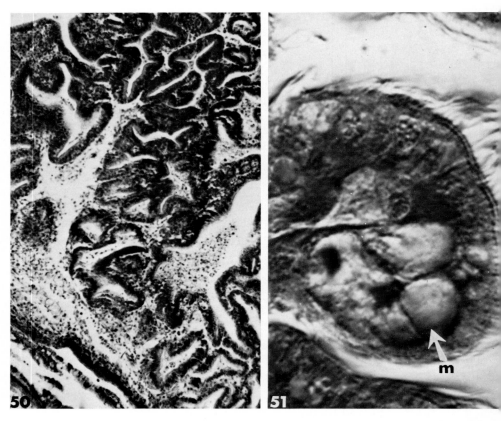

Fig. VII.50. Higher magnification of tumor showing convoluted columnar and cuboidal ep‑
lium near periphery. Note the numerous mucous cells and the leukocytes in the cavities. 3?
(From Wolf, 1969. Courtesy of P. H. Wolf.)

Fig. VII.51. Section from the interior of the lesion showing the ciliated epithelial cells.
the mucous cells (m). Oil immersion; Nomarski interference contrast. 840×. (From Wolf, 1
Courtesy of P. H. Wolf.)

Although access to the Hawkesbury River population is no longer
available, the incidence of tumors (12 in 20,000 oysters examined, or
0.0006%, to date) in the Nowra population is sufficiently high to allow
the types of investigations of tumorgenesis, transplantation experiments,
and epidemiological studies that have provided so much valuable in‑
formation in vertebrate oncology.

As Pauley and Sayce (1968) pointed out, tumors of mollusks are usu‑
ally discovered because they are large, readily visible masses protruding
externally from the body surface of the animal. Internal tumors, not
being grossly obvious and perhaps not being recognized microscopically,

have only rarely been reported. Pauley and Sayce (1968) described an internal lesion that was not grossly visible, but was palpable as a hard mass on the dorsal surface of a Pacific oyster. The growth appeared to consist of a fibrous capsule enclosing a hollow center. It was relatively large, $42 \times 19 \times 23$ mm, and had displaced much of the normal tissue (Fig. VII.56).

Microscopically, the tumor consisted of a stroma of fibrous connective tissue and collagen (Fig. VII.57), with occasional scattered Leydig cells of normal appearance. The nuclei of the fibroblasts were round, but many of the cells were elongated similar to those found in the wound repair process of oysters. There were numerous sinusoids containing leukocytes and numerous tubular gametogenic structures at the edge of the tumor that led the authors to note that it appeared to have originated from undifferentiated gonadal cells (Fig. VII.58). However, they demonstrated by additional study that gonadal origin was unlikely. The tumor cell nuclei were smaller than those of adjacent gametogenic tubules, and their tendency to assume a spindle shape is inconsistent with germ cell origin. The authors noted that the tendency toward a spindle shape with long bipolar cytoplasmic extensions is suggestive of myofibrils, hence a smooth muscle origin. Special stains (Van Gieson, PTAH) support the muscular origin, while the location within the peripheral gonadal tubules supports the gametogenic origin. The lack of any glandular tissue eliminates the possibility of origin from the digestive tubules that were in contact with the tumor in a few locations. The Leydig cells in the tumor contained very little glycogen, while the normal Leydig cells adjacent to the tumor were filled with glycogen. Tumor cells themselves, though they were PAS-positive, contained no glycogen. The hollow center of the growth lacked a limiting membrane and was filled with a clear liquid.

A portion of the tumor that appeared to grossly divide the growth into two lobes was composed of normal Leydig cells heavily infiltrated with leukocytes and contained a moderate increase in collagen. Although this tissue was essentially normal in appearance, it seemed to be a part of the tumor. The border of the neoplasm was well defined, with the tumorous mass apparently pushing aside rather than invading adjacent normal tissue. This is considered a benign growth in view of its apparent noninvasive character, its assumed slow growth in the absence of mitotic figures, and lack of polyploidy.

Another internal tumor was found in a Pacific oyster that also possessed multiple watery cysts. Histological examination revealed an unusual proliferation of tissue infiltrating the area normally occupied by Leydig cells and digestive tubules and virtually replacing the normal

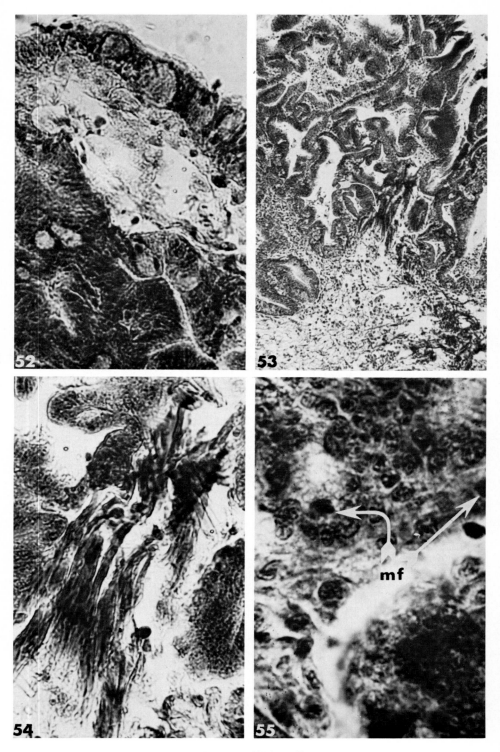

Figs. VII.52–VII.55

tissue (Pauley *et al.*, 1968). The stroma around the watery cysts presented a histological picture of almost complete disorganization. Many areas typically consisting of Leydig cells or digestive tubules with associated Leydig tissue had been replaced by a bizarre proliferation of nervous tissue and collagen (Fig. VII.59), accompanied by edema and a marked leukocytic infiltration. The nerve cells varied considerably in size and arrangement, although most had nerve fibrils obviously leading from them or closely associated with them. Unusually heavy proliferation of collagen was associated with all nervous elements and frequently encapsulated several nerve cells in a thick collagenous capsule (Fig. VII.60). The occurrence of the tumorous tissue may have been directly related to the watery cysts or purely coincidental. It should be noted that the internal tumor occurred in only one of several oysters possessing apparently identical watery cysts. Furthermore, it is difficult to conceive of the tumor formation as simple hyperplasia resulting from stimulation induced by the cause of the watery cysts. Although the marked increase in collagen could be hyperplastic, the nerve cells represent a cell type not found normally in that area, therefore, obviously representing neoplasia.

In addition to being clearly neoplastic, at least in my opinion, the above-described tumor possessed a number of characteristics commonly found in malignant neoplasms of vertebrates. It was invasive, rather than expansive and confined, and the structure of the nervous elements were completely atypical. Additionally, the complete disruption of the normal tissue architecture is suspiciously reminiscent of malignancy. Since the distinction between malignant and benign tumors in mollusks is at least as nebulous as that between neoplasia and hyperplasia, this tumor was described as an invasive tumor, and we ignored the question of malignancy for lack of trenchant criteria (Pauley *et al.*, 1968).

Katkansky (1968) described two small polypoid growths in the intestine of a European oyster (*Ostrea edulis*). They were mushroom-shaped, 0.15 mm in height, stalked, and consisted of slightly modified epithelial cells. There may have been two discrete growths or a single

---

Fig. VII.52. Sydney rock oyster tumor. Fold of proliferating tumorous epithelium. Note the frequent mucous cells and the numerous leukocytes. 480×. (From Wolf, 1969. Courtesy of P. H. Wolf.)

Fig. VII.53. Invasion of mantle tissue by tumor. Note (arrow) invasion and destruction of muscle bundle. 100×. (Courtesy of P. H. Wolf.)

Fig. VII.54. Higher magnification of tumorous epithelium surrounding and invading muscle bundle. 320×. (Courtesy of P. H. Wolf.)

Fig. VII.55. Section of tumor with numerous mitotic figures (mf). Note increased cellularity and hyperchromatism of the neoplastic tissue. 840×. (From Wolf, 1969. Courtesy of P. H. Wolf.)

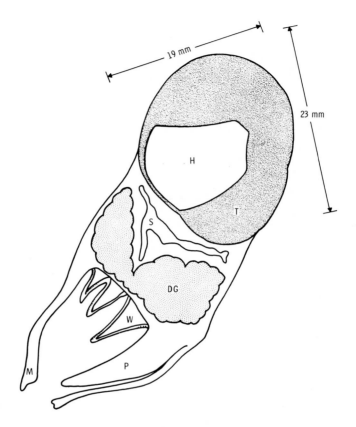

Fig. VII.56. Diagrammatic sketch of oyster (*C. gigas*) with fibrous tumor (T). Through midregion, showing hollow center (H), stomach (S), digestive gland (DG), water tube of gills (W), palp (P), and mantle (M). (From Pauley and Sayce, 1968. Courtesy of G. B. Pauley.)

ridgelike growth approximately 20 mm in length. There is considerable question as to the neoplastic nature of this anomoly.

Some of the earliest descriptions of tumors in pelecypods were in the freshwater mussel, *Anodonta*. Williams (1890) reported that 1 of the 700 freshwater mussels (*A. cygnea*) examined from a pond possessed "a pediculated tumor of about the size of a hazel nut" arising from the mantle. Grossly, Williams was struck by its resemblance to the polyps found in the digestive tract of mammals. Sectioned material stained with hematoxylin and picrocarmine revealed that the growth consisted of gland cells and bundles of muscle fibers. Although his microscopic equipment did not satisfy him, Williams was able to determine that the gland cells, found predominantly toward the broader end of the tumor, were

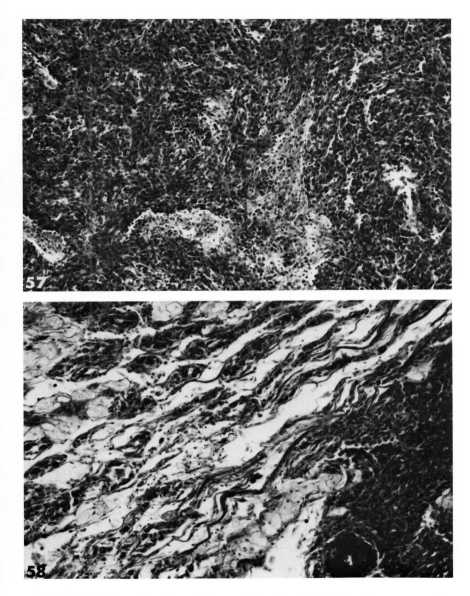

Fig. VII.57. Internal portion of tumor. Note the dark-staining fibrous connective tissue and lighter staining collagen. Mallory's trichrome. 100×. (From Pauley and Sayce, 1968. Courtesy of G. B. Pauley.)

Fig. VII.58. Edge of tumor showing its apparent origin from undifferentiated gonadal cells. Mallory's trichrome. 155×. (From Pauley and Sayce, 1968. Courtesy of G. B. Pauley.)

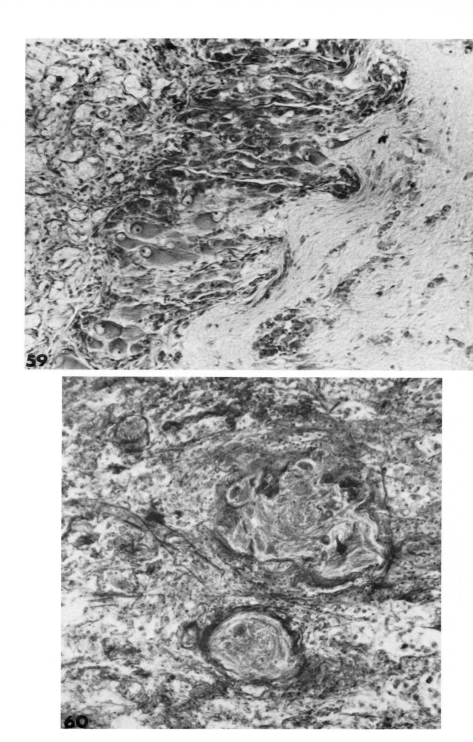

Figs. VII.59 and VII.60

indistinguishable from those of the glandular margin of the mantle. The muscle tissue, though irregularly arranged, apparently formed a core and the stalk. It is difficult not to speculate that the tissues described as muscle by Williams were more likely to have been connective tissue fibers, indistinguishable from muscle tissue with the staining techniques and microscopic equipment available. Unfortunately, this can never be answered because no illustrations were provided and the material was no doubt lost. Despite his lack of sophisticated equipment, the author did not hesitate to classify the tumor, stating that it could be considered an adenomyoma.

A short time later, Collinge (1891) commented that up to the time of reading Williams' paper he had never seen anything similar in freshwater mussels, agreeing with Williams that such growths were certainly uncommon. After reading Williams' paper, Collinge found two or 300 *Anodonta* from a pond had structural abnormalities. One of these, he felt, was similar to the tumor described by Williams, except that it was not stalked.

The other growth, which, interestingly, Collinge attributed to injury, consisted of a "hard, muscular pyriform body about the size of a pea or slightly smaller" located beneath the left lobe of the mantle and near the dorsal siphon. Just above this abnormal structure, at the base of the left mantle lobe and connecting with both the mantle lobe and outer gills, was an elongated mass that he obviously considered a part of the abnormal growth. Collinge remarked on the obvious interference with the mussel's functions as shown by an almost complete lack of nacreous material on the posterior portions of both valves and a loose, vacuolated condition of the calcareous layer of the shell; both of these conditions indicating loss of function of the mantle that deposits the shell. Again, it is unfortunate that no figures were provided and, in this case, no histological description of either of the abnormalities.

A small, firm, tumorlike growth on the foot of another species of *Anodonta* (*A. californiensis*) was reported by Pauley (1967a). The dimensions of the structure were 3 mm in diameter and 2 mm in height. Histologically, the growth consisted of dark, basophilic, glandular cells and muscle fibers with a deeply convoluted epithelial covering (Fig.

Fig. VII.59. Invasive nerve tumor of *C. gigas*. Note normal Leydig tissue on left, the extreme variation in size of tumorous nerve cells and nuclei in center, and abnormal amount of collagenous material on right. Masson's trichrome. Approximately 280×. (From Pauley *et al.*, 1968.)

Fig. VII.60. Tumorous nerve cells encapsulated with heavy collagenous sheaths. Note the collagenous strands, edema, heavy leukocytic infiltrate, and lack of normal Leydig cell architecture. Masson's trichrome. 192×. (From Pauley *et al.*, 1968.)

VII.61). The epithelium had a faded, vacuolated cytoplasm differing markedly from the normal tall columnar epithelium of the foot. The deeply basophilic glandular cells and muscle fibers constituting the stroma of the growth were normal in appearance, though irregularly arranged. Pauley stated that the distribution of glycogen in the lesion exhibited no apparent abnormality, noting that the glycogen deposition in the muscle fibers in the growth was similar to that of normal foot muscles. However, he then said "the basophilic gland cells in the tumor were highly PAS positive, but did not possess glycogen like those beneath normal epithelium."

The lesion was well defined with no evidence of invasion into adjacent tissue, and it was considered to be a benign growth. Pauley first classified it as an adenomyoma, but agreed with Dawe (C. J. Dawe and G. B. Pauley, personal communication) that it was possibly erroneous to assign a name to this growth using terminology developed for use in vertebrate pathology. Pauley went further by pointing out the possibility that the lesion was not a true neoplasm, but a hyperplasia caused by some unknown stimulus of the basophilic gland cells and muscle fibers lying beneath the foot epithelium.

Subsequently, Pauley (1967b) found that four of forty mussels examined possessed pedunculated tumors arising from the foot. Multiple tumors occurred on all mussels, two to six per animal, with a total of fourteen lesions on the four affected mussels (Figs. VII.62, VII.63,

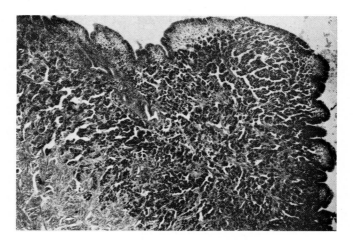

Fig. VII.61. Benign tumorlike growth on the foot of a freshwater mussel, *Anodonta californiensis*. Note the basophilic gland cells comprising the lesion, the deeply convoluted epithelial covering, and the normal muscle tissue at lower left. Hematoxylin and eosine. Approximately 30×. (From Pauley, 1967a.)

VII.64). The tumors varied in size from 1 × 2 × 2 mm to 4 × 4 × 2 mm, were polypoid, and connected to the surface of the foot by a firm but flexible stalk. The growths were all similar in gross appearance and histological architecture. The outer surface consisted of a convoluted tall columnar epithelium (Fig. VII.65), with an abnormally thick and dense layer of basophilic gland cells immediately beneath the epithelium and a central core of compacted muscle fibers. All cells in the lesions were normal in appearance, no mitotic figures were seen, and no inflammatory reactions were noted. The neoplasms were well defined and noninvasive and were classified as pedunculated adenomas by Pauley (1967b); but he pointed out subsequently (Pauley, 1969) that this was probably an erroneous term for the reasons mentioned previously.

An apparently quite different tumor was described from *Anodonta implicata* by Butros (1948). This was a large (2.0 cm × 0.8 cm), nodular growth arising from one of the inner labial palps. The outer layer of the tumor was composed of somewhat abnormal epithelium, smaller and more irregular than that covering the palps, but possessing the normal numerous goblet cells characteristic of palp epithelium. Connective tissue cells were abundant near the periphery of the lesion, but virtually absent near the center. Butros indicated that a few mitotic figures were present, but provided no photomicrographs to substantiate the statement.

Pauley (1968) found tumorlike growths in a population of the freshwater mussels (*Margaritifera margaritifera*) affected by watery cysts, which he designated spongy disease. Five specimens possessed small (2–3 mm in diameter by 2 mm in height) polypoid growths attached to the foot by a small stalk (Fig. VII.66). The growths consisted of basophilic gland cells of normal appearance and a few supporting muscle and collagen fibers covered by a layer of normal ciliated columnar epithelium (Fig. VII.67). There was no inflammatory response associated with any of the lesions, and, since there were no mitotic figures observed and no indications of invasion of the foot, they were obviously benign. Although the cause of the accelerated growth was not known, Pauley thought that they represented hyperplasia of the basophilic gland cells beneath the epithelium of the foot.

Pauley (1969) also found two tumorlike growths on the foot of another freshwater mussel, *Gonidea angulata*. The larger lesion, measuring 5 × 4 × 3 mm, appeared to have originated from the smooth muscle of the foot, but only traces of muscle fibers were present in the growth; the muscle was largely replaced by collagen. The lesion was edematous, had an incomplete epithelial covering, and exhibited hyperplasia of the gland cells. Pauley was uncertain as to whether the growth resulted from chronic irritation or whether the increased collagen resulted from chronic

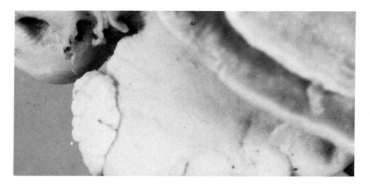

Fig. VII.62. Large fungiform tumor on foot of *A. californiensis*. (From Pauley, 1969.)

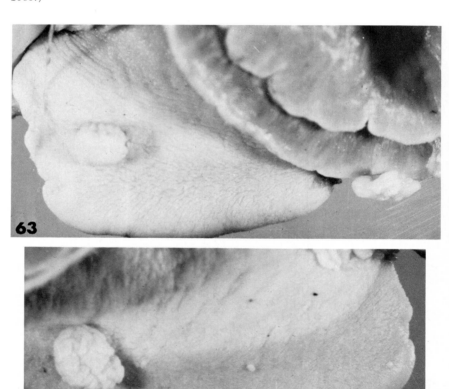

Fig. VII.63. Freshwater mussel (*A. californiensis*) with large polypoid tumor on foot and similar lesion on edge of gill. (From Pauley, 1969.)

Fig. VII.64. Foot of *A. californiensis* with 1 large and several small tumors. (From Pauley, 1969.)

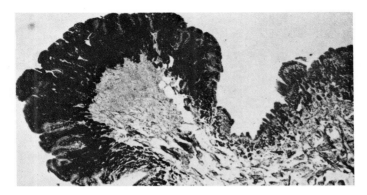

Fig. VII.65. Histological appearance of small polyp such as shown in Figs. VII.62–VII.64. Note deep epithelial crypts on surface, the muscular core, and the abnormally thick, dense layer of basophilic gland cells. (From Pauley, 1967b.)

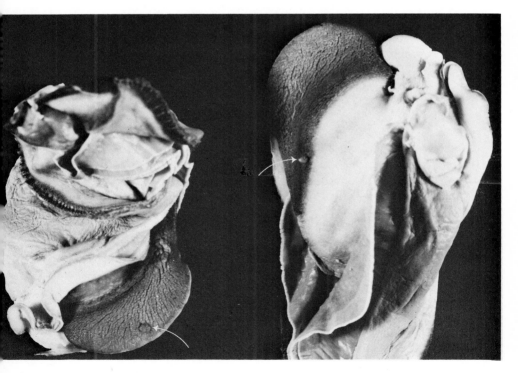

Fig. VII.66. Freshwater mussels. *Margaritifera margaritifera* with small polypoid growths on the ot. (From Pauley, 1968.)

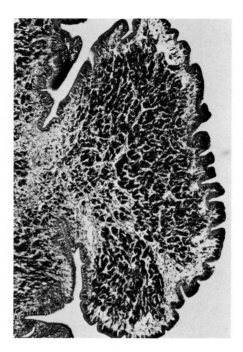

Fig. VII.67. Microscopic appearance of the polypoid growths in *M. margaritifera*.
Note the hyperplasia of the basophilic gland cells. Hematoxylin and eosine. 52×.
(From Pauley, 1968.)

irritation to a polypoid neoplasm arising from an unknown cause. The
smaller lesion (Fig. VII.68), 3 × 2 × 2 mm, also appeared to have originated from the smooth muscle of the foot. It, too, was edematous and
was covered by a highly convoluted epithelium. Neither lesion was invasive nor contained mitotic figures.

Cauliflowerlike papillary tumors around the rectal opening were found
in about 2% of the soft shell clams, *Mya arenaria,* examined by Hueper
(1963). Although excellent photographs of the gross lesions were presented, apparently no microscopic study was undertaken. Hueper also
neglected to state the number of *Mya* he had examined; Cantwell (1967)
examined more than 1000 *M. arenaria* from the same area of Chesapeake
Bay without finding a tumorous individual. However, Potter and Kuff
(1967) did find a growth on the siphon of a soft shell clam. It was firm
and muscular, much like the siphon, and histologically was found to
consist of normal smooth muscle arranged, as in the siphon, in inner
longitudinal and outer circular layers and covered by columnar epithelium. The lesion was solid except for a small diverticulum of the
siphon lumen extending into the base.

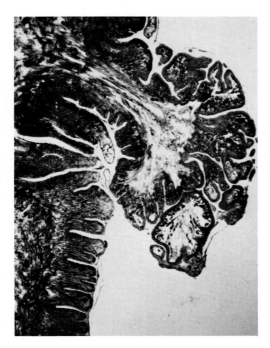

Fig. VII.68. Microscopic appearance of the smaller growth on *Gonidea angulata*. Note highly convoluted epithelial covering and core of smooth muscle. Modified Mallory's trichrome. Approximately 40×. (From Pauley, 1969.)

Polypoid and papillary lesions were found on the foot of a gaper clam, *Tresus* (*Schizothaerus*) *nuttalli* by Taylor and Smith (1966). The large polypoid growth measured 4 × 7 mm, and the 3 papillary lesions were from 1.0 to 1.5 mm in diameter. Grossly, the structures were strongly suggestive of neoplasia. After careful histological study by several independent investigators as well as the authors, it was finally decided that the growths were a hyperplastic response to injury. The highly cellular character of the lesions made the task of differentiating between hyperplasia and neoplasia extremely difficult. Microscopic examination revealed that the firm, fleshy growths consisted primarily of heavy infiltrations of fibroblasts and leukocytes and, thus, represented an unusually intense inflammatory reaction. Microorganisms were observed in some of the phagocytes, but the authors were unable to determine whether they were the initiators of the hyperplasia or secondary invaders.

Tumors are also known to occur in the butter clam, *Saxidomus giganteus*. Pauley (1967c) reported that 1 of 183 butter clams examined had a large (9 mm × 6 mm) polypoid tumor on the foot (Fig. VII.69), and

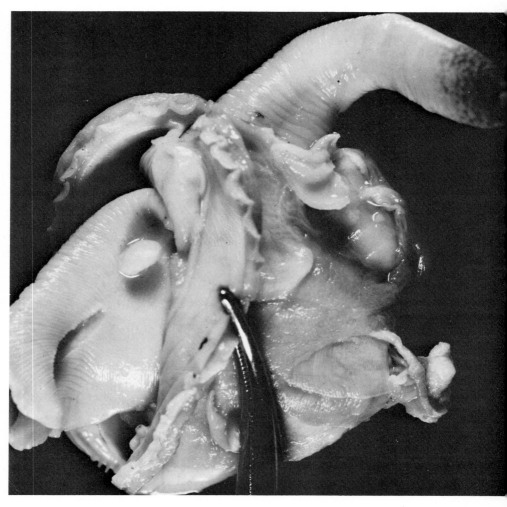

Fig. VII.69. Polypoid tumor on the foot of a butter clam (*Saxidomus giganteus*). (From Pauley 1967c.)

that a similar, but smaller, fingerlike growth was found on 1 of 394 *S. giganteus* subsequently examined (personal communication). Although it is extremely valuable to know the number of animals examined in order to eventually develop some appreciation for the rate of "tumorgenesis" in invertebrates, Pauley's finding of two tumorous individuals among the relatively few specimens examined is probably fortuitous. In studies of the ecology of paralytic shellfish poisoning in both Washington and Alaska, thousands of butter clams were tested for toxicity in my laboratory each year for at least 7 years without a single tumorous individual being recognized. While it may be correctly argued that the clams were opened for bioassay rather than microscopic study, all individuals opening clams were trained shellfish biologists aware of my interest in tumors and other diseases of mollusks. Therefore, it seems unlikely that many, if any, large tumors were overlooked. Incidently, many of the butter clams were opened by me, which perhaps strengthens my opinion that tumors in S. *nuttali* are not common.

The large tumor found in the butter clam by Pauley was firm and similar to the normal portion of the foot in color and texture. Histologically, the bulk of the tumor consisted of a central core of smooth muscle, normal in appearance, which appeared to have originated from the circular layer of smooth muscle located just beneath epithelium of the normal foot. Basophilic gland cells beneath the epithelium were normal in location, numbers, and appearance, and the growth was covered by a tall columnar epithelium, somewhat convoluted, resembling the normal epithelium of the foot. Pauley could find no cause for initiation of the growth, such as microorganisms or foreign bodies. Response to physical trauma was ruled out because of the lack of any inflammatory response. Since the tumor cells were normal in appearance and lacked mitotic figures and since there were no indications of invasion into the adjacent areas of the foot, it was considered to be a benign tumor.

The smaller growth was firm and normal in color and measured 2.5 mm × 1 mm. It was covered by tall columnar epithelium similar to that of the normal foot, but was somewhat hypertrophied and hyperplastic. Like the larger tumor discussed above, the bulk of the tumor consisted of smooth muscle with an apparent origin from the circular layer of muscle, and no inflammatory response or mitotic figures were observed.

Growths have been reported on clam siphons on three occasions (Potter and Kuff, 1967; Pauley and Cheng, 1968; Des Voigne et al., 1970). The first of these, a cylindrical growth 1.3 mm × 4.0 mm, was considered to be a developmental anomaly of the siphon of *Mya arenaria* by the authors. It was covered by normal epithelium and consisted of smooth muscle arranged in inner longitudinal and outer circular layers, again consistent with the normal architecture of the siphon. No epithelium-lined lumen was present in the growth, but a small diverticulum of the lining of the siphon extended into the base of the anomaly.

Pauley and Cheng (1968) described a wrinkled fungiform swelling at the basal portion of the siphons of a soft-shell clam (*Mya arenaria*) (Fig. VII.70) that measured 12 × 9 × 6 mm, with the major portion of the growth located ventrally and along the right side of the siphons.

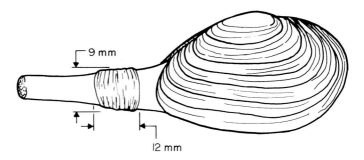

Fig. VII.70. Sketch illustrating the gross appearance of a tumor on the siphons of a soft-shell clam (*Mya arenaria*). (From Pauley and Cheng, 1968.)

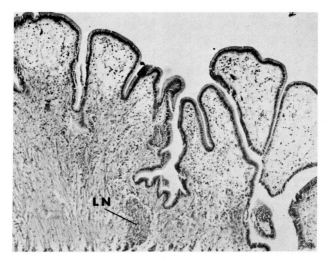

Fig. VII.71. Microscopic appearance of the tumor on the siphons of soft-shell clam. Note the folds in the epithelial covering, the edematous areas beneath the epithelium, the moderate leukocytic infiltration throughout the lesion, and the nest of leukocytes (LN). Hematoxylin and eosine. Approximately 70×. (From Pauley and Cheng, 1968.)

Microscopically, the tumor was covered by a highly convoluted columnar epithelial layer that was deeply folded or cryptlike (Fig. VII.71), but was confluent with the epithelium of the siphon. The stroma of the tumor consisted of smooth muscle that was confluent with the muscle tissue of the siphon. The tumor consisted of two lobes that arose from the ventral side of the siphons by thin stalks, and one lobe grew upward along the right side of the siphons. The lobes were separated by a deep cleft that extended into the lumen of the siphons and was lined by epithelium that was confluent with the epithelial surfaces of the siphons, the tumor, and the lining epithelium of the lumen of the siphon.

The third siphonal tumor was described in the horse clam, *Tresus nuttalli* (Des Voigne *et al.,* 1970). Grossly, the lesion was a cone-shaped, papillomalike growth protruding from the right side of the siphon and was 7 mm in height and 6 mm in diameter at its widest point (Fig. VII.72). It was covered by the dark, horny layer that normally covers the siphon. Over the lesion, however, the layer was loose and easily removed to expose the underlying growth.

Relatively severe autolytic changes occurring prior to fixation caused obvious alterations and made histological study of the growth difficult. The tumor was covered by tall columnar epithelium that was markedly convoluted and consistent in appearance with the epithelium lining the normal siphon (Fig. VII.73). The stroma consisted of muscle fibers similar in appearance to the smooth muscle of the siphon, but they were randomly arranged in circular and longitudinal bands (Fig. VII.74), distinctly dissimilar to the appearance of the siphon in which external circular muscle bundles enclose internal longitudinal bands. Many of the muscle fibers were swollen, separated, and more acidophilic than normal, undoubtedly from the previously mentioned postmortem changes.

Cryptlike structures were present beneath the surface epithelium (Fig. VII.74) and were lined by epithelial cells resembling those covering the external surface of the lesion (Fig. VII.75). These cells were vacuolated and had decreased cytoplasm, presumably because of autolytic changes related to fixation, rather than degenerative effects of the neoplasm. Some of the crypts contained an almost totally autolyzed epithelial lining, and gas-forming bacteria apparently formed gas pockets in the decomposing muscle bundles.

An apparent ovarian tumor in the quahog, *Mercenaria mercenaria,* was described by Yevich and Berry (1969). Three cases were found during the routine histopathological examination of 1300 *M. mercenaria,* all from Narragansett Bay, Rhode Island. None of the affected animals exhibited grossly detectable anomalies. Histologically, it was observed that the normal ovarian follicles (Fig. VII.76) were, in most areas, completely replaced by large polyhedral cells containing vesicular nuclei and clear pink cytoplasm (Figs. VII.77 and VII.78). Smaller cells with smaller hyperchromatic nuclei apparently infiltrated and became scattered throughout the tumor. Mitotic figures were also found, in random fashion, throughout the tumorous tissue. Many of the islands of tumor cells contained a core of deep-pink staining fibrous connective tissue that often appeared necrotic.

Early invasion into the kidney of one tumorous animal occurred, with

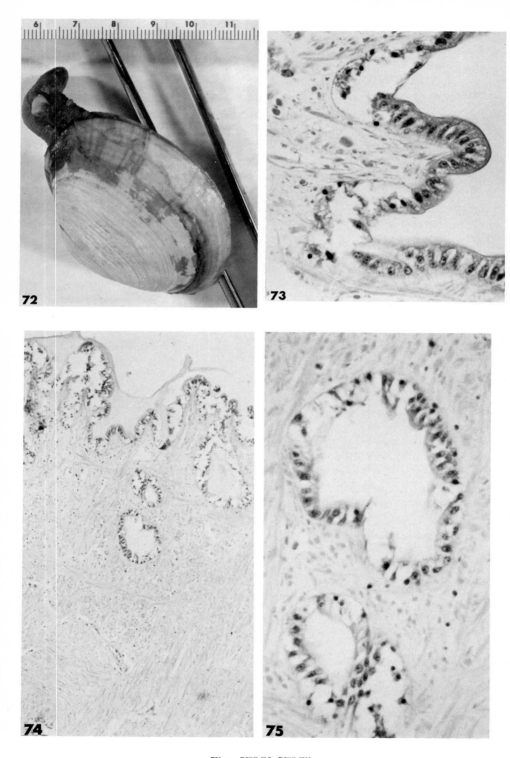

Figs. VII.72–VII.75

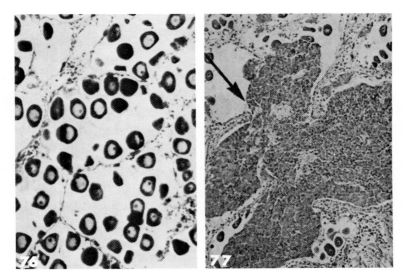

Fig. VII.76. Normal ovarian follicles of the quahog. *Mercenaria mercenaria.* Hematoxylin and eosine. (From Yevich and Berry, 1969.)

Fig. VII.77. Ovarian follicles filled with masses of tumor cells (arrow). Hematoxylin and eosine. 295×. (From Yevich and Berry, 1969.)

nests of tumor cells present in the supporting connective tissue (Fig. VII.79). The authors were correct in their statement that these lesions appear to be the first report of a primary ovarian tumor in a mollusk since Pauley and Sayce (1968) ruled out gametogenic origin in the one Pacific oyster tumor possibly arising in the gonad. Although the authors stated that the ovarian tumor in the quahog is similar histologically to granulosa cell carcinoma of the ovary in humans, it would clearly be premature and highly speculative to either assume that there is a real similarity or to use the vertebrate oncological terminology in describing this anomoly.

Fig. VII.72. Horse clam, *Tresus nuttalli,* with papillomalike lesion on the right side of the siphon. The horny covering has been removed to expose the growth. (From Des Voigne *et al.,* 1970.)

Fig. VII.73. Columnar epithelium covering the surface of the tumor. Note the autolytic changes in muscle fibers and portions of the epithelium. 340×. (From Des Voigne *et al.,* 1970.)

Fig. VII.74. Lower magnification of section through the tumor. Note that the parenchyma consists of irregularly arranged bands of muscle fibers and the epithelial-lined crypts beneath the surface epithelium. 136×. (From Des Voigne *et al.,* 1970.)

Fig. VII.75. Higher magnification of epithelial-lined crypts. 544×. (From Des Voigne *et al.,* 1970.)

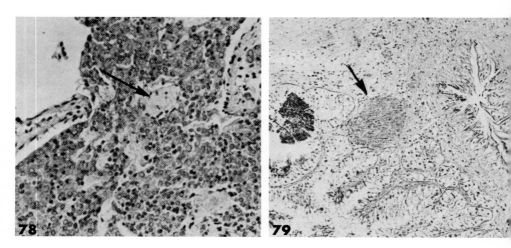

Fig. VII.78. Higher magnification of tumor mass. Note large polyhedral cells containing vesicu
nuclei and infiltration of cells with deeply stained hyperchromatic nuclei. Arrow points out area
necrosis. Hematoxylin and eosine. 713×. (From Yevich and Berry, 1969.)

Fig. VII.79. Mass of tumor cells (arrow) in connective tissue between pericardial funnel a
glandular mass of the kidney. Hematoxylin and eosine. 280×. (From Yevich and Berry, 1969.)

Some of the most intriguing and enigmatic presumed neoplasms occurring in mollusks, or for that matter in all the invertebrates, are a
group of diseases of, apparently, the blood and blood-forming organs of
mussels and oysters. Couch (1969) described a focal, tumorlike lesion
in the right mantle of a gravid female oyster, *Crassostrea virginica*. The
mantle was enlarged transversely because of the lesion, but its vertical
length was unaffected. The tissue within the lesion stained deeply basophilic in comparison to adjacent normal connective tissue. The borders
of the lesion were poorly defined (Fig. VII.80), but it extended from the
proximal edge of the right mantle distally to involve about two-thirds of
the length of the mantle. The inner epithelium of the right mantle lobe
was greatly modified (Fig. VII.81) over the lesion, but the outer epithelium of the lobe was unaffected. There was no indication of connective tissue encapsulation.

A single cell type was predominant in the lesion (Figs. VII.82 and
VII.83). Couch, for convenience, termed the cells "blastoid," because he
believed them to be undifferentiated. The diameter of the blastoid cells
ranged from 4.9 to 12.6 μm, the nuclei were large (3.5–10.5 μm diameter), and the cytoplasm was relatively scanty. The nuclei were round,
oval, or bilobed and possessed a conspicuous nucleolus (Fig. VII.83).
Binucleate blastoid cells occurred rarely, and mitotic figures were numerous (up to seven per oil immersion field) (Fig. VII.84). Leydig cells

(the normal vesicular connective tissue of the mantle) present in the lesion were abnormally small and in many areas were replaced by the blastoid cells.

Blood sinuses in the lesion appeared to be partially occluded by blastoid cells, and large blastoid cells were found in a perivascular focus in a blood sinus of the otherwise normal left mantle. As Couch pointed out, this finding might represent either metastasis or multiple foci of the disease. Since normal leukocytes can wander freely throughout the oyster body, the blastoid cells, if they are of hemic origin, may have the same capacity.

Couch noted that the origin of the blastoid cells was inexplicable. If of leukocytic origin, considerable enlargement of the cell was necessary as well as disproportional enlargement of the nucleus. However, there is little resemblance between the blastoid cells and Leydig cells or any other cells of the oyster. Couch pointed out that the blastoid cells morphologically resemble the cells of certain reticular sarcomas in mice; they have enlarged nuclei, pleomorphic nuclei, occasional binucleate or multinucleate cells, frequent mitoses, and focal rather than diffuse distribution. He carefully stated, however, that he did not imply that there was a similarity in etiology or natural history between the two. The lesion cannot be precisely categorized because its origin is unknown, but it appears to be clearly neoplastic. The cells in the lesion do not resemble host response to injury or parasites, and no recognizable parasites were found. The possibility that the blastoid cells are themselves parasites must be considered, but they certainly do not resemble any known protistan. On the other hand, they do have characteristics of molluscan cells. Although it is speculative at this point, it seems most likely that the blastoid cells are neoplastic cells of leukocytic origin.

A similar disease was reported by Farley (1969a) in the blue mussel, *Mytilus edulis*, from Yaquina Bay, Oregon. Two samples of mussels, forty-three collected in September, 1968 and fifty-seven collected in February, 1969, were examined. No gross lesions were recognizable, but when examined microscopically, three of the forty-three collected in September and seven of the fifty-seven collected in February were found to have neoplasticlike abnormalities characterized by multiple local (Fig. VII.85 and VII.86) or diffuse (Fig. VII.87) lesions; they consisted of abnormally large, undifferentiated cells. Lesions varied from large tumorous masses in the mantle (Fig. VII.88 and VII.89) to small local lesions in the connective tissue between tubules of the digestive diverticula (Fig. VII.86) and in hemolymph spaces of the posterior adductor muscle (Fig. VII.90).

Mussels with no lesions or with only a few of the abnormal cells were

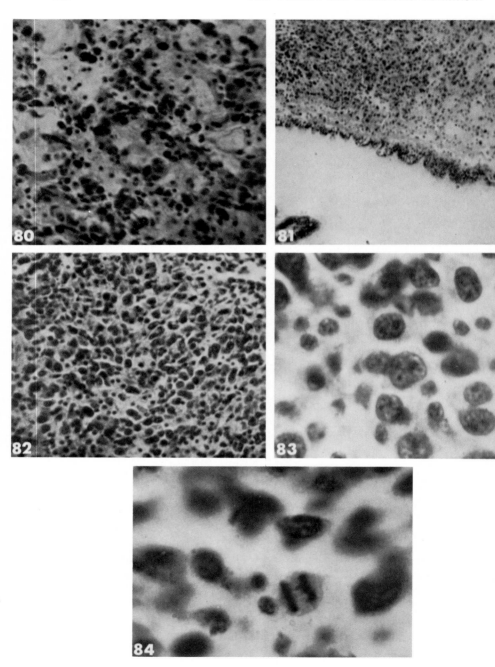

Figs. VII.80–VII.84

fat and exhibited extensive gametogenesis, while those with large lesions or numerous atypical cells were in poor condition, even moribund, and gametogenesis was arrested.

The most striking atypical characteristic of the neoplastic cells is that they are larger than normal, with nuclei 7–13 μm in diameter compared to 3–4 μm for normal leukocyte nuclei. However, they are morphologically similar to normal, undifferentiated molluscan cells, containing a nucleus with a vesicular chromatin pattern and agranular cytoplasm like hyaline hemocytes (leukocytes). In contrast to normal mussel tissue, the atypical cells in this proliferative disease are extremely active mitotically (at least three mitotic figures can be seen in Fig. VII.87), and the mitotic figures are enlarged in comparison with normal mitoses. Considerably more than the normal number of chromosomes (Fig. VII.91) occur in the neoplastic cells, and their division is characterized by numerous atypical features, including displaced chromosomes (Fig. VII.92), tripolar figures (Fig. VII.93), and, perhaps, even tetrapolar stages (Fig. VII.94). Resting cells may also exhibit cytological abnormalities such as multiple nucleoli (Fig. VII.95), binucleate cells (Fig. VII.96), and abnormally shaped nuclei (indented, elongated, irregular, or crenulated) (Fig. VII.97).

Farley (1969b) described six cases of probable neoplastic disease of the hematopoetic system in the genus *Crassostrea*, five in *C. virginica*, and one in *C. gigas*. Subsequently, Farley and Sparks (1970) described additional leukemialike diseases in *C. virginica*, *C. gigas*, and *Ostrea lurida*, reviewed the state of knowledge of these abnormalities in pelecypods, proposed a preliminary classification of cytological types, and reported additional information on the origin of these suspected neoplastic cells.

Although numerous cases of these leukemialike "neoplasms" have been recognized, they are rarely grossly obvious (Fig. VII.98). They are rare in the genus *Crassostrea*, but fairly common in both *Mytilus edulis* and *Ostrea lurida* in Yaquina Bay, Oregon. Pelecypods affected by these dis-

---

Fig. VII.80. Periphery of lesion of "blastoid" cells in connective tissue of proximal end of mantle of *C. virginica*. Note the poor definition of border of the lesion. 409×. (From Couch, 1969.)

Fig. VII.81. Internal epithelium of right mantle. Note modification in structure of epithelium from right to left along the length of the mantle, with normal epithelium at right and abnormally thin and degenerate epithelium overlying the lesion at left. 95×. (From Couch, 1969.)

Fig. VII.82. Center of lesion in right mantle. Note most cells are of "blastoid" type. 409×. (From Couch, 1969.)

Fig. VII.83. Higher magnification of "blastoid" cells in oyster lesion. Note larger size of nuclei compared to adjacent normal cells. 922×. (From Couch, 1969.)

Fig. VII.84. Dividing "blastoid" cell (anaphase). 922×. (From Couch, 1969.)

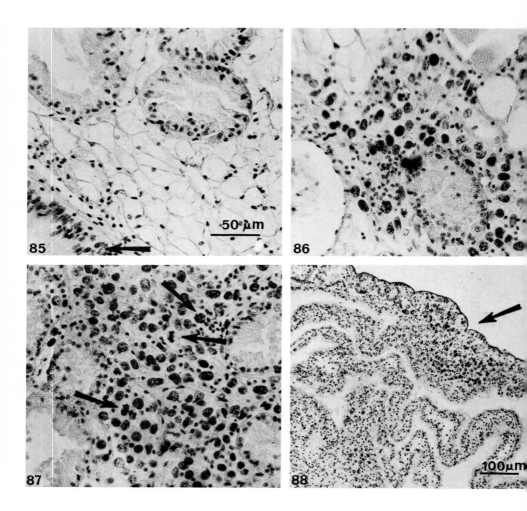

Fig. VII.85. Section of normal mussel (*Mytilus edulis*) tissue. Note alimentary tract epitheli (arrow); Leydig (vesicular) connective tissue; digestive diverticula (at top). Feulgen-picro-met blue. 238×. (From Farley, 1969a. Courtesy of C. A. Farley.)

Fig. VII.86. Local "neoplastic" lesion in *M. edulis* replacing connective tissue between digesti diverticula. Feulgen-picro-methyl blue. 238×. (From Farley, 1969a. Courtesy of C. A. Farley

Fig. VII.87. Diffuse dissemination of "neoplastic" cells in connective tissue of *M. edulis*. N mitoses (arrows). Feulgen-picro-methyl blue. 238×. (From Farley, 1969a. Courtesy of C. Farley.)

Fig. VII.88. Local "neoplastic" lesion in mantle (arrow) of *M. edulis*. Feulgen-picro-methyl bl 95×. (From Farley, 1969a. Courtesy of C. A. Farley.)

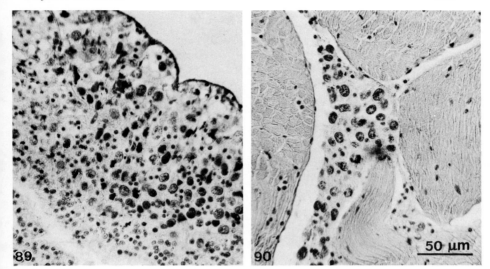

Fig. VII.89. Higher magnification of lesion in Fig. VII.88. Note large nuclei of "neoplastic" cells l dark-staining pycnotic cells. Feulgen-picro-methyl blue. 250×. (From Farley, 1969a. Courtesy C. A. Farley.)

Fig. VII.90. "Neoplastic" lesion in adductor muscle. Feulgen-picro-methyl blue. 250×. (From :ley, 1969a. Courtesy of C. A. Farley.)

orders frequently exhibit one or more gross signs of disease, including poor condition, pale digestive glands, and mantle recession (which are indicative of starvation) with resulting emaciation and signs of impending death. The rarity of grossly detectable signs of the disease is probably related to the nature of the circulatory system and the weak structural development of the connective tissue, both of which are conducive to rapid dissemination of the primary neoplasm before it becomes grossly apparent.

To understand the cytology of these disorders, it is necessary to recall that hemocytes, or leukocytes, of pelecypods occur as two major types: large (15 μm) granular phagocytes with nuclei averaging about 3 μm in diameter and smaller (8 μm) hyaline forms with nuclei that average about 4–5 μm in diameter. The cells involved in these proliferative diseases are of two cytological types. The first and most easily recognized type, which has been found in *C. virginica*, *O. lurida*, and *M. edulis*, is characterized by enlarged cells and nuclei two to four times the size of normal hemocytic cells and nuclei (Figs. VII.99, VII.100, VII.101, and VII.102). These cells frequently exhibit such cytological abnormalities as irregular nuclear shape and contour, binucleated or, rarely, multinucleated forms and increased density of chromatin, diffuse patterns of

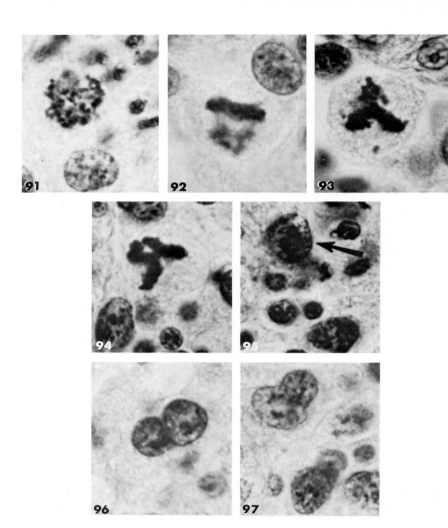

Fig. VII.91. "Neoplastic" cell from *Mytilus edulis,* late prophase. Note the enlarged nucleus compared to adjacent normal nuclei and increased number of chromosomes. Feulgen-picro-methyl blue. 1000×. (From Farley, 1969b. Courtesy of C. A. Farley.)

Fig. VII.92. "Neoplastic" cell, metaphase. Note displaced chromosomes. Feulgen-picro-methyl blue. 1000×. (From Farley, 1969b. Courtesy of C. A. Farley.)

Fig. VII.93. Tripolar figure in "neoplastic" cell. Feulgen-picro-methyl blue. 1000×. (From Farley, 1969b. Courtesy of C. A. Farley.)

Fig. VII.94. Possible tetrapolar figure in "neoplastic" cell. Feulgen-picro-methyl blue. 1000×. (From Farley, 1969b. Courtesy of C. A. Farley.)

Fig. VII.95. "Neoplastic" cell in resting stage with 2 nucleoli (arrow) in nucleus. Feulgen-picro-methyl blue. 1000×. (From Farley, 1969b. Courtesy of C. A. Farley.)

Fig. VII.96. Binucleate "neoplastic" cell. Feulgen-picro-methyl blue. 1000×. (From Farley, 1969b. Courtesy of C. A. Farley.)

Fig. VII.97. Bizarre-shaped nuclei is "neoplastic" cells. Feulgen-picro-methyl blue. 1000×. (From Farley, 1969b. Courtesy of C. A. Farley.)

chromatin and nuclear membranes, multiple nucleoli, and pycnosis and nuclear lysis in advanced lesions. Mitotic abnormalities occur frequently in the form of tripolar figures, chromosomes displaced from the mitotic spindle, and polyploidy. There are some indications that the neoplastic cells in *O. lurida* (Figs. VII.100 and VII.101) are dedifferentiated vesicular connective tissue (Leydig) cells. The enlarged cells forming local lesions in *C. virginica* may also be neoplastic connective tissue cells. The lesion described by Couch (1969) and one described by Farley and Sparks (1970) (Fig. VII.103) appear to have arisen from the endothelial lining of hemolymph vessels, and they consist of enlarged, partially differentiated pleomorphic and spindle-shaped cells. Such an origin and the presence of newly formed hemolymph vessels in the case described by Farley and Sparks is consistent with hemangioendotheliomas of vertebrates.

The second cell type found in pelecypod leukemoid disorders resembles, in size and morphology, hyaline hemocytes (Figs. VII.104 and VII.105). The disease is considered neoplastic because of the highly invasive infiltration of most tissues, especially the vesicular connective tissue, which is destroyed by the invading cells. Mitoses are common in the neoplastic cells, and, although the nuclei tend to be uniform in size, atypical nuclear configurations do occur. Necrosis occurs in massive lesions in the late stages of the disease. It has been found mostly in *C. virginica* (Fig. VII.104) but several cases from *C. gigas* and one from *O. lurida* (Fig. VII.105) have been recognized.

Histopathologically, these diseases are characterized by diffuse accumulations of abnormal, mitotically active cells that invade and replace the vesicular connective tissue (Figs. VII.106–VII.111). Foci are sometimes first evident in hemolymph vessels and sinuses and are of common occurrence in the muscle and byssal tissues of mussels. Although invasion of epithelium is rare, it has been observed in *C. virginica* and *O. lurida*. These diseases appear to originate as focal lesions in either the vesicular connective tissue or the endothelial lining of hemolymph vessels. Early cases are characterized by one or more relatively large microscopic lesions with diffuse disseminations of individual or small collections of neoplastic cells in the hemolymph sinuses, lumina, and endothelial linings of vessels, or in the vesicular connective tissue as widely disseminated cells distributed throughout the hemolymph spaces. Progression of the disease to an intermediate stage is characterized by the development of extensive but localized lesions in the Leydig tissue and gonad. An advanced stage is attained in which there is dense infiltration of the connective tissue and other tissues throughout the organism, and gametogenesis is completely arrested. Gross pathological symptoms of general debility become evident at this stage. Terminal cases are characterized by widespread degenerative changes in the neoplastic and eventually in the remaining normal tissue.

Although these diseases exhibit many of the criteria of neoplasia and even malignancy in vertebrates (including proliferation of single cell

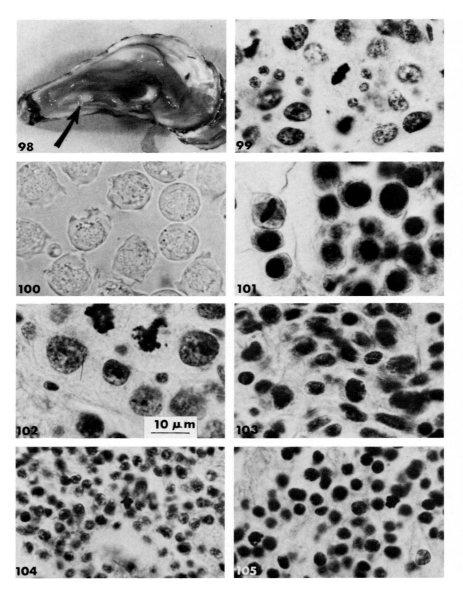

Figs. VII.98–VII.105

lines, unrestricted infiltration, nuclear and mitotic abnormalities, and gross and histological changes indicative of fatal outcome), a definitive interpretation of neoplasia cannot yet be made. Transplantation–transmission experiments and ultrastructural studies in progress have not yielded definitive results. The seasonal nature of the onset of these abnormalities is characteristic of virtually all known oyster epizootics of parasitic causation. The disease in *O. lurida* was studied for several years by me and my associates (Jones and Sparks, 1969). During the early winter months, the first stage of the disease invariably appears in fairly large numbers of native oysters (*O. lurida*) in Yaquina Bay. As the disease progresses through the intermediate, advanced, and terminal stages, increasing numbers of oysters succumb, virtually all of which, when examined microscopically, exhibit the characteristics of the disease described above. The seasonality and progression of onset and the apparent contagious nature of the epizootic were to us, highly suggestive of an infectious disease leading us to suspect parasitic involvement. However, no protistan of known pathogenicity was ever found associated with the diseased oysters. It is, of course, possible that the abnormal cells are themselves a parasite of the mollusk rather than neoplastic cells. Their morphology and mitotic activity are, however, unlike any known parasitic forms, but are strongly suggestive of molluscan cells.

CLASS CEPHALOPODA

There are apparently no reports of naturally occurring tumors in the cephalopods. As mentioned previously, there have been several studies of the effects of carcinogens on a variety of cephalopods. Since tar was found to have no carcinogenic effect and since the reactions to it are the typical inflammatory response to a foreign substance (Jullien, 1940), the

Fig VII.98. Gross tumor (arrow) in mantle of *Ostrea lurida*. (From Farley and Sparks, 1970.)

Fig. VII.99. Mitotic and interphase "hemocytic-type neoplastic cells" in *Crassostrea virginica* disorder. 1000×. (From Farley and Sparks, 1970.)

Fig. VII.100. *In vivo* "vesicular-type neoplastic cells" in *O. lurida*. 1000×. (From Farley and Sparks, 1970.)

Fig. VII.101. Mitotic and interphase "vesicular-type neoplastic cells" in *O. lurida*. 1000×. (From Farley and Sparks, 1970.)

Fig. VII.102. Mitotic and interphase "vesicular-type neoplastic" cells" in *Mytilus edulis*. 1000×. (From Farley and Sparks, 1970.)

Fig. VII.103. "Endothelial-type neoplastic cells" in *C. virginica*. 1000×. (From Farley and Sparks, 1970.)

Fig. VII.104. Mitotic and interphase "hemocytic-type neoplastic cells" in *C. virginica*. 1000×. (From Farley and Sparks, 1970.)

Fig. VII.105. "Hemocytic-type neoplastic cells" in *O. lurida*. (From Farley and Sparks, 1970.)

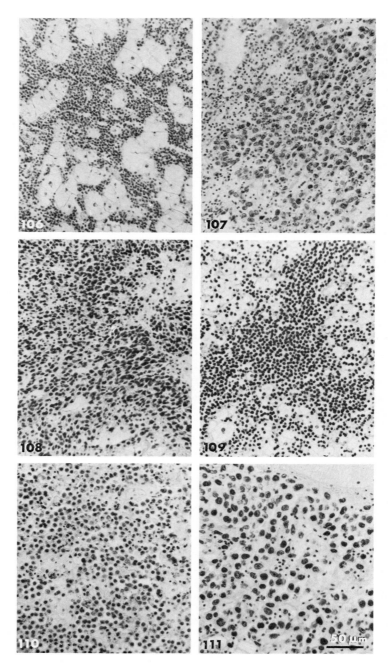

Figs. VII.106–VII.111

tissue responses are discussed in the chapter on reaction to injury and wound repair.

The introduction of crystals of 1:2–5:6-dibenzanthracene, another vertebrate carcinogen, enclosed in Vaseline beneath the epidermis of the cuttlefish, however, evokes a markedly different cellular reaction (Jacquemain *et al.,* 1947). Macroscopically, a rapidly expanding "tumor" develops, having an oval or circular outline and attaining a diameter of several centimeters within 5 days. The lesion is markedly elevated and presents a blanched appearance. It differs from the normal inflammatory reaction by failing to form an elimination tract.

Histologically, as the "tumor zone" is approached, a number of modifications become apparent: the surface epithelium disappears; the continuous band of chromatophores in the superficial part of the dermis is lost with only isolated chromatophores remaining; the dermal connective tissue is considerably thickened, particularly in the central part of the lesion. The remaining cells have become rounded and necrotic, containing homogeneous pycnotic nuclei.

The authors noted that all the cellular elements were those of the normal animal, but had undergone massive autolysis through the entire thickness of the dermis with the superficial part of the underlying muscle layer invaded by a few histocytic cells from the dermis. Since no new tissue is produced in reaction to the irritant, it does not seem appropriate to use the term "tumor" to describe this degenerative lesion.

Phylum Arthropoda

In view of the extensive literature on tumors of insects, especially the fruit fly, *Drosophila,* one might expect tumors to be of frequent occurrence in the remaining classes of the phylum. This, however, is appar-

---

Fig. VII.106. "Hemocytic neoplasm" in vesicular connective tissue of *C. virginica.* 250×. (From Farley and Sparks, 1970.)

Fig. VII.107. "Vesicular cell neoplasm" in gill of *C. virginica.* 250×. (From Farley and Sparks, 1970.)

Fig. VII.108. "Endothelial neoplasm" in *C. virginica.* 250×. (From Farley and Sparks, 1970.)

Fig. VII.109. "Hemocytic neoplasm" in vesicular connective tissue of *O. lurida.* 250×. (From Farley and Sparks, 1970.)

Fig. VII.110. "Vesicular cell neoplasm" in *O. lurida.* 250×. (From Farley and Sparks, 1970.)

Fig. VII.111. "Vesicular cell neoplasm" in *M. edulis.* 250×. (From Farley and Sparks, 1970.)

ently not true, judging from the paucity of reports of neoplasms in non-insect arthropods.

It is outside the purpose of this volume to discuss tumors in insects, but it should be pointed out that a number of reviews of this subject are available, the latest is by Harshbarger and Taylor (1968). Their review was directed primarily toward the literature published since 1950, and they noted that of the more than 400 papers on tumors in insects about 80% are concerned with *Drosophila.*

In contrast to this voluminous literature on the neoplasms of insects, I have been able to find only eight papers concerned with tumors or tumorlike growths in the remaining arthropod groups. Several of these, as will be seen shortly, are of dubious validity or are only briefly mentioned in papers primarily concerned with other information.

A tumorlike structure was reported in the primitive arthropod, *Limulus,* by Hanström (1926). It appeared as a chitinous foreign body near the anterior end of the brain, with a portion of the outer skin between the anterior margin of the carapace and the oral appendages. This structure, as pointed out by the author, resembled "dermoid" cysts of man and other vertebrates in that it was formed by recession of a saclike portion of the ectoderm within the body proper, where it continued to secrete successive chitin layers.

In the class Crustacea, tumors have been reported in a lobster, a shrimp, and a crab. Herrick (1895, 1909) briefly described a large growth on the side of the carapace of a lobster caught in Vineyard Sound, Massachusetts. There was a craterlike depression covered by a membrane over the top of the growth. Herrick thought the growth was a "sore" resulting from a wound that had failed to heal. He noted that a similar case had been mentioned by Rathbun (1887), and that Prince (1897) included a description of a tumorlike growth originating in the wall of the stomach sac of a large lobster that finally perforated the carapace and caused its death. Although I have not seen the latter report, it appears that no photograph, histological studies, or photomicrographs were made of either of these growths. This seems particularly unfortunate in the case of the growth arising in the stomach sac mentioned by Prince.

Savant and Kewalramani (1964), in recording a new host record of the isopod parasite *Bopyrus squillarium,* stated that the parasite caused a 7 mm tumor in the left branchial chamber of the host shrimp, *Paloemon tenuipes.* Since this genus of isopods typically causes a pronounced swelling on the side of the cephalothorax because of the presence of the female in the host's gill chambers, it is suspected that the authors used the term tumor in its broadest sense, and were describing the size of the swelling rather than suggesting neoplasia.

There is no question, however, of Fischer's intentions in his description of tissue resembling cancer in the abdomen of a crab, *Carcinus maenus*, parasitized by *Sacculina carcine*. He also noted that there was tissue resembling adenosarcoma in the eye of the crab (E. Fischer, 1928).

Several abnormalities that, though confusing, could be considered a form of neoplasia or tumor have been observed in ticks. Olenev and Rozhdestvenskaja (1933) found four female ticks (*Hyalomma* sp., *Dermacentor niveus*) in a collection from Turkestan whose bodies were covered with numerous oval, whitish, well-defined elevations with glossy surfaces, which they called papules. One tick had about 1000 papules, equally distributed on the dorsal and ventral surfaces, but confined to the more thinly chitinized parts of the body, thus being absent from the capitulum, scutum, and legs. Sections showed that the papules were composed of chitin, but the cause and significance could not be explained.

Subsequently, Colas-Belcour (1937) found lesions of the integument on two female ticks of the genus *Hyalomma*, which the author noted were similar to those just discussed. Colas-Belcour believed the lesions were probably the result of an infection, but was unable to demonstrate microorganisms in fixed material. He did strongly emphasize the improbability that these obviously related lesions had occurred by chance alone.

Phylum Echinodermata

The only possible neoplasms reported from echinoderms are several pigmented, tumorlike epidermal lesions in the ophiuroid *Ophiocomina nigra* (Fontaine, 1969). The growths were noted incidentally while Fontaine was studying the nature of the dark pigment of *O. nigra*, a widespread, common European brittle star. He ascertained that the pigment has the properties of a phaeomelanin and that the pigment granules are localized in epidermal dendritic melanocytes that are structurally analogous to those found in the skin of amphibians. The tumorlike lesions consist of morphologically abnormal melanocytes containing large numbers of pigment granules and smaller quantities of lipofuscin granules.

Unfortunately, the lesions were not recognized until after Fontaine had left the area where they were collected (Plymouth, England) and had processed them for histological study. However, 7 of the 95 specimens in the collection possessed the tumorlike lesions. Not only was the rate (7.4%) of incidence high, but the frequency of multiple lesions was also high: one specimen contained three; three had two; three had only a single lesion.

Development of the lesions appears to be restricted to certain definite areas of the integument, the epidermis on and especially between the arm spines and the epidermis of the aboral surface of the disc and arms. The most massive growths occur between the arm spines, and, perhaps significantly, the lesions appear most frequently in the most densely pigmented areas and those areas exposed to the greatest illumination under natural conditions.

Grossly, the growths consist of soft, ovoid dark nodules (Figs. VII.112 and VII.113), ranging from 25 to 250 μm in diameter. They appear superficially to be dermal in location, but on stained section are clearly epidermal, lying entirely above the epidermal basement membrane (Fig.

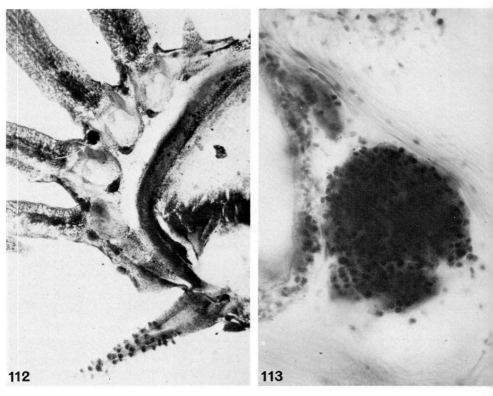

112   **113**

Fig. VII.112. Transverse section, 100 μm thick, of the arm of the ophiuroid, *Ophiocomina n*̶ with an intermediate-stage tumor (arrow) between arm spines. (From Fontaine, 1969. Cour of A. R. Fontaine.)

Fig. VII.113. Tumor in Fig. VII.112. at higher magnification. Note normal, intact exte epithelium and apparent dermal position of the tumor. (From Fontaine, 1969. Courtesy of A Fontaine.)

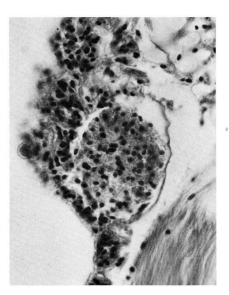

Fig. VII.114. A 10-μm section of an intermediate-stage tumor located between arm spines. Note the basement membrane (to the right of the tumor) proving the epidermal nature of the lesion. (From Fontaine, 1969. Courtesy of A. R. Fontaine.)

VII.114), which delimits the basal portions of the growth. However, the lesion is not encapsulated, since the lateral and superior surfaces are in direct contact with adjacent normal epithelial cells (Figs. VII.115 and VII.116). The external epithelium remains intact and normal in appearance.

Microscopically, the nodules are composed of densely packed cells and are completely devoid of connective tissue elements. Abnormally shaped epidermal melanocytes are the predominant cells making up the nodule; the widely ramifying dendrites characteristic of the normal melanocyte are not present; instead, the cells are attenuated and shaped like mammalian fibroblasts. The dark coloration of the nodule results from numerous pigment granules within the melanocytes. A second cell type commonly present in the lesion contains light-brown, irregular granules consistent in histochemical reaction with lipofuscin. Fontaine interpreted this cell as an abnormal form of the carotenoid-containing lipocyte that is responsible for the red and yellow lipochrome colors in normal skin. A third cell type occurring in the nodules is nonpigmented, agranular, and rounded in shape. Its origin is unknown, but is believed by Fontaine to be an invading coelomocyte and/or a pathologically altered epithelial cell.

Despite the paucity of material, lesions of a wide size range were

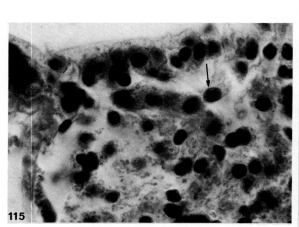
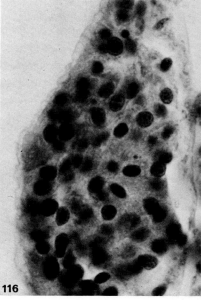

Fig. VII.115. Oil-immersion view of exterior of tumor in *O. nigia* in Fig. VII.114. Pigment g
ules, which have been bleached, are evident in cytoplasm. Note presumptive invasive or migra
cell (arrow). (From Fontaine, 1969. Courtesy of A. R. Fontaine.)

Fig. VII.116. Oil-immersion view of early-stage lesion in *O. nigra*. Note intact external epithel
and basement membrane. Rounded pigment cells are the only recognizable cellular elements wi
the tumor. (From Fontaine, 1969. Courtesy of A. R. Fontaine.)

found. Fontaine arranged them in a sequence of increasing size and found that the cellular pattern clearly demonstrated a developmental sequence in lesion growth that he classified as early, intermediate, and late stages.

The early stage consists of a small, ovoid nodule containing, almost exclusively, abnormal rounded melanocytes. The growth is densely pigmented and causes an inconspicuous bulge in the epithelium (Fig. VII.117). As growth progresses to the intermediate stage, the lesion increases in size, causing the bulge in the intact external epithelium to be more prominent, and the basal portion compresses and tends to distort the underlying dermis (Fig. VII.114). There is, however, no invasion of the dermis, and the basic architecture of the dermis is not altered. Fontaine believed that the basement membrane is an effective barrier to invasion by the tumor cells. Where the tumor occurs over a skeletal plate, the supporting calcite is sufficiently strong to prevent dermal distortion (Fig. VII.118).

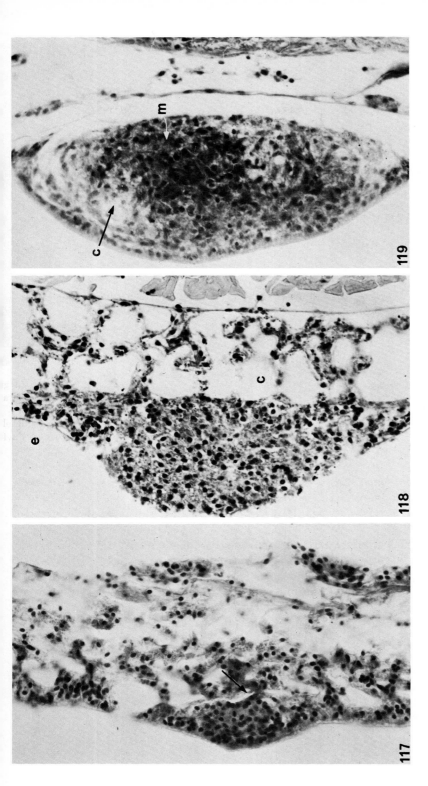

Fig. VII.117. Early-stage tumor in aboral epidermis of an arm. Note obvious epidermal nature and intact basement membrane (arrow). (From Fontaine, 1969. Courtesy A. R. Fontaine.)

Fig. VII.118. Intermediate-stage tumor in epidermis over calcified dermis. Clear spaces occupied in life by calcite, forming skeletal plates of the integument. c, Calcified dermis; e, epidermis. (From Fontaine, 1969. Courtesy of A. R. Fontaine.)

Fig. VII.119. Late-stage tumor showing cytological zonation into cortical (c) and medullary (m) regions. (From Fontaine, 1969. Courtesy of A. R. Fontaine.)

Cytologically, abnormal melanocytes, lipocytes, and clear cells comprise the tumor at the intermediate stage, and the dark pigmentation is less marked. Numerous pigmented fibroblastlike cells appear at this stage; they are morphologically identical to embryonic pigment cells that appear to have become detached from the tumor mass. Their location and orientation is consistent with a pattern of outward migration between the basement membrane and the epidermal epithelium to invade tissues adjacent to the original lesion (Fig. VII.115). Small, satellite tumor foci, presumably originating by migration of the embryonic pigment cells, are frequently observed a short distance from the main tumor mass.

Architecturally, the late-stage lesion (Fig. VII.119) may be roughly divided into cortical and medullarly regions. Fontaine states that "the medullary portion is composed of cells which look highly undifferentiated and fibroblastlike," they are "completely devoid of pigment and show frequent mitotic figures." It seems more likely to me that the cells are highly dedifferentiated, or anaplastic, a common characteristic of malignant tumors in vertebrates. The cortical region is pigmented and is composed of the same cell types and patterns. The lack of pigment in the medullary portion of late-stage tumors causes the lesion to be very lightly pigmented. Although mitotic figures are common in all three stages, they are most frequent in the medullary region of the late-stage lesion.

Fontaine's interpretation of the developmental sequence of the growths appears to be correct. He believed the origin of the growth to be an abnormality and subsequent hypertrophy of a locus of epidermal melanocytes, with the melanocytes dedifferentiating to an embryonic condition. As growth continues, caretenoid-containing lipocytes and possibly epithelial cells and/or wandering coelomocytes are incorporated and involved in the neoplastic proliferation. During this intermediate stage, migrant or invasive cells are budded off, and peripheral or satellite tumors are established. Dedifferentiation proceeds in later stages, with the resulting melanocytes losing their melanogenetic properties, thus producing a central or medullary zone with cells that look like fibroblasts or embryonic cells in tissue culture. As Fontaine correctly points out, the limited amount of material and the static, observational nature of the data precludes unequivocal diagnosis of these growths as neoplasms. Three salient points noted by Fontaine, however, strongly indicate that they are neoplastic in nature, and, although his scientific conservatism restrained him from so stating, possibly malignant. "First, the earliest detectable stage has its focus or germinal origin in a cluster of abnormal pigment cells." Since these cells are consistent in appearance with echinoderm embryonic melanocytes, the lesion probably originates from defective epidermal cells. "Secondly, migrant cells occur peripheral to the main cellular masses and appear to move outward to establish satellite colonies at a distance." Fontaine notes that this characteristic is suggestive of metastasis or, more probably, epidermal spread (invasion) commonly observed in junctional nevi. "Thirdly, the nodule grows by frequent cell divisions." Since mitoses are of infrequent occurrence in normal, adult echinoderm tissues, this point is of considerable significance, suggesting that the mitotic activity within the lesion is, at least, of an order of magnitude greater than normal.

Because of the involvement with and apparent origin in defective pigment cells, Fontaine tentatively diagnosed the growth as a melanoma or, due to the lack of evidence of detrimental effect on the host, more probably a pigmented junctional nevus. Further study is needed to establish the possible malignancy of this growth and to determine which of these diagnoses is correct. There is no question, however, of the importance of this lesion in invertebrate oncology. It is the first record of an apparent neoplasm in an echinoderm, and it presents evidence of metastasis in an invertebrate tumor. This report reemphasizes the point made earlier of the difficulty of recognition of invertebrate neoplasms. Only someone like Fontaine, familiar with the histology and embryology of echinoderms, would realize the lesions were abnormal and would be able to work out the probable origin and development.

A small amount of work on the effects of vertebrate carcinogens on echinoderms does appear in the literature. According to Harshbarger (1967), Kaiser reported that crystals of 3,4-benzopyrene implanted in the starfish, *Asterias rubens,* initiated a general toxic reaction and a foreign-body reaction, but apparently nothing resembling neoplasia. In contrast, Tchakhotine (1938) reported wild mesenchymal proliferation in the eggs of sea urchins of several species, induced by treatment with sodium monobromo- or monoiodoacetate. The atypical proliferation filled the blastulae, preventing gastrulation, and thus killing the developing embryo. Tchakhotine believed the proliferation to be neoplastic, an opinion with which Gersch (1950), who felt the experiments demonstrated nothing resembling cancerous-type cell proliferation, disagreed. Since Tchakhotine provided no photomicrographs of his material, it is difficult to evaluate his report.

Phylum Chordata—Subphylum Urochordata

The only tumor or tumorlike growth reported from the ascidians is of parasitic origin. A gregarine, *Monocystis ascidae,* spends the greatest part of its developmental period in an epithelial cell of the gut of a tunicate,

Ciona intestinalis. Siedlecki (1901) noted that infection of an intestinal cell initiated hypertrophy, beginning with a slight enlargement of the cell and a distinct hypertrophy of the nucleus and condensation of the chromatin. As the gregarine grows, the hypertrophy of the cell is accentuated; the protoplasm lacks the density and homogeneity of normal cells and is studded with clear vacuoles, while the still enlarging nucleus is displaced into a corner of the cell. In subsequent stages, the cells continue to enlarge, attaining ten to twenty times their normal size, and the nuclei, after becoming smaller and very chromatic, disintegrate.

The response of the tunicate's intestinal cells is, of course, a clear case of cellular hypertrophy, as Scharrer and Lochhead (1950) pointed out. However, Siedlecki reported that neighboring uninfected epithelial cells also proliferated, and, in some cases, proliferation of surrounding connective tissue contributed to the formation of an abnormal growth, which Siedlecki called a tumor similar to liver adenoma of rabbits. As Siedlecki noted, if one examines the advanced stage of the abnormal growth, the original cause (the gregarine) is missing, and it thus appears as a spontaneous neoplastic growth.

Siedlecki felt that the irritation initiating the hypertrophy was chemical rather than mechanical in nature. He theorized that, when the chemical effect was weak, only the parasitized cell would react, but when it was stronger, it would initiate proliferation of neighboring cells resulting in tumor formation.

CARCINOGENS

Responses of invertebrates to vertebrate carcinogens, although of great importance, have not been considered in a specific section; instead, where experimentally induced neoplastic growth was claimed by authors, their findings were discussed in the section dealing with tumors of their taxonomic group. Harshbarger (1967) reviewed the responses of invertebrates to vertebrate carcinogens dividing the review into the effects of carcinogenic viruses, carcinogenic chemicals, and ionizing radiation, and the reader is referred to his article for more detailed discussion and an excellent bibliography. It appears worthwhile, however, to note several of Harshbarger's conclusions. He remarked on the lack of cohesion of the work on carcinogens, pointing out that no one has systematically investigated the effects of any individual carcinogen or group of carcinogens on representatives from various invertebrate phyla. There has been little effort to relate work to that of others, and where such efforts have been made, basic parameters such as dosage and method of application have been largely ignored. Finally, there has never been an in-depth study of the effects of a single carcinogen on any one animal. Despite the inconsistency of data, however, Harshbarger felt some general observations were germane. One relevant observation was that a carcinogenic vertebrate virus (Rous sarcoma virus) has been reported to cause chromosomal aberrations in an insect. As Harshbarger pointed out, this, if corroborated, indicates a probable area of fruitful research in other invertebrates.

Although some invertebrates have failed to produce a detectable response to externally applied carcinogens or have responded with typical foreign body reactions to internally introduced substances, these results will not prove that the chemicals are inactive carcinogenically until all variables of time of application, method of application, and means of detection are understood. Numerous studies have shown that vertebrate carcinogens generally stimulate regenerative cells in invertebrates. And, interestingly, virtually all chemical carcinogens, including heavy metals, polycylic compounds, hydrocarbons, azo dyes, alkylating agents, and aromatic amines, tested for their cytogenetic action on the fruit fly, *Drosophila,* have proved to be mutagenic. Radiation pathologies may involve damage to the endocrine system in addition to direct cellular damage.

Finally, Harshbarger notes that "the fact that invertebrate and vertebrate responses to certain carcinogens may be quite different should stimulate a further research with the invertebrates; significant contributions to the understanding of carcinogenesis may be discovered in the simpler system."

References

Boettger, C. R. (1956). Ueber einen Fall von pathologischer Gestaltveränderung bei einer Wegschnecke der Art *Arion ater* (L.). *Biol. Zentralbl.* **75**, 257–267.

Boveri, T. (1914). "Zur Frage der entstehung malignen Tumoren." Fischer, Jena.

Butros, J. (1948). A tumor in a fresh-water mussel. *Cancer Res.* **8**, 270–272.

Cantwell, G. E. (1967). *In* "Activities Report af the Registry of Tumors in Lower Animals for the Period September 1, 1966 to March 31, 1967," p. 3. Smithsonian Institute, Washington, D. C.

Colas-Belcour, J. (1937). Observations de lésions pathologiques chez des tiques du genre *Hyalomma. Bull. Soc. Pathol. Exot.* **30**, 876–878.

Collinge, W. E. (1891). Note on a tumour in *Anadonta cygnaea* Linn. *J. Anat. Physiol. Norm. Pathol. Homme Anim.* **25**, 154.

Cooper, E. L. (1968). Multinucleate giant cells, granulomata, and "myoblastomas" in annelid worms. *J. Invertebr. Pathol.* **11**, 123–131.

Cooper, E. L. (1969). Neoplasia and transplantation immunity in annelids. *In*

"Neoplasms and Related Disorders of Invertebrate and Lower Vertebrate Animals." *Nat. Cancer Inst., Monogr.* 31, 655–669.

Couch, J. A. (1969). An unusual lesion in the mantle of the American oyster, *Crassostrea virginica*. In "Neoplasms and Related Disorders of Invertebrate and Lower Vertebrate Animals." *Nat. Cancer Inst., Monogr.* 31, 557–562.

Dawe, C. J. (1968). Invertebrate animals in cancer research. *Exp. Anim. Cancer Res., Gann Monogr.* 5, 45–55.

Dawe, C. J. (1969). Phylogeny and oncogeny. In "Neoplasms and Related Disorders of Invertebrate and Lower Vertebrate Animals." *Nat. Cancer Inst., Monogr.* 31, 1–40.

Des Voigne, D. M., Mix, M. C., and Pauley, G. B. (1970). A papillomalike growth on the siphon of the horse clam, *Tresus nuttali*. *J. Invertebr. Pathol.* 15, 262–270.

Dubois, F. (1949). Contribution à l'étude de la migration des cellules de régénération chez les planaires dulcicoles. *Bull. Biol. Fr. Belg.* 83, 215–283.

Farley, C. A. (1969a). Sarcomatid proliferative disease in a wild population of blue mussels (*Mytilus edulis*). *J. Nat. Cancer Inst.* 43, 509–516.

Farley, C. A. (1969b). Probable neoplastic disease of the hematopoietic system in oysters, *Crassostrea virginica* and *Crassostrea gigas*. In "Neoplasms and Related Disorders of Invertebrate and Lower Vertebrate Animals." *Nat. Cancer Inst., Monogr.* 31, 541–555.

Farley, C. A., and Sparks, A. K. (1970). Proliferative diseases of hemocytes, endothelial cells, and connective tissue cells in mollusks. In "Comparative Leukemia Research 1969" (R. M. Dutcher, ed.), Bibl. Haematol. No. 36, pp. 610–617. Karger, Basel.

Fischer, E. (1928). Association chez le crabe d'un tissu parasite et d'une trame conjunctive analogue à certains processus tumoraux. *Bull. Ass. Fr. Etude Cancer* 17, 468–470.

Fischer, P. H. (1954). Tumeur fibreuse chez un Pleurobranche. *J. Conchyliol.* 94, 99–101.

Fontaine, A. R. (1969). Pigmented tumor-like lesions in an ophiuroid echinoderm. In "Neoplasms and Related Disorders in Invertebrate and Lower Vertebrate Animals." *Nat. Cancer Inst., Monogr.* 31, 225–261.

Foster, J. A. (1963). Induction of neoplasms in planarians with carcinogens. *Cancer Res.* 23, 300–303.

Foster, J. A. (1969). Malformations and lethal growths in planaria treated with carcinogens. In "Neoplasms and Related Disorders in Invertebrate and Lower Vertebrate Animals." *Nat. Cancer Inst., Monogr.* 31, 683–691.

Frömming, E., Peter, H., and Reichmuth, W. (1961). Beitrag zur Frage der pathologischen Gestaltsveränderung und der Geschwülste bei unserer Nacktschnecken. *Zool. Anz.* 166, 139–147.

Gersch, M. (1950). Ueber Zellwucherungen und Geschwulstbildung in der Lunge von *Helix*. *Biol. Zentralbl.* 69, 500–507.

Gersch, M. (1954). Einwirkung von kanzerogenen Kohlenwasserstoffen auf die Haut von Regenwürmern. *Naturwissenchaften* 41, 337.

Goldsmith, E. D. (1939). Spontaneous outgrowths in *Dugesia tigrina* (Syn. *Planaria maculata*). *Anat. Rec.* 75, Suppl., 158–159.

Hancock, R. L. (1961a). Giant nuclei in the earthworm *Lumbricus*. *Nature* (*London*) 189, 685.

Hancock, R. L. (1961b). Neoplasms in *Lumbricus terrestris* L. *Experientia* 17, 547–549.

Hancock, R. L. (1965). Irradiation induced neoplastic and giant cells in earthworms. *Experientia* **21**, 33–34.

Hanström, B. (1926). Ueber einen Fall von pathologischer chitinbildung in Inneren des Körpers von *Limulus polyphemus*. *Zool. Anz.* **66**, 213–219.

Harshbarger, J. C. (1967). Responses of invertebrates to vertebrate carcinogens. *Fed. Proc., Fed. Amer. Soc. Exp. Biol.* **26**, 1693–1697.

Harshbarger, J. C., and Taylor, R. L. (1968). Neoplasms of insects. *Annu. Rev. Entomol.* **13**, 159–190.

Henderson, T. R., and Eakin, R. H. (1961). Irreversible alteration of differentiated tissues in planaria by purine analogs. *J. Exp. Zool.* **146**, 253–264.

Herrick, F. H. (1895). The American lobster: A study of its habits and development. *Bull. U. S. Fish Comm.* **15**, 1–252.

Herrick, F. H. (1909). Natural history of the American lobster. *Bull. U. S. Fish Bur.* **29**, 149–408.

Hérubel, M. A. (1906). Sur une tumeur chez un invertébré (*Sipunculus nudus*). *C. R. Acad. Sci.* **143**, 979–981.

Hueper, W. C. (1963). Environmental carcinogenesis in man and animals. *Ann. N. Y. Acad. Sci.* **108**, 963–991, and 1028–1038.

Jacquemain, R., Jullien, A., and Noel, R. (1947). Sur l'action de certains corps cancérigènes chez les céphalopodes. *C. R. Acad. Sci.* **225**, 441–443.

Jones, E. J., and Sparks, A. K. (1969). Histopathology of an unusual histopathological condition of *Ostrea lurida* from Yaquina Bay, Oregon. *Proc. Nat. Shellfish. Ass.* **59**, 11.

Jullien, A. (1940). Sur les réactions des Mollusques Céphalopodes aux injections de goudron. *C. R. Acad. Sci.* **210**, 608–610.

Katkansky, S. C. (1968). Intestinal growths in the European flat oyster. *Calif. Fish Game* **54**, 203–206.

Korschelt, E. (1924). "Lebensdauer, Alten und Tod," 3rd ed. Fischer, Jena.

Ladreyt, F. (1922). Sur une tumeur cancéreuse de Siponcle (*Sipunculus nudus* L.). *Bull. Inst. Oceanogr.* **405**, 1–8.

Lange, C. S. (1966). Observations on some tumours found in two species of planaria —*Dugesia etrusca* and *D. ilvana*. *J. Embryol. Exp. Morphol.* **15**, 125–130.

Lenhoff, H. M., Rutherford, C., and Heath, H. D. (1969). Anamolies of growth and form in hydra: Polarity, gradients, and a neoplasia analog. *In* "Neoplasms and Related Disorders of Invertebrate and Lower Vertebrate Animals." *Nat. Cancer Inst., Monogr.* **31**, 709–750.

Metcalf, M. M. (1928). Cancer (?) in certain protozoa. *Amer. J. Trop. Med.* **8**, 545–557.

Moewus, F. (1959). Stimulation of mitotic activity by benzedine and kinetin in *Polytoma revella*. *Trans. Amer. Microsc. Soc.* **78**, 295–304.

Mottram, J. C. (1940). 3:4-Benzypyrene, *Paramecium* and the production of tumors. *Nature* (*London*) **145**, 184–185.

Nolte, A. (1962). Eine Geschwulstbildung bei *Helix pomatia* L. *Zt. Zellforsch. Mikrosk. Anat.* **56**, 149–156.

Olenev, N. O., and Rozhdestvenskaja, V. S. (1933). A pathological condition observed in ticks (Ixodidae). *Parasitology* **25**, 478–479.

Pauley, G. B. (1967a). A tumorlike growth on the foot of a freshwater mussel (*Anodonta californiensis*). *J. Fish. Res. Bd. Can.* **24**, 679–682.

Pauley, G. B. (1967b). Four freshwater mussels (*Anodonta californiensis*) with pedunculated adenomas arising from the foot. *J. Invertebr. Pathol.* **9**, 459–466.

Pauley, G. B. (1967c). A butter clam (*Saxidomus giganteus*) with a polypoid-tumor on the foot. *J. Invertebr. Pathol.* 9, 577–579.

Pauley, G. B. (1968). A disease in freshwater mussels (*Margaritifera margaritifera*). *J. Invertebr. Pathol.* 12, 321–328.

Pauley, G. B. (1969). A critical review of neoplasia and tumor-like lesions in mollusks. *In* "Neoplasms and Related Disorders in Invertebrate and Lower Vertebrate Animals." *Nat. Cancer Inst., Monogr.* 31, 509–539.

Pauley, G. B., and Cheng, T. C. (1968). A tumor on the siphons of a soft-shell clam (*Mya arenaria*). *J. Invertebr. Pathol.* 11, 504–506.

Pauley, G. B., and Sayce, C. S. (1967). Descriptions of some abnormal oysters (*Crassostrea gigas*) from Willapa Bay, Washington. *Northwest Sci.* 41, 155–159.

Pauley, G. B., and Sayce, C. S. (1968). An internal fibrous tumor in a Pacific oyster, *Crassostea gigas. J. Invertebr. Pathol.* 10, 1–8.

Pauley, G. B., Sparks, A. K., and Sayce, C. S. (1968). An unusual internal growth associated with multiple watery cysts in a Pacific oyster (*Crassostrea gigas*). *J. Invertebr. Pathol.* 11, 398–405.

Potter, M., and Kuff, E. (1967). A developmental anomaly of the siphon of the soft-shell clam, *Mya arenaria,* from Chesapeake Bay. *In* "Activities Report of the Registry of Tumors in Lower Animals for the period Sept. 1, 1966 to March 31, 1967," p. 11. Smithsonian Institute, Washington, D. C.

Prince, E. E. (1897). Special report on the natural history of the lobster. *29th Annu. Rep., Dep. Mar. Fish. Can.* Suppl. 1, Parts I–IV, pp. 1–36.

Rathbun, R. (1887). Part V. Crustaceans, worms, radiates, and sponges. *In* "Natural History of the Useful Aquatic Animals" (by G. B. Goode), Sect. I, pp. 759–850.

Robbins, S. L. (1959). "Textbook of Pathology" Saunders, Philadelphia, Pennsylvania.

Ryder, J. A. (1887). On a tumor in the oyster. *Proc. Acad. Natur. Sci. Philadelphia* 39, 25–27.

Savant, K. B., and Kewalramani, H. G. (1964). On a new record of host species of isopod parasite, *Bopyrus. Cur. Sci.* 33, 217.

Scharrer, B., and Lochhead, M. S. (1950). Tumors in the invertebrates: A review. *Cancer Res.* 10, 403–419.

Siedlecki, M. (1901). Sur les rapports des Gregarines et de l'epithelium intestinal. *C. R. Acad. Sci.* 132, 218–220.

Simroth, H. (1905). Ueber zwei seltene Missbildungen an Nachtschnecken. *Z. Wiss. Zool.* 82, 494–522.

Smith, G. M. (1934). A mesenchymal tumor in an oyster (*Ostrea virginica*). *Amer. J. Cancer* 22, 838–841.

Sonneborn, T. M. (1954). Problems of tumors in relation to studies on *Paramecium. Proc. Nat. Cancer Conf., 2nd, 1952* pp. 1139–1147.

Soule, J. D. (1965). Abnormal corallites. *Science* 150, 78.

Sparks, A. K. (1969). Review of tumors and tumor-like conditions in Protozoa, Coelenterata, Platyhelminthes, Annelida, Sipunculida, and Arthropods, excluding insects. *In* "Neoplasms and Related Disorders of Invertebrate and Lower Vertebrate Animals." *Nat. Cancer Inst., Monogr.* 31, 671–682.

Sparks, A. K. (1970). The future of invertebrate pathology—one man's opinion. (Presidential address presented at the Second Annual Meeting of the Society for Invertebrate Pathology, Burlington, Vermont, August 17–22, 1969.) *J. Invertebr. Pathol.* 15, i–iv.

Sparks, A. K., Pauley, G. B., Bates, R. R., and Sayce, C. S. (1964a). A mensenchymal tumor in a Pacific oyster, *Crassostrea gigas* (Thunberg). *J. Insect Pathol.* **6**, 448–452.

Sparks, A. K., Pauley, G. B., Bates, R. R., and Sayce, C. S. (1964b). A tumor-like fecal impaction in a Pacific oyster, *Crassostrea gigas* (Thunberg). *J. Insect Pathol.* **6**, 453–456.

Sparks, A. K., Pauley, G. B., and Chew, K. K. (1969). A second mesenchymal tumor from a Pacific oyster (*Crassostrea gigas*). *Proc. Nat. Shellfish. Ass.* **59**, 35–39.

Spencer, R. R., and Melroy, M. B. (1940). Effect of carcinogens on small free-living organisms. II. Survival value of methylcholanthrene adapted *Paramecium. J. Nat. Cancer Inst.* **1**, 343–348.

Squires, D. F. (1965a). Neoplasia in a coral? *Science* **148**, 503–505.

Squires, D. F. (1965b). Abnormal corallites. *Science* **150**, 78.

Stéphan, F. (1960). Tumeurs spontanées chez la Planaire *Dugesia tigrina. C. R. Soc. Biol.* **156**, 920–922.

Stolk, A. (1961a). Enzymatic activity (three dehydrogenase systems) of the pharyngeal tumors in the earthworm *Lumbricus terrestris. Experientia* **17**, 306.

Stolk, A. (1961b). Occurrence of giant nuclei and pharyngeal tumor, in the earthworm *Lumbricus terrestris. Experientia* **17**, 306.

Szabó, I., and Szabó, M. (1934). Epitheliale Geschwulstbildung bei einem wirbellosen Tier *Limax flavus* L. *Z. Krebsforsch* **40**, 540–545.

Taylor, R. L., and Smith, A. C. (1966). Polypoid and papillary lesions in the foot of the gaper clam, *Tresus nuttalli. J. Invertebr. Pathol.* **8**, 264–266.

Tchakhotine, S. (1938). Cancérisation expérimentale des éléments embryonnaires, obtenue sur des larves d'Oursins. *C. R. Soc. Biol.* **127**, 1195–1197.

Thomas, J.-A. (1930). Etude d'un processus néoplasique chez *Nereis diversicolor* O.F.M. due à la dégénérescence des oocytes et quelquefois des soies. *Arch. Anat. Microsc.* **26**, 251–333.

Tittler, I. A. (1948). An investigation of the effects of carcinogens on *Tetrahymena geleii. J. Exp. Zool.* **108**, 309–325.

van Wagtendonk, W. J. (1969). Neoplastic equivalents of Protozoa. *In* "Neoplasms and Related Disorders of Invertebrate and Lower Vertebrate Animals." *Nat. Cancer Inst., Monogr.* **31**, 751–768.

von Hansemann, D. (1890). Uber asymetrische zellteilung in Epithelkrebsen und deren biologische Bedentung. *Virchons Arch. Pathol. Anat. Physiol.* **119**, 229–326.

White, P. R. (1965). Abnormal corallites. *Science* **150**, 77–78.

Williams, J. W. (1890). A tumor in the fresh water mussel. *Anadonta cygnea* Linn. *J. Anat. Physiol. Norm. Pathol. Homme Anim.* **24**, 307–308.

Wolf, P. H. (1969). Neoplastic growth in two Sydney rock oysters, *Crassostrea commercialis* (Iredale and Roughley). *In* "Neoplasms and Related Disorders of Invertebrate and Lower Vertebrate Animals." *Nat. Cancer Inst., Monogr.* **31**, 563–573.

Wolf, P. H. (1971). Unusually large tumor in a Sydney rock oyster. *J. Nat. Cancer Inst.* **46**, 1079–1084.

Wolman, M. (1939). A proliferative effect of carcinogenic hydrocarbons upon multiplication of paramecia. *Growth* **3**, 387–396.

Yevich, P. P., and Berry, M. M. (1969). Ovarian tumors in the quahog *Mercenaria mercenaria. J. Invertebr. Pathol.* **14**, 266–267.

Author Index

Numbers in italics refer to the pages on which the complete references are listed.

A

Abbott, D. P., 205, *227*
Abbott, W., 142, *161*
Abbud, L., 213, *226*
Abeloos, M., 218, *224*
Abeloos, R., 218, *224*
Adams, E., 157, 158, 159, 160, *162*
Akiya, R., 223, *225*
Aldrich, D. V., 177, *200*, 217, 218, *227*, *228*
Alexander, P., 234, *267*
Allen, D. M., 190, 191, *200*
Allgen, C. A., 45, *127*
Anderson, A. H., 31, *129*
Anderson, J. M., 117, 118, *128*
Ando, Y., 223, *225*
Andrew, W., 126, *128*
Andrews, G., 232, *268*
Angelovic, J. W., 243, *270*
Ansell, A., 184, *202*
Antoine, S., 157, 158, 159, 160, *162*
Applegate, V. C., 195, *199*
Archambeau, J. O., 230, 231, 232, *267*

Arendt, J. A., 126, *128*
Asano, M., 209, *225*

B

Back, A., 233, *267*
Bacq, Z. M., 220, *225*, 234, *267*
Baetjer, F. H., 236, *267*
Balbiani, E. G., 22, *128*
Baltzer, F., 221, *225*
Bang, B. C., 55, 57, *128*
Bang, F. B., 55, 57, 63, 108, 117, *128*
Bardeen, C. R., 236, *267*
Barnes, R. D., 112, *128*
Bates, R. R., 310, 311, 312, 313, 314, 317, *370, 371*
Bearden, C. M., 192, *199*
Beaven, G. F., 198, *199, 201*
Beaver, P., 38, 39, *128*
Beckett, G. E., 198, *199*
Bennetts, M. J., 103, *129*
Bernard, F. J., 178, *199*
Berry, M. M., 343, 345, 346, *371*
Bisignani, L. A., 63, 65, *133*
Blackman, R. B., 187, 188, 189, *201*

373

Subject Index

A

Acanthaster planci, venom apparatus, 212
Acanthocephala, reaction to injury, 43
Acellus, thoracic limb regeneration in, 111
Acetylcholine content of cockroach nerve
  cords, 136
Actenosphaerium, reaction to injury, 22
Adamsia palliata toxin, effects, 205
Alcohols, toxic effects, 166
Aldrin, toxic effects, 191
Algae, toxigenic, 214
Algor mortis, 6
Amoebocytes in inflammatory response of
  mollusks, 87–98
Amphiporine, 220
Anemone toxin, effects, 205
Annelids
  effects of heat, 147
  effects of irradiation, 241
  reaction to injury, 46
  toxic, 220
  tumors, 295
  venomous, 206
Anodonta

  reaction to injury, 65
  tumors in, 330–338
Anodonta oregonensis, wound repair in,
  99
Anoxia, necrosis due to, 2
Anthozoa toxin, effects, 205
Anticoma procera, wound repair in, 45
Aplysia toxins, 222
Appendicularia, postmortem changes in,
  7
Arbacia, reaction to injury, 119
Arcella, reaction to injury, 22
Archeocytes, 25
Arion, tumors in, 305, 308
Arion ater, metaldehyde poisoning, 168
Arsenic poisoning, effects, 174
Arthropods
  effects of ionizing radiation, 265
  myriapodous, reaction to injected ma-
    terial, 112
  pathological effects of high temperature
    on, 154
  poisonous, 222
  postmortem changes in, 6